# Advanced Models for Project Management

## INTERNATIONAL SERIES IN OPERATIONS RESEARCH & MANAGEMENT SCIENCE

**Frederick S. Hillier, Series Editor**
Stanford University

Saigal, Romesh
*LINEAR PROGRAMMING: A Modern Integrated Analysis*

Nagurney, Anna/ Zhang, Ding
*PROJECTED DYNAMICAL SYSTEMS AND VARIATIONAL INEQUALITIES WITH APPLICATIONS*

Padberg, Manfred/ Rijal, Minendra P.
*LOCATION, SCHEDULING, DESIGN AND INTEGER PROGRAMMING*

Vanderbei, Robert J.
*LINEAR PROGRAMMING: Foundations and Extensions*

Jaiswal, N.K.
*MILITARY OPERATIONS RESEARCH: Quantitative Decision Making*

Gal, Tomas / Greenberg, Harvey J.
*ADVANCES IN SENSITIVITY ANALYSIS AND PARAMETRIC PROGRAMMING*

Prabhu, N.U.
*FOUNDATIONS OF QUEUEING THEORY*

Fang, S.-C./Rajasekera, J.R./ Tsao, H.-S.J.
*ENTROPY OPTIMIZATION AND MATHEMATICAL PROGRAMMING*

Yu, Gang
*OPERATIONS RESEARCH IN THE AIRLINE INDUSTRY*

Ho, Teck-Hua / Tang, Christopher S.
*PRODUCT VARIETY MANAGEMENT*

El-Taha, Muhammad / Stidham, Jr., Shaler
*SAMPLE-PATH ANALYSIS OF QUEUEING SYSTEMS*

Miettinen, Kaisa M.
*NONLINEAR MULTIOBJECTIVE OPTIMIZATION*

Chao, Hung-po & Huntington, Hillard G.
*DESIGNING COMPETITIVE ELECTRICITY MARKETS*

Weglarz, Jan
*PROJECT SCHEDULING: Recent Models, Algorithms & Applications*

Sahin, Izzet & Polatoglu, Hakan
*Quality, Warranty and Preventive Maintenance*

# Advanced Models for Project Management

by

Prof. L. Valadares Tavares
Technical University of Lisbon
Instituto Superior Técnico
Lisbon, Portugal

KLUWER ACADEMIC PUBLISHERS
BOSTON/DORDRECHT/LONDON

**Distributors for North, Central and South America:**
Kluwer Academic Publishers
101 Philip Drive
Assinippi Park
Norwell, Massachusetts 02061 USA
Telephone (781) 871-6600
Fax (781) 871-6528
E-Mail <kluwer@wkap.com>

**Distributors for all other countries:**
Kluwer Academic Publishers Group
Distribution Centre
Post Office Box 322
3300 AH Dordrecht, THE NETHERLANDS
Telephone 31 78 6392 392
Fax 31 78 6546 474
E-Mail <orderdept@wkap.nl>

Electronic Services <http://www.wkap.nl>

**Library of Congress Cataloging-in-Publication Data**

Tavares, L. Valadares (Luís Valadares), 1946-
Advanced models for project management / by L. Valadares Tavares.
p. cm. -- (International series in operations research & management science ; 16)
Includes bibliographical references and index.
ISBN 0-7923-8222-6
1. Industrial project management. 2. Network analysis (Planning)
I. Title. II. Series.
T56.8.T38 1998
658.5--dc21 98-39321
CIP



*Printed on acid-free paper.*

Printed in the United States of America

## TABLE OF CONTENTS

*Foreword*

More than forty years have passed since the early attempts to model projects. A large domain of theoretical developments has grown producing a high number of analytical and numerical results, but it seems that the main model is still the same: the concept of **project network**. This concept has come to represent the two major features underlying the notion of a project: the **sequential** and the **competitive** nature of its components, the project's **activities**. Actually, the sequential property defines the structure of the project and the competitive nature stems from the use of common resources (facilities, goods, equipment, management, etc.) to carry out the different activities.

However, significant advances have been achieved in project modelling, allowing the production of much more powerful results:

A. the concept of precedence and the description of activities has been generalized to produce a wide range of realistic representation of projects.

B. the stochastic study of the features of projects such as the duration and cost of their activities is carried out by several analytical and numerical models, allowing experimental and forecasting analyses.

C. the allocation of resources can be now studied for more complex situations and restrictions.

D. the financial description of projects is more accurately studied and its optimization is thoroughly pursued.

E. the assessment and the evaluation of projects now can be studied within the framework of multicriteria decision theory considering multiple perspectives and supporting the project manager to select the most appropriate compromises between risk, time and expected gains.

F. the decisions concerning the scheduling of a project can be supported by new models based on synthetic indicators, helping the manager to select the most convenient solutions.

All these developments can be implemented using the modern world of computing facilities which seems so far way from the tools adopted ten or twenty years ago.

In this book, the author has followed these six lines of progress and analyzes the major achievements of each:

A - on modelling and structural analysis → Chapter 2 and 3

B - on simulation and stochastic risk analysis → Chapters 4 and 5

C and D - on scheduling → Chapter 6

E - on assessment and evaluation → Chapter 7

F - on synthetic support to decision making → Chapter 8

Several original advances are presented, namely in the areas of:

- assessing the complexity and hardness of a project network
- describing the network's morphology
- proposing a new approach to simulate project networks
- developing a model based on continuous variables to optimize project schedules
- developing a three dimension model (MACMODEL) to assess and to evaluate projects
- developing a new synthetic indicator to support the process of scheduling

Some of these developments produce new results such as the generation of a test set of networks including instances with up to 1000 activities and the ability to use GAMS-MINOS to solve problems with more 100 activities.

Several software products were developed to help the project manager using these new tools such as RISKNET and MACMODEL.

Project Management has become a key approach to understand and to lead organizations as is discussed in Chapter 1 and therefore the author believes that the developments presented in this book can be useful not just for a post-graduation course on Operational Research, Systems Engineering, Management Sciences or Project Management, but also as a guide to help project managers make better use of the potential of this flourishing scientific domain.

*Acknowledgements*

The work presented in this book stems from my research, teaching and consulting work carried out at Instituto Superior Técnico as full professor of Operational Research of the Department of Civil Engineering and as head of research a the Centro de Sistemas Urbanos e Regionais during more than 20 years.

Also, activities as invited professor at the Faculty of Economics and Management (Faculdade de Ciências Económicas e Empresariais) of the Universidade Católica Portuguesa have contributed greatly to this book due to the interaction with my colleagues and to the participation in research projects carried out in the area of Management Sciences at this University

The contacts and meetings within the European Working Group on Project Management and Scheduling of the European Association of O. R. Societies (EURO) have been a major source of influence while compiling this book, and periodic contacts with colleagues such as S. Elmaghraby, J. Weglarz and Burton Dean proved useful when developing the major lines of research.

Most of the work was discussed with my colleague Prof. José Álvaro Antunes Ferreira who is also co-author of several research papers quoted and used in this book. The contribution of our colleague José Pedro Silva Coelho was also particularly important in the area of software production.

The research presented in this book was partially supported by the project 3/3.1/CEG/2677/95 - PRAXIS XXI (**FUNDAÇÃO DE CIÊNCIA E TECNOLOGIA**, Lisbon, Portugal) and cooperation with american collegues also was supported by the **FUNDAÇÃO LUSO - AMERICANA PARA O DESENVOLVIMENTO** (Lisbon, Portugal).

The help and support of my wife, Maria Manuel, and of my two children, Ana Teresa and Paulo Tiago, has been a key condition to the development of my scientific career and of this book. A final word of thanks is due to Patrícia Nunes who efficiently and patiently typed several versions of this manuscript.

L. VALADARES TAVARES,
LISBOA, INSTITUTO SUPERIOR TÉCNICO, 1998
CESUR - IST, AV. ROVISCO PAIS
1000 LISBOA - PORTUGAL
(lavt@alfa.ist.utl.pt)

**About the author**

Luís Valadares Tavares is full professor of Operational Research at Instituto Superior Técnico (Civil Engineering Department) and invited professor at the Faculty of Economics and Management Sciences of the Portuguese Catholic University.

He is author of more than 60 research papers published by major international journals and of six other scientific books.

He was the founder of the portuguese OR Society (APDIO) and currently he is the president of APDIO.

He has performed a long list of research and consulting activities for international institutions (OECD, UNDP, WHO, IFORS, European Union, etc.) and was visiting professor in more than 20 universities in Europe, US and Asia.

The author is president of the Working Group on Project Scheduling of the European Association of the OR societies (EURO).

He is married, has two children and lives in Lisbon

# 1

# A SYSTEMIC INTRODUCTION TO PROJECT MANAGEMENT

## *1. Projects and organizations*

Initially, the concept of project was used to describe the **accomplishment** of a **physical system** such as a new building or a new industrial product. This explains why this term was borne within the Engineering literature (Battersby, 1967).

However, since the early fifties, the study of the management of organizations is giving growing importance to the concept of project as a central notion to **understand**, to **structure** and to **lead organizations**.

Any organization, either public or private, with or without profit aims, is affected by **external factors** (such as, market fluctuations, legal changes, or technological innovations) and by its **internal features**, its **ethics** and **culture**, its **organizational structure** and **available resources** (Simon, 1960).

The development of an organization demands a continuous effort of change to preserve its **identity and mission** under the turbulent environment of the conditioning factors. The growing rate and extension of these changes require a more flexible and systemic process of development which should be based on a **general cycle** of development including the stages presented in Figure 1.1, following the principles successively presented and discussed since the classical works on management by Fayol (1949) and Taylor (1947).

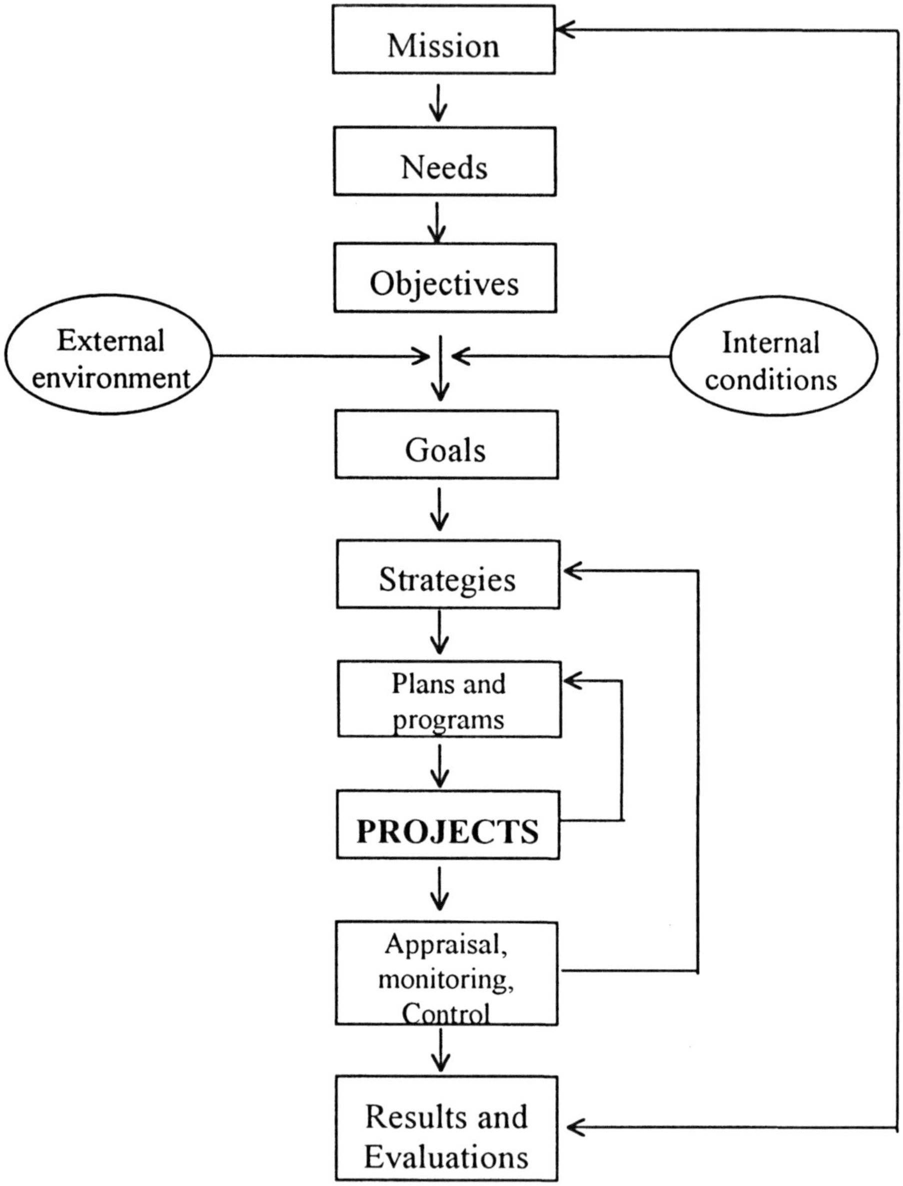


**FIGURE 1.1**
**THE CYCLE OF DEVELOPMENT OF AN ORGANIZATION**

Defining objectives and analyzing internal and external conditions allows one to set the **goals** to be reached within a given period of time.

The following stage establishes a **strategy** that should allow the organization to profit from its competitive advantages, use any favorable external factors, compensate or surpass any negative influences and increase the potential of the resources to be used.

Any strategy must be made according to plans, defined as broad guidelines for the development of the organization, which must include **projects** as their core units.

The planning and implementation of plans must be monitored, evaluated and controlled. This will create iterative cycles which allow for the adjustment of goals, correction of strategies and updating of objectives according to real life experiences.

This means that the idea of **change** within the framework of each organization is fundamental and that the concept of **project** is the core unit to develop the process of changes required to achieve the pre-defined goals (Drucker, 1954).

Therefore, the definition of a project implies:

- clear definition of the objectives to be attained which have to be translated into goals;
- choice of technologies and of the appropriate people or organizations to deal with them;
- analysis of the changes and activities to be undertaken;
- definition of the resources to be acquired in order to implement the project;
- planning, programming and management of the project, so that it will be developed in the best way possible;
- monitoring, evaluation and control.

In some cases, the **changes** that are acquired are largely external to the organization (as it the case with, say, a construction company that builds a new estate); in others, the changes to be made are at the **very core of the organization** (for example, creating a new career planning scheme within the organization or developing a new management support information system).

Sometimes, the objectives will be to position the organization differently in the market that it serves (creating a new product or making sales promotions, for example), while in other cases objectives correspond to internal changes, such as converting a production technology or the distribution system.

The advantage of identifying the concept of project and of making all its requirements explicitly clear is that it allows for the **rational management of change** within the organization. One therefore concludes that the concept of project is a fundamental notion, both structurally and instrumentally, in organization management, which in turn should be designed and implemented as a function of given objectives.

**Goals** must be defined thoroughly and in terms of the stated objectives, so that dubious situations are avoided. Thus, if a project is concerned with a real estate development, it should be defined previously (by the so-called "program") in terms of all the indicators and attributes of the development (Areas, quality of finishing, comfort levels, etc). If the purpose of the project is the development of new product, its properties must be defined beforehand so that its design, prototype and production are accomplished in the best possible conditions.

A clear and visible definition of goals assumes a period of thought and discussion within the organization; this is usually regarded as one of the first (or main) benefits that a project induces, since it is always most useful in encouraging an organization to rethink its course and destination.

A more clear focus on goals, their formulation, measurement and evaluation has a wide range of advantages for the optimization of the strategies, namely the reduction of internal entropic politics (Jones, 1990).

Therefore, within each organization, Project Management becomes the basic approach to implement the general principles of **MBO-Management by Objectives** (Drucker, 1970).

After knowing the objectives and the goals, a project must be designed carefully, by means of a **backwards approach**. Or, in other words, starting from the specified goals, one must imagine, identify and characterize all **activities** that are essential to its fulfilment.

An **activity** is therefore each **component** of the project. An activity includes a sub-group of closely linked tasks that must be accomplished to each of the pre-established goals.

Defining an activity includes:

- Describing the work it encompasses;
- Analyzing the resources that are needed;
- Defining the group of other activities that must be completed before the current task is begun (**precedence conditions**) or other types of precedence conditions;
- Defining the responsibility of those who will execute, monitor and control its development.

Specifying each activity usually includes the definition of several actions, with diverse resources and characteristics.

This means that designing and implementing a project will require:

- **Breaking down** the project into its components (**activities**) and choosing the most appropriate technologies, since any activity can be carried out by alternative means;
- Selecting the resources, the people or the services that will be used for each activity; They can belong to the organization (departments, teams, etc.) or not and this last alternative corresponds to **outsourcing solutions** which are becoming increasingly popular and successful.

Thus, the concept of project helps the organization to concentrate its energy on the identification of **needs** which will be translated into **objectives**. These objectives correspond to goals which will be pursued by a **structured system** of **activities**. The project management approach allows the analysis and control of the whole range of operations of each organization in terms of activities ("**activity - based management**" - ABM - see, e. g., Kearney, 1993). Therefore, this approach can support the development of benchmarking measures and any process of **business reengineering** (See Hammer and Stanton, 1995).

## 2. Project management and model building

The professional activity of designing, structuring, scheduling, organizing, managing and controlling projects is usually called **Project Management**. Nowadays, there is a growing number of professionals with an initial background in Engineering, Sciences, Economics or Management who have developed specialized skills in this area and work as project managers. Project Management is clearly an interdisciplinary field requiring a scientific methodology and appropriate technical instruments.

The main methodological contributions stem from **Management Sciences** and **Operational Research**, and the major instruments are based on **Statistics, Information Systems** and, of course, **Software Development**.

The scientific approach to Project Management is very much based on a **systemic model** of the process of development required by the project and therefore the progresses achieved by **OR** in **modelling** (Ackoff, 1967) have contributed to substantial advances in Project Management.

A wider use of **information systems** in most sectors is providing larger and more significant volumes of data allowing the development of statistical analyses which can support the activity of Project Management.

The recent spectacular progresses of **computer systems** and of **specialized software** are providing portable and economic tools with very powerful abilities supporting the professional activities of Project Management.

These new tools allow the implementation of more ambitious and complex models requiring the adequate representation of the problem and their solution. These recent advances explain the growing interest of Project Managers in Project Management Modelling which is the subject of this book.

The wide diversity of models being used by Project Management can be understood and developed if studied within the framework of the **Systems Theory**.

The use of more advanced models in any professional area, from Medicine to Engineering, implies the fulfilment of multiple requirements:

- the adopted models have to be sufficiently representative of the studied reality in order that they don't "cut corners" which will be important for the required support;
- these models cannot be too complex because if this happens then their estimation and use can be hardly implemented;
- the vality of the models should be tested (**Model Validation**) and they should be transparent making the conclusions understandable by the professionals (**Model Transparency**);
- the models to be built should be oriented to support the process of decision making and training in each professional area. Training is no less important because it is actually a crucial way to improve the know-how of professionals and is also an essential condition to gain the required **trust** and general acceptability concerning each model (**Model Legitimization**).

Summing up, the development of a model should correspond to an harmonic equilibrium between the desired levels of **feasibility** for its **estimation** and **implementation,** of **transparency** for **understanding** and **trust** by the professionals, as well as of **representativeness** to describe the important features of the **studied system** (Fig. 1.2 ):

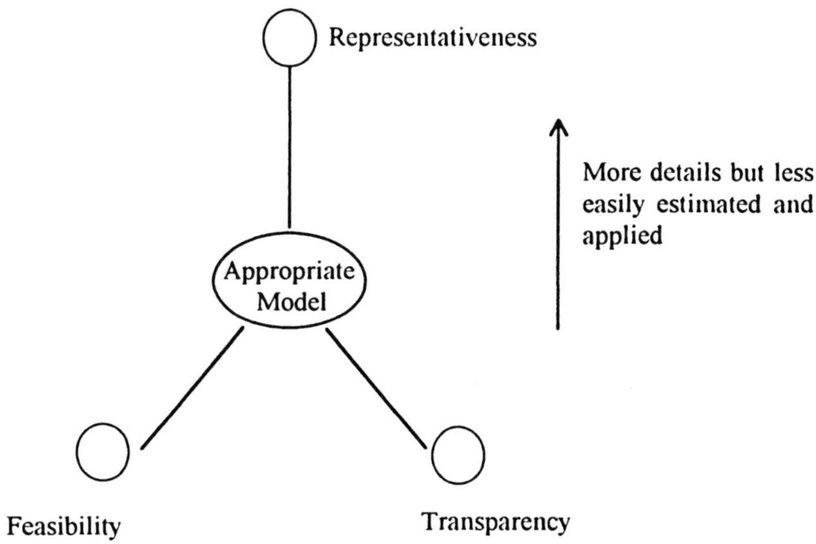


**FIGURE 1.2**
**MODEL BUILDING COMPROMISE**

In Project Management, human and organizational features often have been ignored by OR models which is a serious drawback. However, the development of Organizational Sciences as well as of the Information and Computer Technologies allows the **development** of more balanced models pursuing the well known paradigm of $H^3$ for any recommended model:

$$H^3\begin{cases} H- & \textbf{human} \\ H- & \textbf{harmonic} \\ H- & \textbf{honest} \end{cases}$$

A better understanding and a more coherent and unified perspective on PM models can be achieved through the general framework of Systems Theory. The presentation of its application to Project Management is the subject of next sections.

## *3. The general systemic model*

The most general model of **Systems Theory** (Ashby, 1956), which is particularly useful to understand organizations (Handy, 1979), is based on the following scheme (Fig.1.3) where:

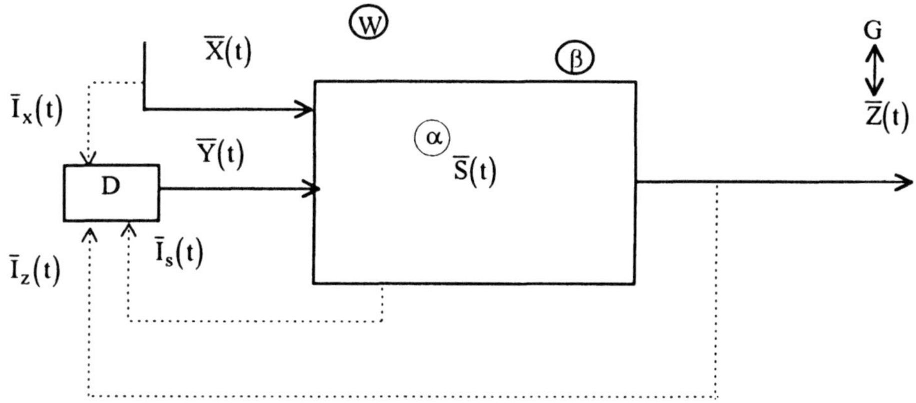


**FIGURE 1.3**
**THE GENERAL MODEL OF SYSTEM'S THEORY**

- Ⓐ- Studied system
- Ⓦ- Context where the studied system is defined
- Ⓑ- Border of the system
- $\overline{S}(t) = \{S_1(t), \ldots, S_p(t)\}$ defines the system's state for time t with t=0, ..., H where H is the studied horizon. $\{\overline{S}(t)\}\tau$ denotes $\overline{S}(t)$ with t=0, ...,τ.

- $\overline{X}(t)$ is the vector of uncontrolled variables affecting the system: $\overline{X}(t) = \{X_1(t),...,X_N(t)\}$ with t=0, ..., H.

- $\overline{Y}(t)$ is the vector of decision variables controlled by the decision-maker, D:
  $\overline{Y}(t) = \{Y, (t), ..., Y_M (t)\}$ with t=0, ..., H.

- $\overline{Z}(t)$ is the vector containing the output or response function
  $\overline{Z}(t) = \{Z_1 (t), ..., Z_Q (t)\}$ with t = 0, ..., H

- $\overline{I}(t) = \{\overline{I}_X(t), \overline{I}_s(t), \overline{I}_z(t)\}$ contains the information vectors about $\overline{X}(t)$, $\{I_1^X,(t),...,I_N^X(t)\}$, about $\overline{S}(t)$, $\{I_1^s,(t), ..., I_p^s(t)\}$ and about $\overline{Z}(t)$, $\{I_1^Z,(t),...,I_Q^Z(t)\}$, with t = 0,...,H.

- Response function, F, is the function defining $\overline{Z}(t)$ in terms of the system's state and of the acting variables:

$$\overline{Z}(t) = F\{\overline{S}(t), \overline{X}(t), \overline{Y}(t)\}$$

- G - Goals of the system. They can be defined by a scalar G, or by a vector, $\overline{G}$, with a specific value to be maximized (or minimized). Also G or $\overline{G}$ can be a function of the time, G (t) or $\overline{G}$ (t).

- State function g determines the evolution of $\overline{S}(t)$ on time:

$$\overline{S}(\tau) = g\left\{\{\overline{S}(t)\}_{\tau-1,} \overline{X}(\tau-1), \overline{Y}(\tau-1)\right\} \text{ with } \tau = 1,...,H.$$

Usually, the Markov hypothesis is adopted and so

$$\overline{S}(\tau) = g\{\overline{S}(\tau-1), \overline{X}(\tau-1), \overline{Y}(\tau-1)\} \text{ with } \tau = 1,...,H.$$

The system's state, $\overline{S}(t)$, has to belong to its feasibility domain, Σ.

The general problem of Decision Theory is determining $\overline{Y}(\tau)$ in terms of $\{\overline{X}(t)\}_\tau$ and $\overline{S}(\tau)$ in order that $\overline{Z}(t)$ will be optimized or, at least, achieving a minimal level of satisfaction.

**Linear Programming (LP)**

The canonical formulation of a LP problem is:

$$\underset{\text{Min}}{\overset{\text{Max}}{}}\{F(Y)\} \text{ with } g_m(Y) \leq b_m \text{ and } Y \geq 0,\ m = 1, \ldots, M.$$

F is a scalar and linear function and the restrictions $g_m$ are also linear.

The application of the presented model can be done making:

a) $Y = \{\overline{Y}(t)\}_H$
b) $\{\overline{X}(t)\}_H$ are given by the coefficients and the independent terms of the restrictions
c) the state vector is given by $\{g_m\}_{m=1,\ldots,M}$ and so the restrictions define the feasibility domain $\Sigma$.

**Dynamic Programming**

In this case, F is also a scalar function and the Bellman (Bellman et al, 1962) equation is given by:

$$F_t[\overline{S}(t), \overline{Y}(t)] = U[\overline{S}(t), \overline{Y}(t)] + f. \sum_{i=1}^{N} P(\overline{X}(t) = i).\ F_{t+1}^{*}[\overline{S}(t+1)]$$

where $U[\overline{S}(t), \overline{Y}(t)]$ is the short term utility generated by $\overline{Y}(t)$ when the system is in state $\overline{S}(t)$ and where f is the appropriate discount factor. The optimal value of F from (t+1) until the end of the studied process is denoted by $F_{t+1}^{*}[\overline{S}(t+1)]$.

The variable i=1, ..., N is a discrete description of the domain of $\overline{X}(t)$ and P(i) is the probability of occurrence of i. The state function is given by $\overline{S}(t+1) = g\{\overline{S}(t), \overline{X}(t), \overline{Y}(t)\}$.

The optimal solution is given by:

$$F_t^*\left[\overline{S}(t)\right] = \underset{\overline{Y}(t)}{Max}\left\{U\left[\overline{S}(t), \overline{Y}(t)\right] + f \cdot \sum_{i=1}^{N} P(i).F_{t+1}^*\left[\overline{S}(t+1\right]\right\}$$

**Non-linear optimization and optimization heuristics**

The previous model now can be generalized assuming a non-linear objective function and non-linear restrictions (Non-linear Optimization).

Another generalization is particularly important and corresponds to the assumption that F and $g_m$ with m = 1, ..., M can be computed for each of their domains but without knowing their analytical expressions.

Many optimization heuristics have been proposed to study this problem.

**Simulation**

This method adopts alternative choices of $\{\overline{Y}(t)\}$ which will be described by the index k=1,...,K and it is developed to estimate $\{\overline{Z}(t)\}$ in terms of random generation of $\{\overline{X}(t)\}$. At each time, t, the system's state, $\{\overline{S}(t)\}$, is computed in terms of $[\overline{X}(t), \overline{Y}(t)]$ and the output, $\overline{Z}(t)$ is defined as a function of $\overline{S}(t), \overline{X}(t)$ and $\overline{Y}(t)$.

After estimating $\{\overline{Z}(t)\}_H$ for k=1,...,K, the most convenient alternative is selected, k=k*.

## *4. The systemic model of project management*

The application of the presented systemic model to Project Management can be carried out through the identification of its major components.

**A - System, $(\alpha)$, and its border $(\beta)$.**

The considered system, $(\alpha)$, should include all the reality affected by the changes associated to the project and all the organizations, instruments and resources required by the project. The border $(\beta)$ may have to be extended or contracted due to the nature of the project or to special conditions affecting the system.

Consider two examples:

- Construction of a bridge

The system $(\alpha)$ should include the construction area, all the materials, equipment and human resources as well as the financial and organizational structures.

However, a more thorough analysis can identify environment problems due, for instance, to the use of an adjacent area of this portion of the river by bird species to locate their nests, and so this area should be also considered within the border $(\beta)$ as it will be part of the induced changes.

- Launching a new cosmetic

This project requires a system $(\alpha)$ including the market segment as well as the technological and production facilities associated to the new product.

Quite often, this type of development induces positive or negative effects in other products due to synergetic or cannibalization reasons and if so, they should be also included by $(\alpha)$.

**B - Uncontrolled variables, $\overline{X}$ (t)**

The uncontrolled variables can include an wide range of:

a) external factors such as geophysical conditions, social, political legal or economic changes;

b) unforeseen needs to accomplish the project goals;

c) changes of the internal features of $\textcircled{\alpha}$ not resulting from the evolution of the project (those coming up in terms of such evolution are included within $\overline{S}(t)$ ).

In Table 1.1, examples are given for the two previous cases.

**Table 1.1**
**Examples of uncontrolled variables**

| Building a Bridge | Launching a new cosmetic |
|---|---|
| a) Rainfall, wind and tidal conditions;<br>legal and union restrictions;<br>prices of raw materials;<br>exchange rates. | a) Weather, namely, average temperature;<br>consumer habits;<br>competition. |
| b) additional works due to geological problems | b) additional market to be penetrated |
| c) social turbulence, general strikes | c) inefficiency of the distribution channels |

**C - Decision variables, $\overline{Y}(t)$**

The decision variables are the key instruments of the project manager to lead the project.

The following areas of decision can be identified:

- **Design**

  This choice concerns the solutions conceived and designed to achieve the project goals. Usually, alternative designs can be adopted to implement the same project. For instance, there is usually a wide range of alternative constructive solutions to build the same bridge and a similar situation applies to launch a new product.

- **Technology**

This variable is related to the previous one and nowadays the degree of success of any project is becoming strongly dependent on this choice.

It seems that often insufficient attention is given to the selection and management of technologies originating the well known problems of inadequate technology integration.

- **Organization**

Project management has to be implemented through an organizational structure which implies selecting its most appropriate features (e.g., larger and self-contained organizations or smaller unit depending heavily on outsourcing, hierarchical or matrix model, etc).

- **Resources**

These decisions concern the critical variables of financial, human and material resources and their choice has to be made in terms of the goals and available conditions to manage the project.

- **Schedule**

This decision corresponds to the time location of the whole process of the development of the project. This area of decision may be considered as the initial "seed" justifying the development of basic models of project management, answering questions like: Which calendar? When can the project be finished?

- **Goals, G**

The goals should describe the objectives of the project.

- **Output,** $\overline{Z}(t)$

These variables should describe the project outcomes in terms of the resources already spent ($R_1$ (t)) and of the generated value ($R_2$ (t)). This achievement can be described by a measure of the value already generated by the project.

This concept of generated value is quite useful as it can be used to estimate a measure of the project **efficiency** $R_2$ (t) / ($R_1$ (t) and of the project effectiveness ($R_2$ (t)/G (where G is a scalar describing the value associated to the project goal).

The generated value also can be related to key stages of the project ($t_k = t_1, ... t_K$) and the difference $[R_2(t_k) - R_1(t_k)]$ can help us understand the chain of actions responsible for the project value ("**value chain**").

Obviously, the description of $R_2$ (t) can include many other features related to the goals of the project such as time, quality, environmental equilibrium, social acceptance, etc.

In the previous examples, one can illustrate $R_1$, $R_2$, and it is shown in Table 1.2.

**Table 1.2**
**Examples of G, $R_1$, $R_2$**

| | Building a bridge | Launching a new cosmetic |
|---|---|---|
| $R_1$ | Cost | Cost |
| $R_2$ | Dimension of the construction | Dimension of the market campaign |
| G | Average reduction of the travel time per commuter | Market share for the new product |

- **System's state ,** $\overline{S}(t)$

This vector should contain all the relevant information to support the decision process of the project manager and be able to apply the state equation. Therefore, this vector usually includes:

- Most of the components of $\overline{Z}(t)$, which describe how much was spent and was achieved;

- The relevant internal features (available resources, level of motivation, etc.).

For the two discussed examples, $\overline{S}(t)$ includes the magnitudes given in Table 1.3.:

**Table 1.3**
**$\overline{S}$ (t) examples**

| Building a bridge | Launching a new cosmetic |
|---|---|
| • Dimension of the construction carried out | • Advertizing campaign |
| • Available equipment | • Capacity of the promotion network |
| • Moral and motivation of the team | • Level of sales |

- **Information, $\overline{I}(t)$**

  The information sub-system should provide the decision maker with all the relevant data about $\overline{X}(t)$ (such as the information concerning a general strike or an exchange rate), about $\overline{S}(t)$ (such as a description of that describing the enthusiasm of the team or the survey was already done) and about $\overline{Z}(t)$ (such as, the variables describing how close or far we are from the project goals).

Unfortunately, the development of models of Project Management has not been oriented to all the decision problems already mentioned with the same depth, but rather they have been mainly concerned with the following major areas:

**Project modelling** as an interconnected structure of activities;

**Allocation of resources** to the development of the activities and costing analysis;

**Scheduling of activities**;

**Project appraisal.**

The author has developed new models and obtained further results in the areas of:

- structural **modelling** and analysis of the **morphology** of the project network;
- **simulation** of project networks and estimation of their duration;
- continuous models for **project scheduling**;
- **multi-criteria evaluation** of projects and risk analysis;
- synthetic **support** to **decision making** on project scheduling.

All these topics are covered in the next seven chapters of this book.

# 2

# BASIC MODELS FOR PROJECT MANAGEMENT

## *1. Project networking*

After presenting the general systemic model for project management, special attention should be given to the basic model of a project which includes a model of the system (α) and of the criteria to compare alternative decisions.

The basic modelling of a project can be developed in terms of an interconnected set of activities which implies:

a) the decomposition of the project into a **set of activities**;

b) the definition of all the **relations** connecting such activities;

c) the definition of the criteria to evaluate **alternative decisions**.

This model is called a **project network** as the interconnected set of activities define a network.

The first problem a) requires understanding which are the most appropriate criteria to decompose the project.

According to the usual principles of System's Theory, this decomposition should produce units as self-contained as possible, which means minimizing their interdependence. However, the decomposition into a very large number of parts may be very inconvenient as the analysis and the study of the project become too hard. Unfortunately, this has happened quite often, obtaining too complex representations with very little value for supporting the decisions of project managers. (A common joke within the "milieu" of project managers says that the main utility of network diagrams with 10000 or 20000 of activities is the decoration of walls behind their desks...).

A better approach can be followed by adopting multiple levels of decomposition and keeping the number of components (so-called activities) in each level under reasonable limits.

This corresponds to an hierarchical approach, (**hierarchical decomposition approach**) as it is shown in the following Figure 2.1.

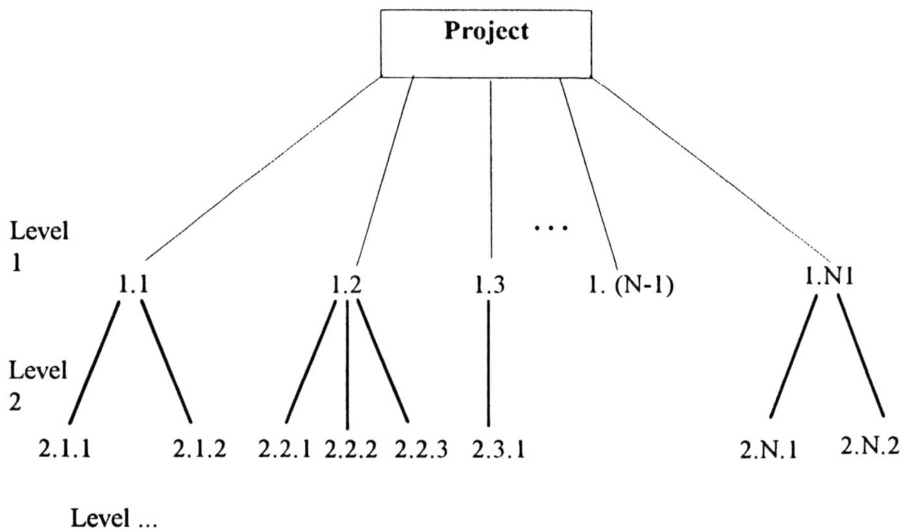


**FIGURE 2.1**
**AN HIERARCHICAL DECOMPOSITION OF THE PROJECT INTO ACTIVITIES**

Obviously, this method provides a "telescopic" view of the project allowing the "zooming" of any specific activity and describing it as a **sub-project** composed by sub-activities.

A few authors have considered this formulation but most of the research has not given sufficient attention to this approach, namely to the study of interconnections between activities belonging to different branches.

In some classes of applications, the project of the first level is called "**program**", the activities of the next level are called "**projects**", those of level 3 receive the label of "**sub-projects**" (or "super-activities") and the activities of the next levels are simply called "**activities**".

Whether or not the hierarchical decomposition approach is used, the definition of each activity is usually based on one or more of these reasons:

- presenting a strong physical or technical unity (for instance, the construction of a component of a bridge or the production of a questionnaire).

- being subject to sub-contracting.

- being its completion correspondent to important events like milestones or stages of the project. The construction of the priority part of a sewage network or carrying out the campaign to advertize a new product are examples of this situation.

The nature of activities can be very heterogeneous as they may have to describe actions so different as building a basement, obtaining a permission, announcing a decision or selecting a new staff member. In any case, the activities are the result of the decomposition of the project.

Therefore, a project can be analysed in terms of its activities and the following elements are essential to its definition through the concept of a **project network**:

a) the **discrete and finite set of activities** $\Omega = \{A_i : i = 1, ..., N\}$ being mutually exclusive and giving an exhaustive description of the project (N is the number of activities);

b) a set of **precedence conditions**, L, with $L = \{J_i: i = 1, ..., N\}$ where $J_i$ is the set of activities immediately preceding $A_i$, $J_i$ can be defined by:
$J_i = \{k \varepsilon J_i') \cap (k \varepsilon J'_{m)} = \emptyset$ for any $m \varepsilon J_i'\}$ where $J_i'$ or $J'_m$ is the whole set of activities which have to be completed before starting $A_i$ or before starting $A_m$, respectively (**precedent activities**);

c) a **discrete and finite set of attributes** $\{B_1 (i), ..., B_Q (i)\}$ with $Q \geq 1$ defined for each activity and describing its properties relevant for project management such as duration, cost, consumption required of each resource ( $q = 1, ..., Q$);

d) a **discrete and finite set of criteria** ( $F_p$ with p = 1, ..., P) which should express the **values** and the **preferences** of the project manager to compare alternative decisions concerning the management of the project. The most common criteria are the **total duration**, the **total cost**, a **cost- benefit function**, a **measure of the project risk, and the net present value** (NPV).

Obviously, in b) other relationships can be added such as the **simultaneous incompatibility between the activities**.

However, the most general and important type of relationship within Project Management concerns the **sequentiality** presented in b) which deserves special attention.

## 2. Sequentiality

The precedence conditions already defined in b) should describe a feasible project which means that:

for any $A_j \in J'_i$ one should have $A_i \notin J'_j$ for any activity $A_i \in \Omega$.

This property is called **acyclicity condition** and implies that there are one or more activities, $A_i$

- with an empty $J_i$("**initial activities**", as they can start as soon as the project starts)
- not belonging to $J_k$ of any activity $A_k \in \Omega$ ("**final activities**", as the project will be finished as soon as they will be ended).

The acyclicity condition also implies that if there is a direct precedence link between j and i, $j \rightarrow i$, then one cannot have $i \rightarrow j$, which means that the pairwise direct precedence relation is not symmetric.

This type of network is called a **directed acyclic graph**.

In this type of graph, the concept of **path** can be quite useful to study several properties or functions and it can be easily defined by any sequence, L, of activities $\{A_1, ..., A_M\} \in L$ satisfying the following condition:

$$A_m \rightarrow A_{m+1} \text{ for } m = 1, ..., M-1$$

which means that there is a direct precedence link from activity $A_m$ to activity $A_{m+1}$.

The acyclicity property means that no path can be constructed with $A_1 = A_M$ and it can be checked by several procedures. One of them was suggested by Elmaghraby, 1977 and it is based on the so-called Adjacency Matrix $\{A_{ij}\}$, with i, j=1, ..., N and:

$$A_{ij} = \begin{cases} 1 \text{ if there is a direct precedence link between i and j} \\ 0 \text{ otherwise} \end{cases}$$

This matrix should have at least one line of zeros (initial activity) and at least one column of zeros (final activity).

This procedure starts by deleting any line and/or column of zeros as well as the corresponding activities from $\Omega$. This means that a new line or column of zeros is generated and the same deleting procedure is iteratively applied until the matrix is empty (acyclic graph), or no further deletion is possible (cyclic graph).

An example of a project network is given in Figure 2.2 where each circle (**node**) represents an activity and where each direct precedence link is represented by an **arrow**. Obviously, the arrow ⓧ is redundant and therefore it can be deleted. This representation is called the AoN ("**Activity on Node**") and, usually, two other activities are added to represent the start of the project (Node S) and the end of the project (Node E) as it is shown in Figure 2.2. Here, the sets $J'_i$ and $J_i$ for a generic activity i (i=6) are also described.

Alternatively, another representation can be adopted AoA ("**Activity on Arc**") corresponding each activity to an arc and representing by a node the event of completion of the arcs converging into it. The previous example is now presented in Figure 2.3 using AoA. The start and the end are now represented by the initial and dummy nodes (S, E) and redundant precedence links cannot exist.

However, now representing the full set of activities with the appropriate precedence links may be more difficult and it may require the introduction of additional and dummy activities as is shown in the example (Fig. 2.3) using dotted lines.

In any case, the project network will be represented by a finite and discrete set of nodes and by a set of **direct links** defining a set of connections between nodes.

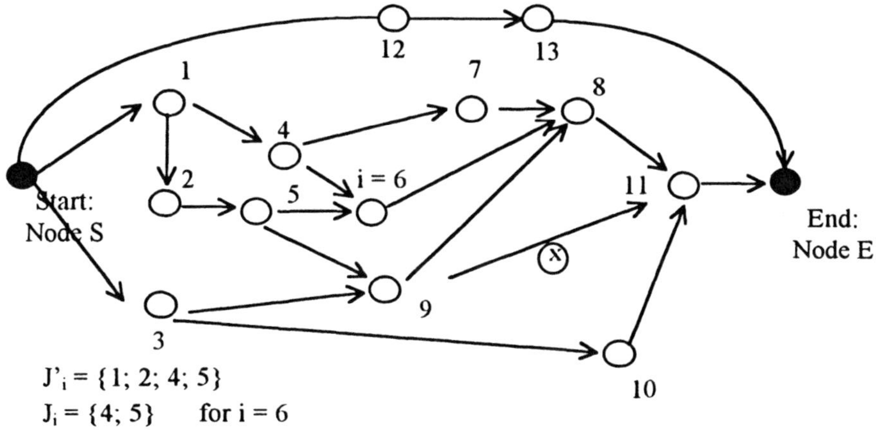


**FIGURE 2.2**
**EXAMPLE OF A PROJECT NETWORK USING AoN**

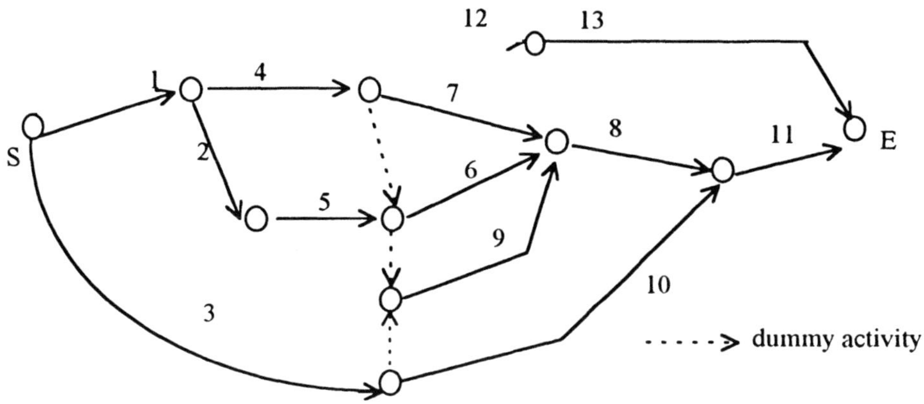


**FIGURE 2.3**
**EXAMPLE OF THE PROJECT NETWORK OF FIGURE 2.2 USING NOW AoA**

The study of the precedence conditions also can be developed in terms of the description of the activities which can be carried out just after completing each activity i, $K'_i$ (**succeedent activities**). The set of immediately succeedent activities can be defined for any activity, $A_i$, by $K_i = \{k : (k \in K'_i) \cap (k \in K'_m) = \emptyset$ with any $m \in K'_i\}$.

The sets $K'_i$ and $K_i$ for the studied example given and with i=6 are given by:

$K'_i = \{8; 11\}$
$K_i = \{8\}$

For any network, (either with AoA or AoN representation) the number of arcs directed to (or from) a node is called **in-degree**, I(i) (or **out-degree** O(i)).

A **series-network** (Fig. 2.4) is defined by having I (i) = O (i) = 1 for any node, excepting the Start and the End (I (S) = 0; O (S) =1; I (E) = 1; O (E) = 0).

A **parallel-network** (Fig. 2.4) is defined by having just two nodes, S and E, with I(S) = O(E) = 0 and O(S) = I(E)>1.

Series-network

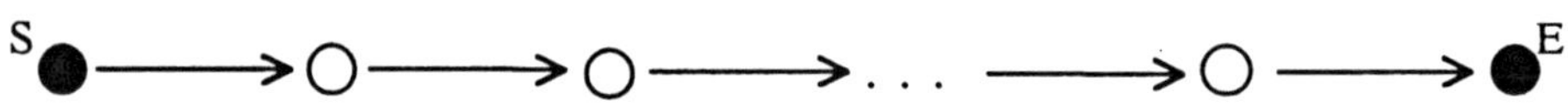


Parallel-network

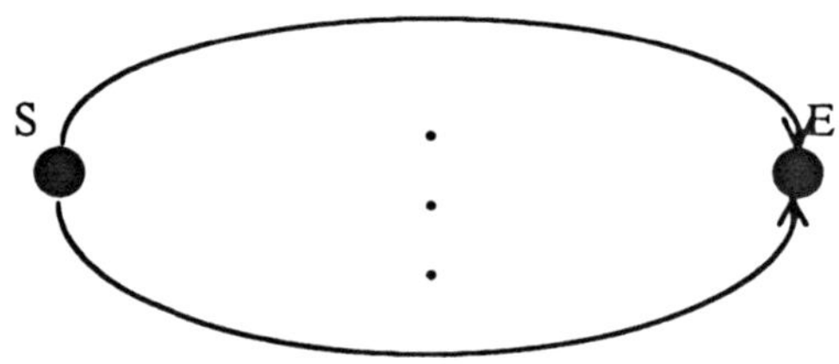


**FIGURE 2.4**
**EXAMPLES OF SERIES-NETWORK AND OF PARALLEL-NETWORK**

The concept of precedence adopted so far implies that the **start** of i is conditioned by the **end** j. (Type ①). However, this concept can be generalized to cover three other situations:

Type ② End of j → End of i

Type ③ Start of j → Start of i

Type ④ Start of j → End of i

These new types are now easily represented adopting AoN and decomposing each activity into its start, its development and its end (Fig. 2.5)

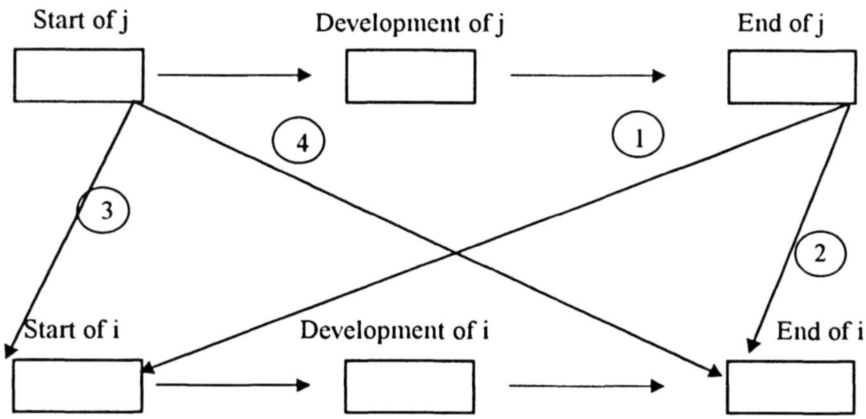


**FIGURE 2.5**
**THE FOUR TYPES OF PRECEDENCE**

However, in many practical problems, the explicit consideration of these four types of links (or of some of them) can help the decision maker to understand the project and to compare alternatives. Then, the representation AON is preferable and four different connections should represent the four different cases.

Obviously, the acyclicity property has now to be revised because even within the same pair of activities (j, i) there can be links in opposite directions as it is shown in Figure 2.6:

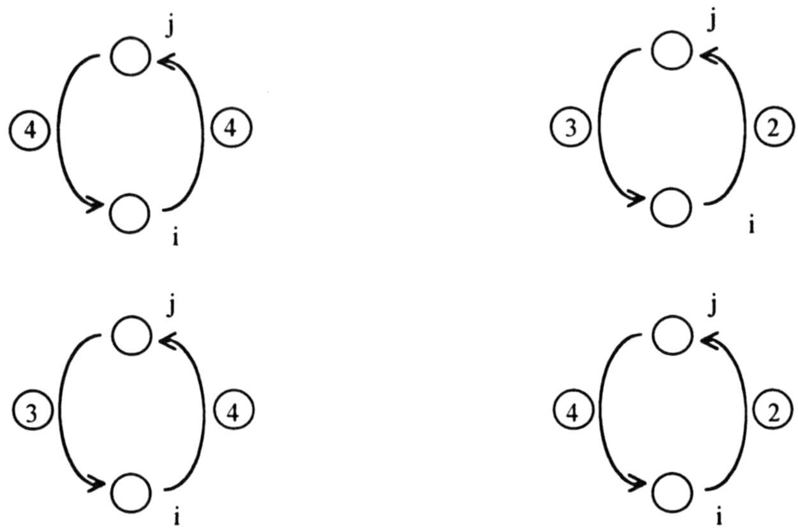


**FIGURE 2.6**
**CYCLICITY WITH DIFFERENT TYPES OF PRECEDENCE**

Therefore, the cyclicity testing should be done after transforming each activity into the sequence of stages already presented and adopting the usual procedures applied to the precedence links connecting the stages of the activities.

Other types of precedence links can describe different relations such as:

a) **probabilistic precedence links**

In this case, a set of alternative links start from one or more nodes (either representing events or activities) of the network, and a probability is associated to the occurrence of each link ($t_k$) as is shown in Figure 2.7.

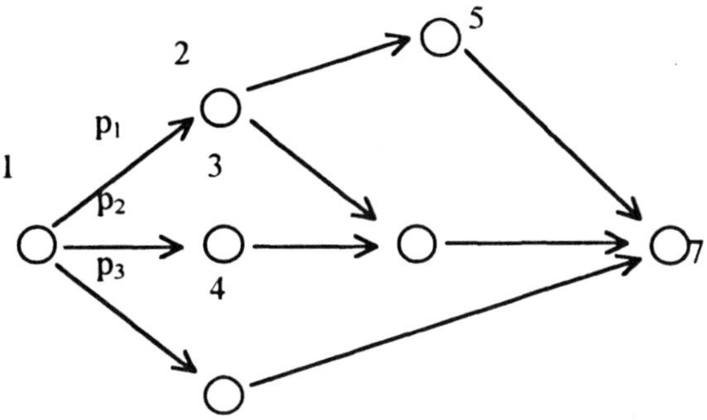


Where $p_1 + p_2 + p_3 = 1$

**FIGURE 2.7**
**EXAMPLE OF PROBABILISTIC NETWORK**

b) modelling of **logical conditions** between activities

A first example was already given by the **simultaneous incompatibility between activities** but other examples can be given using the logical operation of intersection or union. Elmaghraby 1977 suggested the following cases (represented in Fig. 2.8 using AoA):

- **Intersection** → the node d just occurs if a, b and c occur.
- **Inclusive union** → the node d occurs if a, b, c or if one or more nodes occur.
- **Exclusive union** → the node d just occurs if one and just one of the nodes a, b, c occurs.

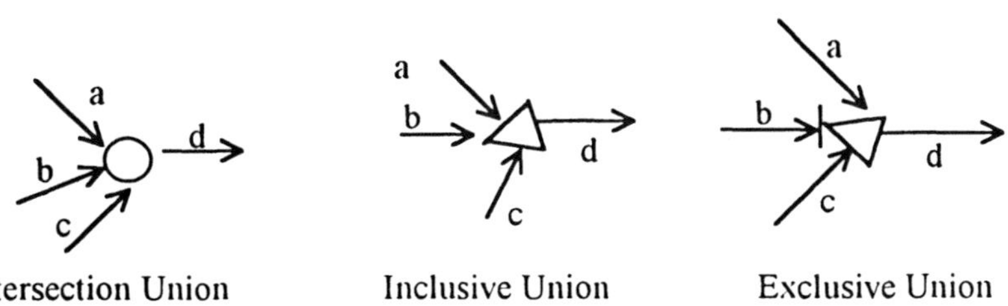


**FIGURE 2.8**
**EXAMPLES OF INTERSECTION, EXCLUSIVE UNION AND EXCLUSIVE UNION**

c) modelling of **overlapping** or **delayed** activities

The possibility of overlapping between two or more activities can be exemplified by two construction operations like building a wall (activity 1) and painting it (activity 2), where a minimal lag of time should be respected between the progress of 1 and of 2. A way of modelling this case can be shown by using:

- the precedence of type ③ with a minimal lag Δ between 1 and 2 means that the start of 2 should not occur before the start of 1 plus Δ;

- the precedence of type 2 between 1 and 2 with the same lag.

The concept of delayed activity (with a delay Δ) means that i can be just started after the completion of the precedent activity, j, plus Δ time units.

Therefore, the generalization concerning overlapping or delayed activities justifies an extension of the concept of precedence link: the direct precedence link with a finite duration, Δ.

Obviously, this relation also should be considered as a **dummy activity** without any cost or resources consumption.

## 3. Resources

In job-shop scheduling, the concept of activity concerns a well defined job which has to be performed using a set of specific resources. The available choices concern the scheduling and the selection of a technology which determines the required resources and the job duration.

The problems of Project Management have a different nature as each activity corresponds usually to a sub-project or, at least, to a large set of actions. In this context, even for the same technology, there is always a wide range of choices about the amounts of resources allocated to each activity and their allocation on time (**intensity of use**). Obviously, the duration of the activity depends directly on such decisions. Therefore, any project manager tends to consider the allocated resources as independent variables and the durations of the activities as the dependent variables.

Unfortunately, the study of Project Management has been influenced too much by the models of job shop scheduling which may explain why this distinction has not been clarified in literature. Therefore, in this book, the study of the resources or cost allocated to each activity precedes the study of the durations of the activities and of the network.

Any activity implies the use of resources which can be classified in terms of different criteria.

These criteria can describe the nature of the resources (capital, equipment, manpower, raw materials, etc.) or their functional features. According to such features, the resources can be classified into two major groups, **renewable and non-renewable resources. The non-renewable resources** such as capital, energy, concrete, bricks, etc. are consumed by the activity following the well known **principle of mass conservation** and a certain rate of consumption on time, r (t).

For instance, an activity i can require a constant consumption per time unit equal to $\alpha$ as is shown in Figure 2.9 for an activity with duration $D_i$.

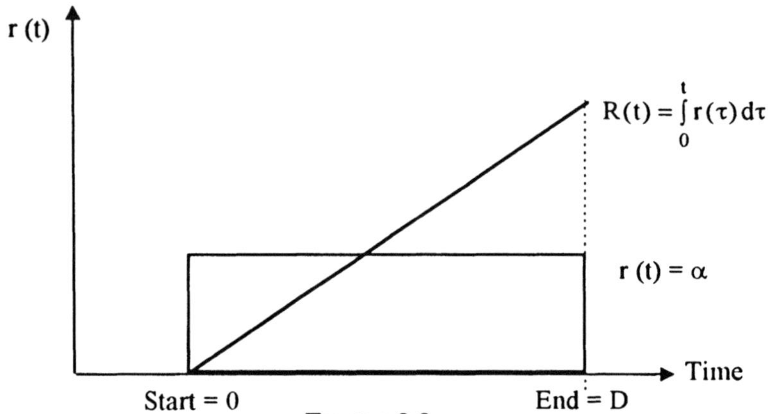


**FIGURE 2.9**
**EXAMPLE OF r(t) AND R(t)**

More often, this consumption rate is not constant but rather shows a peak near the middle duration as is shown in Figure 2.10.

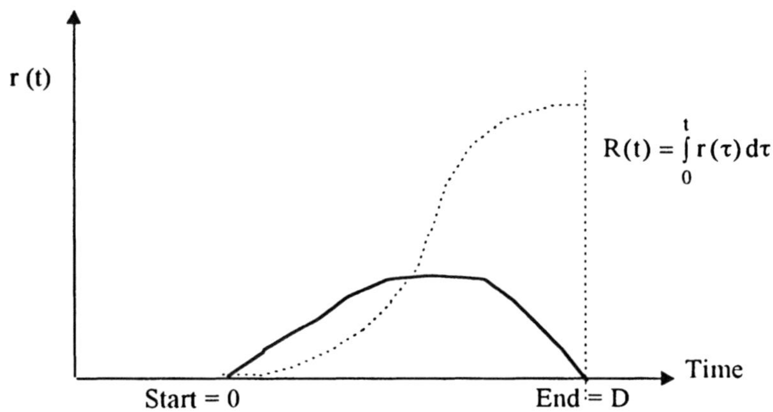


**FIGURE 2.10**
**EXAMPLE OF NON-UNIFORM r(t)**

The cumulative function R (t) can be easily computed from r (t) and it represents the required stock of this resource until time t.

Alternatively, the consumption of a non-renewable resource can be located on specific instants of time, for instance at the beginning of an activity as is the case of the initial payment due to a sub-contractor hired to carry out an activity (Figure 2.11):

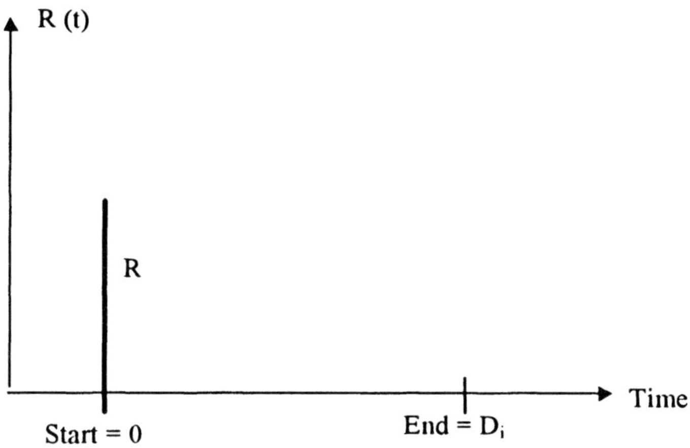


**FIGURE 2.11**
**EXAMPLE OF $R_i$**

Obviously this analysis can be generalized to the whole set of activities of a project and so the cumulative consumption of a non-renewable resource can be studied as is shown in Figure 2.12.

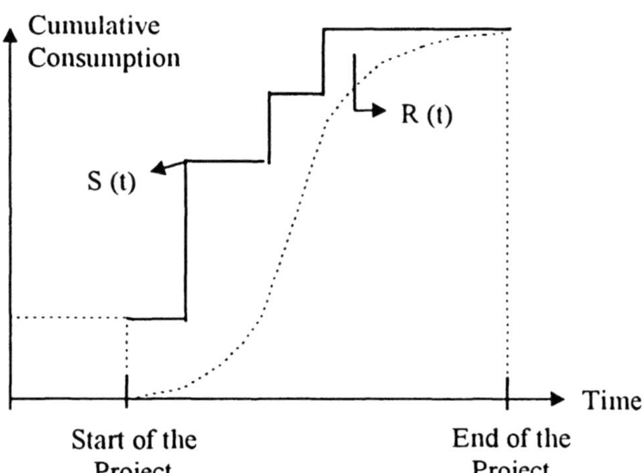


**FIGURE 2.12**
**EXAMPLE OF R(t) AND S(t)**

Therefore, the feasibility of this project implies a cumulative supply curve, S (t), above R (t) to avoid any delays due to a shortage of the resource, as is also shown in the previous Figure.

The management of this resource is based largely on the models developed for **Inventory Control** to optimize an overall function of the cost of orders and of the holding cost of stock.

The total cost per time unit, F, can be easily deduced assuming periodic orders, constant consumption rate, and that no shortage is acceptable (Fig. 2.13):

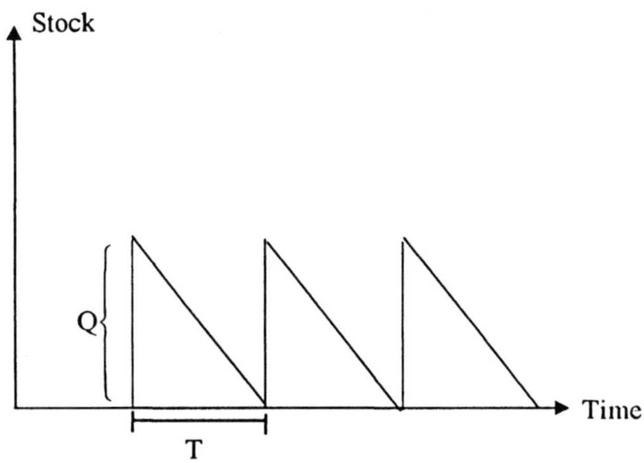


FIGURE 2. 13
STOCK IN TERMS OF TIME

$$F = \frac{1}{T}\left[K + cQ + \frac{h.r.T^2}{2}\right]$$

where T is the order period, (K + c Q) is the ordering cost, r is the constant consumption rate (Q= rT) and h is the holding cost per time unit and stocked unit.

The optimal solution is then given by:

$$Q^* = \sqrt{\frac{2Kr}{h}}$$

Often, r(t) is not constant along the development of a project and so this optimization can be adapted either sub-dividing the duration into sub-periods with an approximate constant consumption rate or developing a dynamic solution in terms of time. The latter is usually studied by simulation. The uncertainty concerning r(t) can be significant and then a random or stochastic model should be adopted. Usually, a safety stock is then adopted to reduce the probability of shortage.

The area of inventory control can play a crucial role to improve the management of most projects, although avoiding delays and reducing their costing is hardly mentioned by most books on Project Management.

The **renewable resources** describe a capacity or a maximal level of availability such as voltage, storage space, etc., but its use is not expressed by consumption rates.

This constraint of maximal availability, C(t), can be constant or time varying and it is described by the so called **capacity curve** (Fig. 2.14).

The inventory analysis is meaningless for this type of resources as they are not additive but the **efficiency** level of use can be computed by $\frac{A_1}{A_0}$ as the project has usually to pay for $A_0$ even if it is just using $A_1$, where $A_0$ is the area defined by C(t) and $A_1$ is the area defined by the availability required at each time.

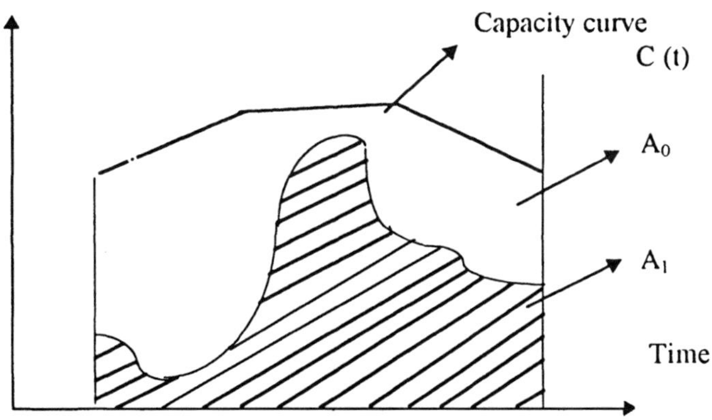


**FIGURE 2. 14**
**CAPACITY CURVE**

Therefore, the achievement of a uniform level of use of renewable resources is much more efficient than having a consumption curve with strong fluctuations. This objective is called "**resource levelling**".

There are also resources **doubly constrained** which means that there are restrictions concerning their accumulated function, R (t), and also their capacity constraint, C(t).

The most common example is **human resources** with R (t) restricted in terms of (man.day) of work and C (t) expressed by the maximal number of workers during the time unit t. Another example concerns energy as R (t) is defined by the consumption in (kwh) and C(t) is given by the installed power expressed in kw.

Actually, in theoretical terms, any non-renewable resource is also constrained by C(t) because the consumption or the use of any good cannot be made at an infinite rate, and so there is a **maximal availability constraint** (for instance, the use of $m^3$ of concrete cannot be done above the maximal flow of output of the adopted equipment). However, in most cases this restriction is less significant and just the cumulative function R (t) is studied (Non-renewable resource).

Nowadays, most economies are adopting (or pretending to adopt) the market model and therefore most (if not all!) requirements for a project are subject to trade which means that they can be bought if **capital** is available.

This explains why capital is being used more often as the resource synthetizing the whole spectrum of resources.

Capital is a non-renewable resource although in theory the expenditure per time unit also will be bounded by the feasibility of making proper deals.

Therefore, money can be considered as a typical example of a non-renewable resource and the **interest rate** can be considered as the holding cost of this good called **capital**.

The capital allocated to each activity is called its **cost** and it may be not associated to an **expenditure**. For instance, the use of a piece of equipment of the company in charge of the project to carry out an activity may not generate an expenditure but it certainly implies a cost. Similarly, any value generated by a project is a **benefit** but it may not be associated to an **income** (e.g., an improvement of environmental conditions).

In Economics, a value can be considered capital if more value can be generated from its use along time ("**capital reproduction**").

Therefore, now having 1 unit of capital is equivalent to having (1+$\alpha$) units in one year's time (measured in terms of the same monetary units), where $\alpha$ is the additional generated capital. This means that:

| Time t | | Time (t+1) |
|---|---|---|
| 1 | $\longrightarrow$ | $1+\alpha$ |
| f | $\longleftarrow$ | 1 |

and so having one unit on (t+1) is equivalent now to having $f = \frac{1}{1+\alpha}$ (f is called the **discounting factor** and $\alpha$ is the **discounting rate**). Therefore, one cannot add values of capital associated to different instants of time without using the discounting conversion. For instance, the present cost (or expenditure), PC, for t=0 of a time-series of values, C(t), with t=0, ..., H will be given by:

$$PC = \sum_{t=0}^{H} C(t)\, f^{t}$$

Similarly, the present value of a series of benefits, **present benefit,** (or of income present income), PB, generated by a project, B(t) with t=0, ....., H will be determined by:

$$PB = \sum_{t=0}^{H} B(t)\, f^{t}$$

Often, the **net present value**, NPV, of a project is considered a good indicator of its utility and so can be easily computed by:

$$NPV = \sum_{t=0}^{H} (B(t) - C(t))\, f^{t}.$$

Usually, the time-series of benefits is delayed much more than that of costs (crops come after sowing!) as is shown in Figure 2.15:

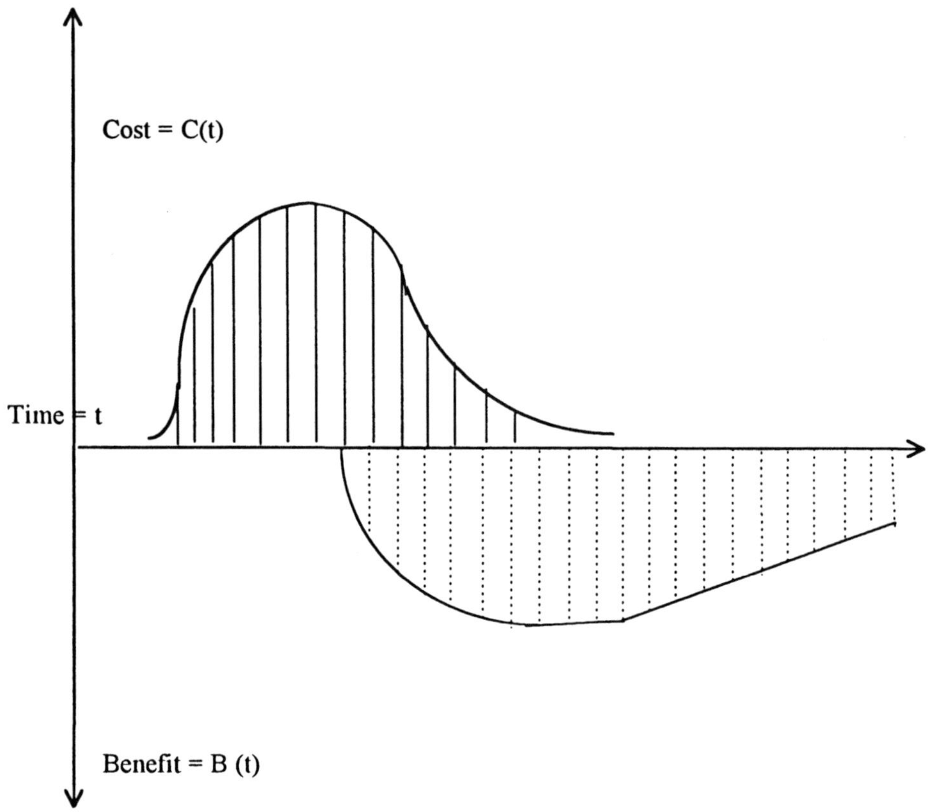


**FIGURE 2.15**
**AN EXAMPLE OF C(T) AND B(T)**

Other economic and financial indicators can be used to synthetize these two time-series:

a) IRR - Internal rate of return, $\alpha^o$, which is the minimal $\alpha$ for which NPV $\geq 0$.

b) $T_r$ - Return period which is defined by the minimal period, H, for which NPV$\geq$0.

c) RPB - Relative present benefit which is defined by:

$$RPB = \sum_{t=0}^{H} B(t)f^t \Big/ \sum_{t=0}^{H} C(t)f^t .$$

d) RNPV - Relative net present value which is defined by: $RNPV = \dfrac{NPV}{\sum_{t=0}^{H} C(t)f^t}$.

All these indicators are very sensitive to the selection of $\alpha$ which can be defined in terms of the value generated by one unit of capital allocated during one year on the best alternative project if this capital is owned by the owner of the project. This rate corresponds to the **marginal rate of return** for the studied economy and so $\alpha$ can be considered as an opportunity cost (how much was the value not generated during one year by that unit of capital which was spent on the project).

Alternatively, the owner of the project may need to borrow the required capital and then $\alpha$ will be defined in terms of the corresponding interest rate adopted for this type of project.

For instance, assuming $\alpha = 0.10$, the terms corresponding to $t > 25$ have a negligible impact on the present value of the project ( $f^{25} = 0.09$ ). This means that long term effects are poorly represented by financial analysis tending to ignore impacts above 20 or 25 years. This myopic nature is particularly serious if $\alpha$ is high. This tends to happen in more profitable economies because the opportunity cost there is much higher (less care is given to long term impacts within richer economies!).

Thus, the assessment or evaluation of projects should be carried out within a more comprehensive and consistent framework, the financial perspective being just one dimension of such analysis. This topic is treated in Chapter VII.

A very convenient strategy to increase the NPV of a project can be based on the anticipation of benefits and on the delay of costs. This delay can be achieved by post-poning more expensive activities until their latest starting times and the anticipation of benefits can be carried out by sub-dividing the project into several phases if benefits can be generated after each phase (**Project phasing**). This strategy tends to delay activities and then all activities will start at their latest starting times which means that all of them will be critical.

Unfortunately, other criteria are deteriorated by this approach as is the case of the risk of not completing the project on schedule.

Thus, a compromise between NPV, risk and levelling of resources has to be pursued (Chapter 7).

The presented analysis has assumed constant monetary units (**constant prices**). However the monetary units are permanently subject to change due to **inflation** (or **deflation** ) dynamics, but the equivalence between the study in terms of **current prices** and the constant prices analysis can be easily deduced:

$$\begin{array}{lll} \text{1 unit at time} & t \longleftrightarrow & (1+d) \text{ at time } (t+1) \\ & g \longleftrightarrow & 1 \end{array}$$

where d is the devaluation rate assuming that 1 unit now (t) can purchase the same as (1+d) on time (t+1). The correction factor, g, is given by $\frac{1}{1+d}$ and therefore current prices easily can be transformed into constant prices:

$$1 \longleftrightarrow \frac{1+\alpha}{1+d}$$

$$f \longleftrightarrow 1$$

where $f = \frac{1+d}{1+\alpha}$, f being the discount factor computed in terms of current prices and corrected for constant prices.

The discount rate at constant prices, α', can be determined in terms of α (current prices):

$$1+\alpha' = \frac{1+\alpha}{1+d} \quad \text{or} \quad \alpha' = \frac{\alpha - d}{1+d}$$

showing that the common idea of getting the corrected constant by subtracting the inflation from the current rate is not exact because that subtraction has to be divided by (1+d).

For instance, with α= 0.10 and d= 0.05, one should have α'= 0.0476 instead of α'= 0.0500.

The assessment and planning of a project should be carried out using constant prices. However, the study of cash flows may be done in current prices to describe the forecasted evolution and then such flows should be converted into constant prices before any further studies.

## *4. Duration*

The two attributes of the set $\{B_1\ (i), \ldots, B_p\ (i)\}$ which have received most of the attention are the **cost** and the **duration** of each activity, C(i) and D(i), as they allow the study of the two major criteria: **total duration of the project**, T and its **total cost**.

The costs have already been studied in the last sub-chapter.

The study of the total duration in terms of $D = \{D(i): i = 1,\ldots,N\}$ has been developed since the early papers of Project Management (See, e. g., Kelly, 1961 and Clark, 1962).

The simplest case assumes that D is deterministically known and then T can be computed by the **Critical Path Method**:

a) The **earliest start time** of any activity, $A_i$, is given by

$$t_i = \max_{j \in J_i} \left[ t_j + D(j) \right]$$

assuming that $t_i = 0$ for activities with $J_i = \emptyset$.

b) The minimal T can be determined by:

$$T^* = \max_i \{ t_i + D_i \}$$

c) The **latest start time** of any activity, $A_i$, compatible with $T^*$ is given by:

$$t_i' = \min_{k \in K_i} \left\{ t_k' - D(k) \right\}$$

making

$$t_i' = T - D(i)$$

for any activity $A_i$ with $K(i) = \emptyset$.

This method also allows the identification of the **critical activities**. A **critical activity** is an activity such that any increase of its duration implies an increase of the total duration of the project. This implies that

$$t'_i = t_i$$

and the set of these activities is called the **critical path**.

In Figure 2.16 and 2.17, the critical path is shown for the presented network in Fig. 2.3 using AoA and AoN and assuming the durations given in Table 2.1 .

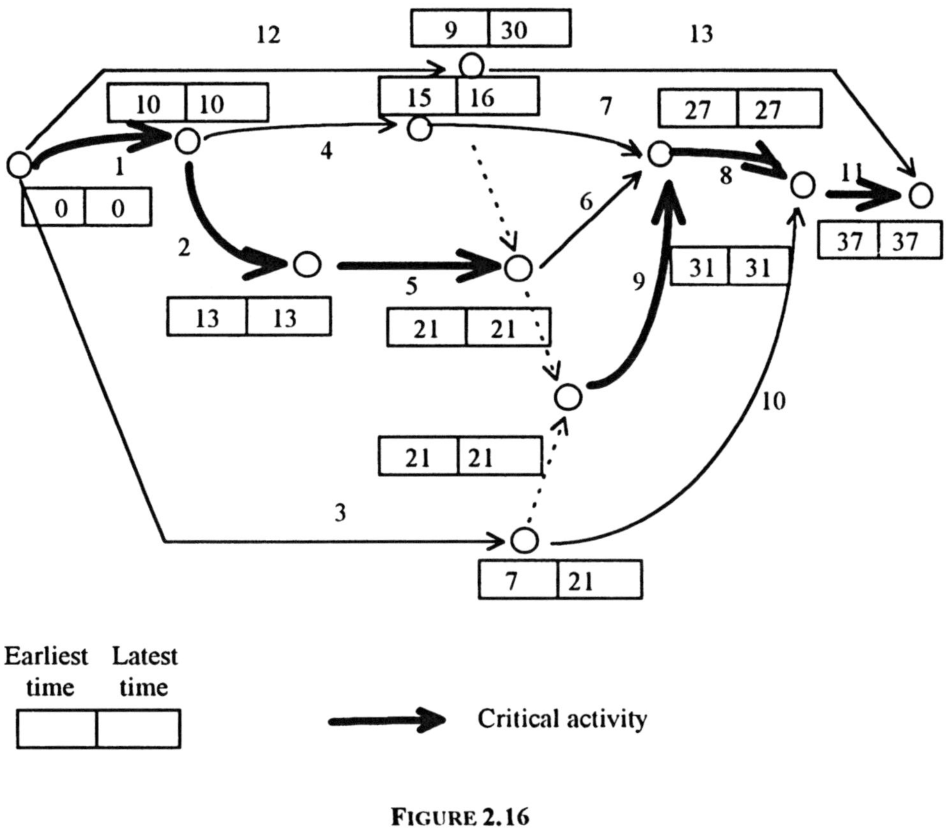


**FIGURE 2.16**
**EARLIEST AND LATEST OCCURRENCE TIME OF THE NODES OF THE NETWORK PRESENTED IN FIG. 2.3 (AOA)**

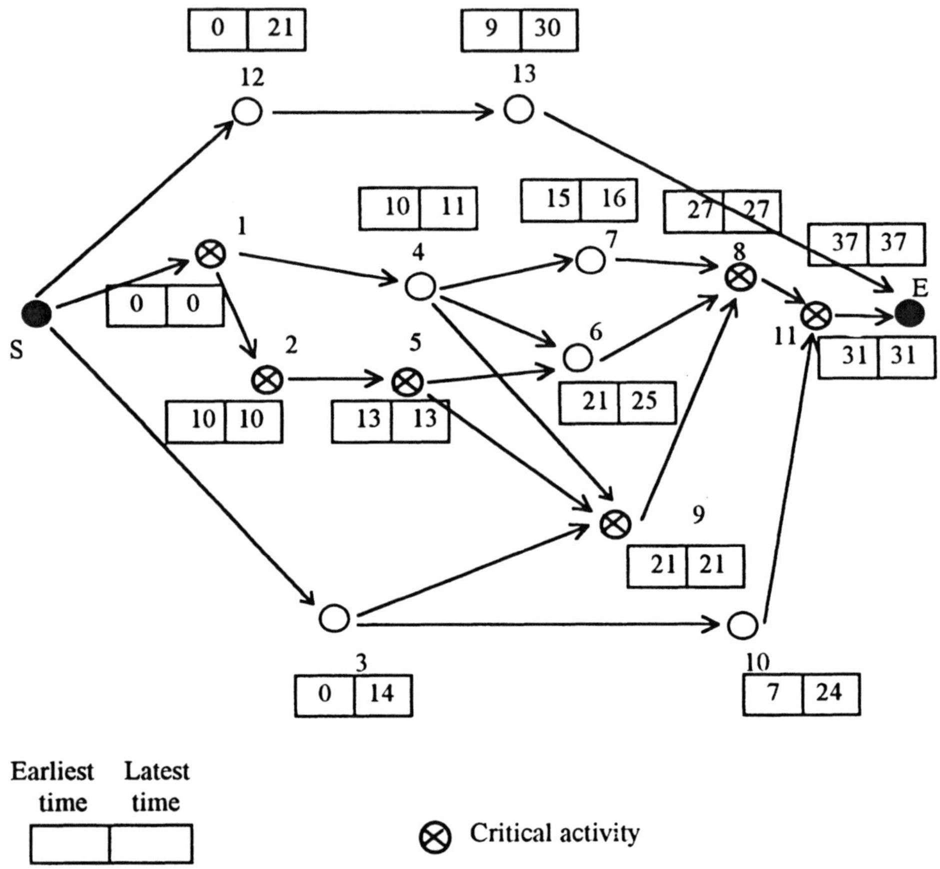


**FIGURE 2.17**
**EARLIEST AND LATEST STARTING TIMES OF THE ACTIVITIES OF THE NETWORK PRESENTED IN FIG. 2.3 (AON)**

**Table 2.1**

| Activity | Duration |
|---|---|
| 1 | 10 |
| 2 | 3 |
| 3 | 7 |
| 4 | 5 |
| 5 | 8 |
| 6 | 2 |
| 7 | 11 |
| 8 | 4 |
| 9 | 6 |
| 10 | 7 |
| 11 | 6 |
| 12 | 9 |
| 13 | 7 |

Obviously, the definition of a critical activity also can be given in terms of the earliest and latest times of the nodes using AoA:

$A_i$ is critical if:

a) $t_k = t'_k$ and $t_m = t'_m$, k being the node where $A_i$ starts and m the node where $A_i$ ends.
b) $D_i = (t_m - t_k)$

It should be noted that the condition a) is just a necessary condition.

The difference $S_k = (t'_k - t_k)$ is called the **node float** (or **slack**) and four other definitions of float can be given for a generic activity $A_i$ connecting two generic nodes, k, m (Fig. 2.18):

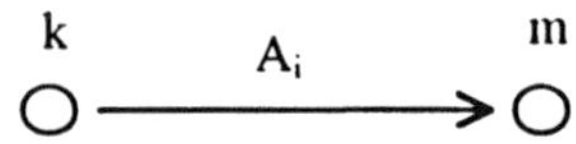


**FIGURE 2.18**
**EXAMPLE OF FLOATS**

$$S(k,m)_M = t'_m - t_k - D_i$$
$$S(k,m)_L = t'_m - t'_k - D_i$$
$$S(k,m)_E = t_m - t_k - D_i$$
$$S(k,m)_F = \max[t_m - t'_k - D_i;0]$$

designated by **Total Float** (actually, the correct name should be **Maximal Float**), **Safety Float, Free Float** and **Interfered Float**. S (k, m)$_M$ corresponds to a maximal float as it assumes that k occurs as early as possible and m as late as possible. The meaning of S (k, m)$_L$ and S (k, M)$_E$ is also obvious as the former corresponds to the latest times and the latter to the earliest times of occurrence. The value of S(k, m)$_F$ can be zero as an earliest calendar for m and a latest one for k may be infeasible.

If there are several activities directly connecting k and m, (Fig. 2.19),

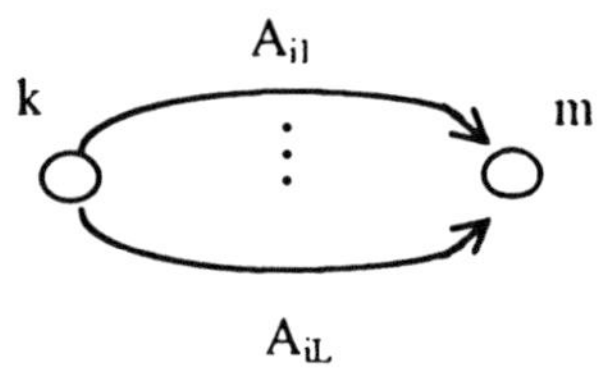


**FIGURE 2.19**
**EXAMPLE OF MULTIPLE ACTIVITIES BETWEEN A PAIR OF NODES**

the previous definitions can still be applied substituting $A_i$ by $\left[\max_{i=i_1,\dots,i_L} A_i\right]$ where $A_{i1}$, ..., $A_{iL}$ are the activities connecting k to m.

The planning and management of a project described by a project network with $B = \{B_1(i) = D_i\}$ implies setting up a calendar for the project which corresponds to define $\{t_i : i = 1,...,N\}$.

This calendar is called the **project schedule** and this definition is called **project scheduling**.

There is no degree of freedom to select $t_i$ for $i \in CP$ (**critical path**) if T should be equal to its minimal, T*, but the schedule of any other activity has to be selected within its earliest and latest bounds. The latest bounds can be determined just in terms of T* but the earliest bounds have to be computed sequentially using the basic equation:

$$\min t_i^s = \max_{j \in J_i} \left( t_j^s + D_j \right)$$

and

$$\min t_i^s \le t_i^s \le t'_i$$

where $t_i^s$ is the adopted starting time of i in this schedule.

This means that if later times are adopted to start activities in the beginning of the project, most of the floats then can be "consumed" and less flexibility exists to schedule later activities.

The extreme case of scheduling all activities at their earliest starting times is called **Earliest Schedule** (ES) and the opposite one is named **Latest Schedule** (LS).

Of course, the set of critical activities is a minimal set with ES but includes all the activities of the network if LS is adopted.

The critical path will be empty if T > T* and if ES is adopted.

For each T, the comparison between different schedules should be done in terms of multiple criteria based on cost, risk, etc.

The concept of **criticality** can be extended to the definition of a **Δ - critical activity** which is an activity that just generates an increase of the total duration of the project if its duration is increased by more than Δ time units. However, this definition depends on the adopted schedule and usually the definition of $\Delta$ - critical activity is done assuming ES.

A software was developed to model the project schedule - RISKNET modules 1 and 2 - which is presented in Annex 1.

Often, the duration of the activities is not known deterministically and therefore the modelling of their durations in terms of random variables has been developed since the early papers on Project Management.

The general approach to treat this problem has assumed that every time a project is carried out, the duration of each activity is assumed to be sampled randomly from a given distribution.

Even assuming the statistical independence between the duration of different activities, the study of the project's total duration, T, is a very difficult problem because the cumulative distribution function of T, $F_T$ (X) is given by:

$$F_T(X) = \int_{D_1,\ldots,D_L} \underset{k}{MAX} = \left[ D_{L_k}(D_1,\ldots,D_N) \right] \le X \Big\} f(D_1)\cdot f(D_2)\ldots f(D_N) \, dD_1 \ldots dD_N$$

where $L_k$ with k=1,..., K describes all paths from the start until the end of the project, $D_{L_k}(D_1,\ldots,D_N)$ is the duration of $L_k$ given the duration of the activities, $D_i$, and f ($D_i$) is its probability density function, with i=1, ..., N.

The method PERT has simplified drastically this problem substituting $\underset{k}{\mathrm{Max}}\left[D_{L_k}(D_1,\ldots,D_N)\right]$ by $D_L(D_1,\ldots,D_N)$ where L is the Critical Path obtained by assuming that the duration of each activity is deterministically equal to its mean, $\mu_i$. Then $D_L = \sum_{i \in L} D_i$ and so $\mu(D_L) = \sum_{i \in L} \mu_i, \sigma^2(D_L) = \sum_{i \in L} \sigma_i^2$ .

The distribution of $D_L$ can be usually assumed to be gaussian either because $D_i$ is normal or because of the number of activities included in L is sufficiently large **(Theorem of Central Limit)**.

Obviously, $D_L(D_1,...,D_N) \leq \underset{k}{Max}\left[D_k(D_1,...,D_N)\right]$ and therefore there is a systematic error of underestimation of T due to the PERT assumption.

However, the use of this method is extremely simple and the produced results can provide interesting lower bounds for T.

## 5. Optimal scheduling

Any activity, i, can be done faster or slower depending on the technology and the resources allocated to i which means that its cost, C (i), should be considered as a function of its duration, D (i), C (i) = f (D(i)).

Usually, this relationship follows the shape presented in Fig. 2.20.

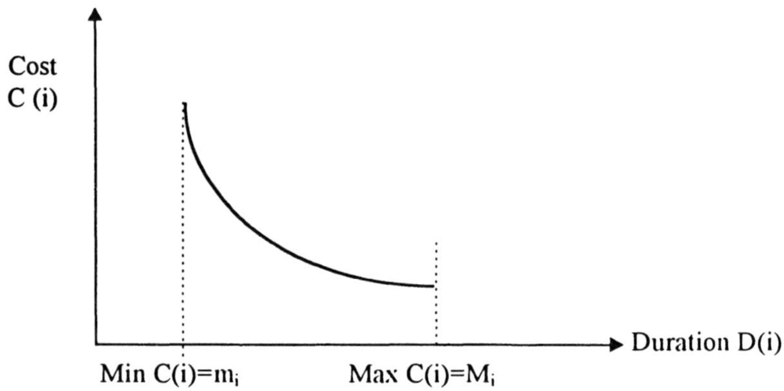


**FIGURE 2.20**
**C(I) IN TERMS OF D(I)**

which can be adjusted by a piece-wise linear function.

Traditional methods of **Project Management** consider that the durations of the activities are given and that the first analysis to be carried out should be to schedule the project in terms of such durations. The study of the required resources is then presented as a second analysis to complement the previous one.

However, in real life the sequence may be the opposite one as the project manager should have *a priori* information about resources and so the durations of the activities will be selected in terms of the acceptable levels of resource consumption using a relationship of the type presented in the previous figure. This means that the durations are not "data" but a consequence of decisions also considering the resource constraints.

Unfortunately, this more realistic formulation requires analytic methods much more complex than the standard algorithms.

However, a very simple and well known method allows the minimization of the total cost of a project, $C = \sum C_i(D_i)$, obtaining a total duration not higher than a maximal bound, $L_T$, the **Crash Duration Method**. This method is based on the following iterative procedure:

1 - Schedule the project with $D(i) = M_i$. If the total duration, T, is lower than $L_T$, this is the optimal schedule. If not, go to the next step.

2 - Identify the activity k which belongs to the critical path and with the lowest $f_k = \left|\frac{\Delta C(k)}{\Delta D(k)}\right|$. Then reduce D(k) until:

ⓐ- $L_T$ is achieved or
ⓑ- k leaves the critical path or
ⓒ- $f_k$ is no longer the minimum.

Whenever ⓑ or ⓒ happens one should go back to step 2.

This procedure is repeated until ⓐ occurs.

It should be noted that at each step, the reduction of T may imply the reduction of the duration of more than one activity (e. g., two parallel activities belonging to the critical path) and then the cost should include the sum of the increase of the cost of each of these activities.

Obviously, a similar algorithm can be proposed to minimize the total duration under the restriction of not exceeding a maximal cost, $L_C$:

1 - Schedule the project with $D(i)=m_i$. If the total cost, C, is lower than $L_C$ this is the optimal schedule. If not, go to the next step.

2 - Identify the activity k which belongs to the critical path and with the lowest $f_k = \left|\frac{\Delta C(k)}{\Delta D(k)}\right|$. Then increase D(k) until:

(a)- $L_C$ is achieved or
(b)- k leaves the critical path or
(c)- $f_k$ is no longer the minimum.

Whenever b or c happens one should go back to 2 until (a) occurs.

An equivalent problem can be formulated if it is assumed that for each activity a set of alternatives discrete modes **[Multi-mode assumption]** is available instead of a continuous function C(i) = f (D(i)):

$$[C(i), D(i)]_k \text{ with } k=1, \ldots, K(i).$$

The general problem of project scheduling now can be formulated as a decision making problem with:

a) $t_i^s$ and k (i) for i = 1, ..., N and k (i) = 1, ..., K (i) as decision variables.
b) T and C as major criteria for minimization.

This means that the major decisions of project scheduling concern:

c) the selection of resources to each activity $\{ t_i^s \}$;
d) the allocation of resources to each activity $\{C_i\}$.

The first problem corresponds to the ordering of N objects (starting times) and the second one to an allocation problem to distribute a total cost, C, by N components.

A very long list of heuristics has been developed to determine its optimal solution as will be studied in Chapter 6.

# 3

# STRUCTURAL MODELLING OF PROJECTS NETWORKS

## *1. Hardness and complexity*

The project network describes the restrictions of sequentiality to be respected and it represents one of the major problems faced by project managers. Obviously, there are networks much more easily managed than others and the manager's understanding of the difficulties and the pitfalls of each project network can be a very useful contribution to its management.

This is such a key issue in Project Management that one could expect that a large number of scientific results would have been published to support this assessment. Unfortunately the reality is quite the opposite as this problem has been hardly studied.

This situation stems from the lack of results to describe the morphology of the network and from the confusion between the difficulty of managing a project and the complexity of applying specific algorithms (like those oriented to optimize the project schedule). Actually, this is why two different concepts are proposed - -**project hardness** and **project complexity**.

**Project hardness** measures the difficulty of managing a project in terms of its project schedule.

**Project complexity** for a specific problem measures the difficulty to solve that problem in terms of its project schedule.

The difficulty of managing a project tends to grow with the restrictions imposed upon the manager assigned to run its activities.

This reduction of freedom concerns particularly two types of decisions already pointed out: the selection of $\left\{t_i^s\right\}$ and of $\left\{C_i\right\}$.

Obviously, the simplest case corresponds to the maximal freedom to select these decisions and it is the case of the **parallel - network** (Fig. 3.1):

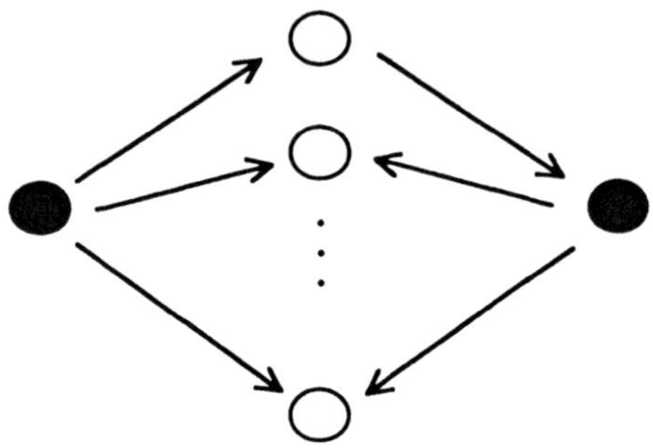

**FIGURE 3.1**
**EXAMPLE OF THE PARALLEL - NETWORK**

The worst case corresponds to the "**series network**" (Fig. 3.2):

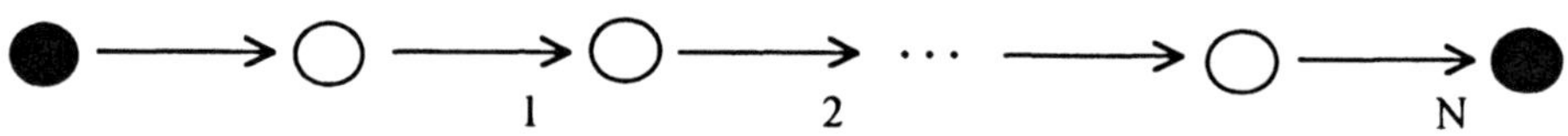


**FIGURE 3.2**
**EXAMPLE OF THE SERIES NETWORK**

as in this case $t_i^s \geq t_{i-1}^s + D(i-1)$ with $i = 2,\dots,N$.

It should be noted that the parallel case is also the easiest one because any implementation problem concerning one activity does not affect the scheduling of the other activities and the serial case is the other extreme case.

This means that an indicator to measure this type of difficulty (**project hardness**) can be proposed:

$$H = \frac{1}{N}\sum_{i=1}^{N} N(i)$$

where N(i) is the total number of activities preceding i.

Then, H represents the average number of precedent activities for each activity of the project network.

Obviously, H will be zero for the parallel - network and its maximal value is equal to $\frac{N-1}{2}$ which corresponds to the series - network. Therefore, an adimensional indicator can be proposed ranging from 0 to 1 (**Hardness indicator**):

$$h = 2H / (N - 1)$$

The concept of complexity has been discussed but its definition depends on the specific problem under study. Two major problems have been considered:

a) the study of the distribution of the total duration assuming that each activity has a random duration (Problem 1).

b) the optimal scheduling of the activities (Problem 2).

A few suggestions have been given:

| | |
|---|---|
| A/N | (Pascoe, 1966) |
| 2 (A-N+1) / (N-1) (N-2) | (Davies, 1974) |
| $A^2/N$ | (Kaimann, 1974) |

where A is the number of arcs and N the number of nodes of the network. However, it has been shown that there is usually a low correlation between these indicators and the level of complexity "expressed by the relative prolongation of the project-durations due to limited availability of resources" (Elmaghraby and Herroelen, 1980).

Elmaghraby and Herroelen, (1980) have discussed this topic but they have not clarified the distinction between hardness and complexity. These authors studied the concept of complexity in terms of the computation of the critical path (Problem O) and of the two presented problems (Problems 1 and 2). They conclude that the complexity for Problems 0, 1 and 2 should be defined, respectively, in terms of

- $(A-e_1)$; (A-N+1) where $e_1$ is the number of arcs out of the initial node;

- m, v, q which are the number of required multiplications, convolutions and integrations;

- S which is the number of alternative sequences of activities, but they do not propose any practical indicators. Actually, they conclude that the complexity function for each case has to be estimated for each specific network. This conclusion has a very reduced practical utility, also, due to the difficulty of computing m, v, q and S.

However, several general considerations can be made concerning the degree of complexity. This degree tends to increase with:

- the difficulty of estimating objective-functions or criteria for each studied solutions. Usually, these functions are computed for each path and this difficulty increases with the number of paths and, particularly, with the number of interdependent paths.

  A classical example is the total duration for a network with activities with random durations;

- the number of alternative solutions

  The search to reach the optimum is much longer if the domain of search is larger. Therefore, this perspective can be responsible for a negative correlation between complexity and hardness as more freedom can help the manager but it also can require a longer optimization process. This means that **a parallel network is less hard but it can be more complex than a series-one to optimize.**

- the possibility of decomposing the decisional problem into sub-problems. This is a general strategy adopted in any multi-variable optimization problems as it reduces the size of the problem. For instance, assuming that one has N decisional variables (e.g., starting times of activities or resources allocated to the activities) and that each variable has K alternative values, the maximal total number of solutions to be considered is $K^N$. However, if a decomposition into M independent sub-problems is possible, one just has $\sum_{m=1}^{M} K^{N_m}$ with $N = \sum_{m=1}^{M} N_m$, which is usually much smaller than $K^N$.

This decomposition is easier to be carried out if the network can be considered as a sequence of sub-networks because Dynamic Programming then can be used or, in more general terms, a decomposition technique can be successfully applied. These techniques will be studied in next section.

The accurate estimation of the degree of complexity then can be carried out by simulation. The simulation of networks is becoming a central issue in Project

Management and this is why a specific chapter of this book is devoted to this methodology (Chapter 4).

## 2. Modular decomposition

Modular decomposition has been developed to subdivide the project network into subsets of activities named **modules** (Muller and Spinrad, 1989), which can be substituted by a "macro-activity".

Two examples are given in Figure 3.4 to illustrate this concept using AoN. In the first network, each sub-set, ⓐ or ⓑ can be easily substituted by a macro-activity and a second level decomposition allows the aggregation of [5; 6] and of [9; 10].

**Network I**

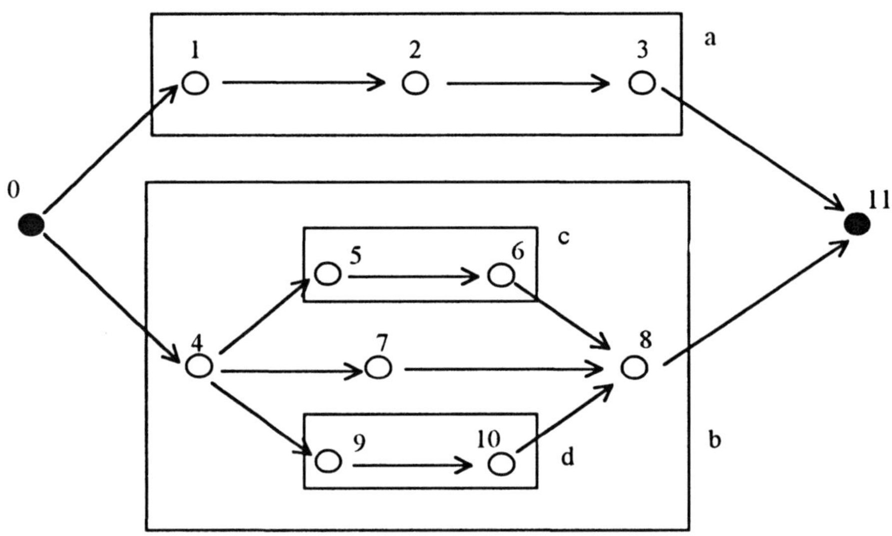


FIGURE 3.4 (a)
EXAMPLES OF MODULAR DECOMPOSITION
(NETWORK I)

**Network II**

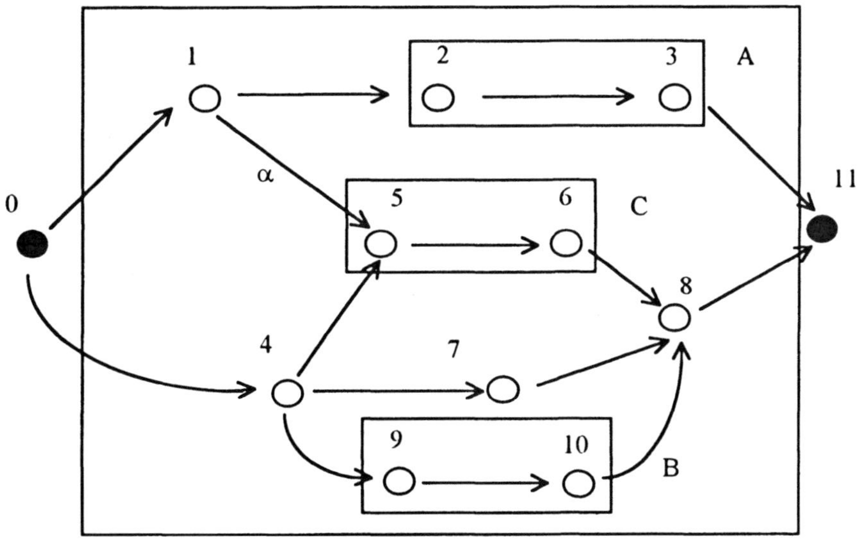


**FIGURE 3.4 (b)**
**EXAMPLES OF MODULAR DECOMPOSITION**
**(NETWORK II)**

However, the introduction of an additional precedence link (α) shown in the next network destroys the "independence" between a and b. Therefore, the decomposition will allow just at this stage the substitution of A, B and C by macro-activities.

In each module, the starting ($S_M$) and the end nodes ($E_M$) have the usual definition:

$$J_{S_M}(M) = \emptyset$$
$$K_{E_M}(M) = \emptyset$$

where J. (M) or K. (M) represents the set of predecessors or successors of . belonging to M.

For instance, in the first network of Fig. 3.4 the node $S_M$ and $E_M$ for a (or b) are 1 and 3 (or 4 and 8).

Obviously, if there are more than one node satisfying the starting (or finishing) condition, an additional node can be introduced as the starting or the finishing node (Fig. 3.5).

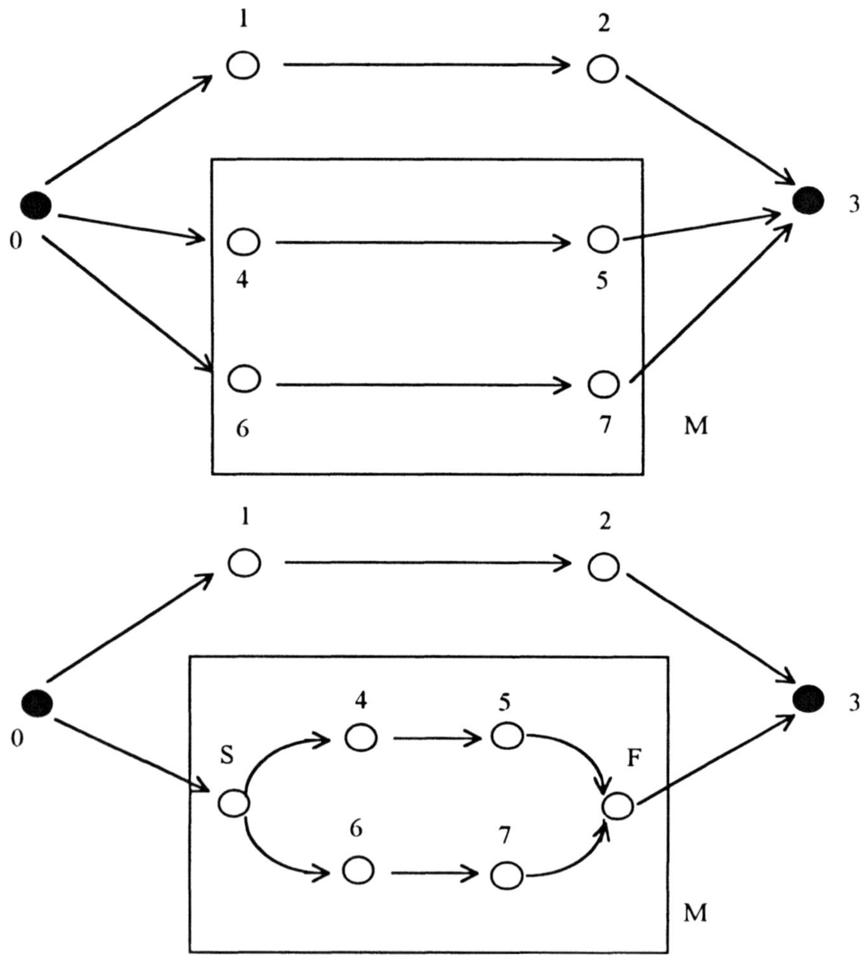


**FIGURE 3.5**
**INTRODUCTION OF A STARTING AND FINISHING NODE ON MODULE M**

In mathematical terms, the concept of module can be defined by:

M is a module if and only if for any activities, i, k ∈ M, one has

$$J_i(\overline{M}) = J_k(\overline{M})$$

and

$$K_i(\overline{M}) = K_k(\overline{M})$$

where $\overline{M}$ represents the set of nodes of the network not belonging to M.

This definition corresponds to say that any node of M has the same set of predecessors or successors outside M.

This means that any precedence link flowing into M flows to $S_M$ and any precedence link flowing from M comes from $E_M$.

The operation of decomposition also can be successfully applied to any module and individual nodes are the simplest modules to be defined. Therefore, the two networks of Fig. 3.4 can be decomposed according to the trees presented in Fig. 3.6 and obtaining the modules included in Fig. 3.7.

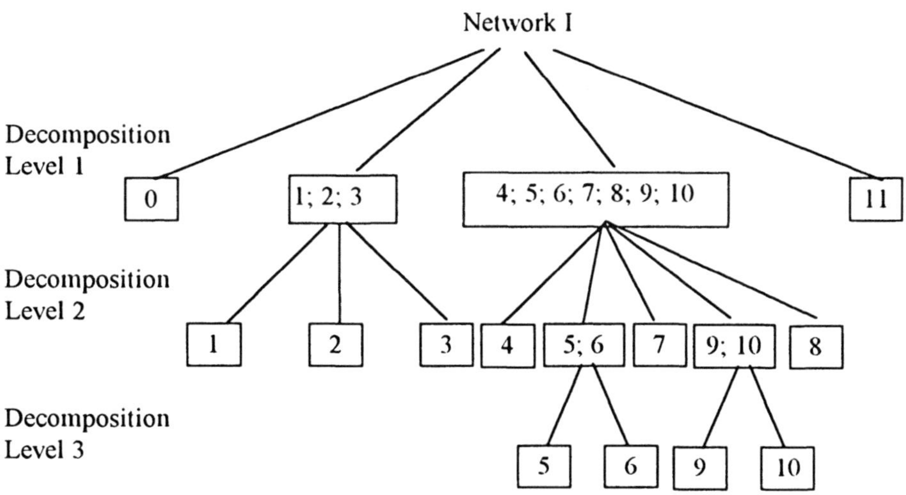


**FIGURE 3.6 (a)**
**HIERARCHICAL DECOMPOSITION NETWORKS I AND II**
**(NETWORK I)**

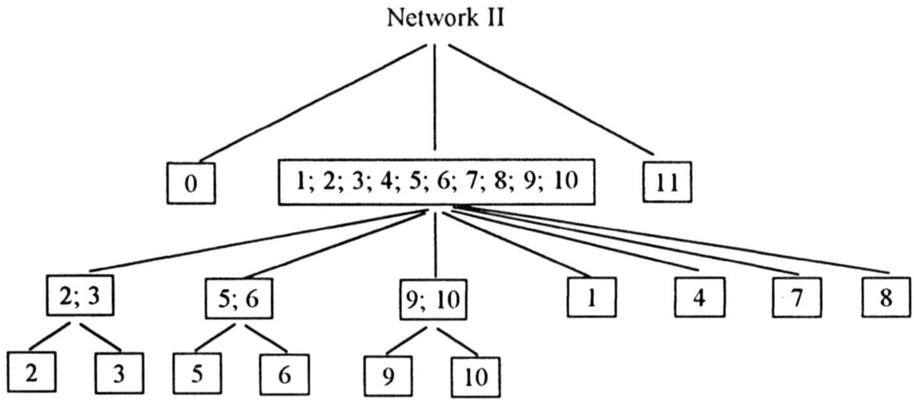


**FIGURE 3.6 (b)**
**HIERARCHICAL DECOMPOSITION OF NETWORKS I AND II**
**(NETWORK II)**

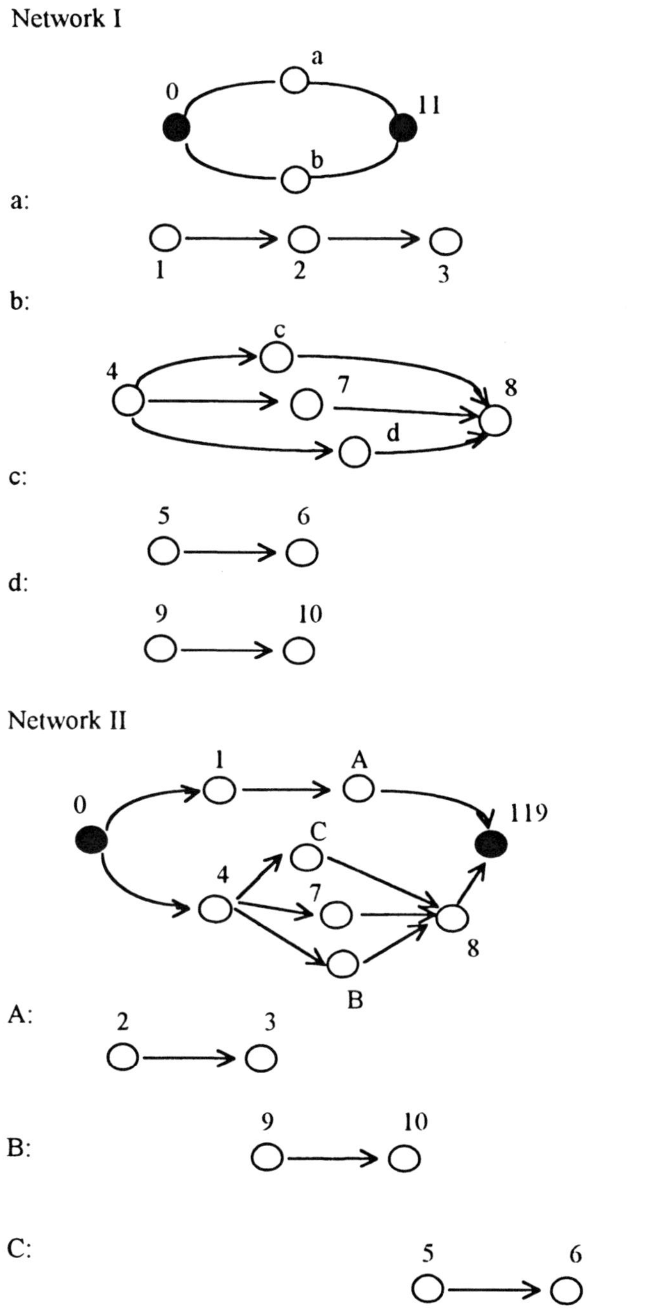


**FIGURE 3.7**
**MODULES OF NETWORKS I AND II**

Several algorithms have been proposed to decompose a network into modules until reducing the network to simple nodes (see Muller and Spinrad, 1989) and the computational time is proportional to $N^2$ which is quite convenient.

Obviously, the purpose of analysing and understanding the network is helped by the identification of macro-activities which implies that this method is less useful if the achieved decomposition does not identify such modules.

An example is given in Fig. 3.8 where, after excluding from the network the starting and the end nodes (0 and 6), one can just identify individual nodes as modules within A.

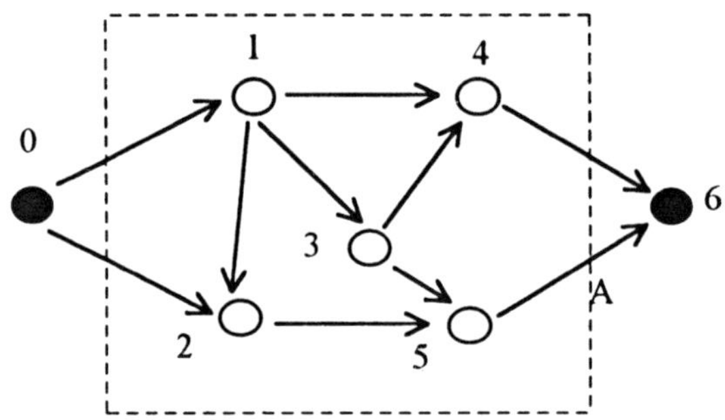


**FIGURE 3.8**
**AN EXAMPLE OF DECOMPOSITION**

Fortunately, this is not the general case and so this approach can help to apply the hierarchical definition of activities presented in 2.1 and also to simplify any network under study.

## 3. Network reduction

Another approach to decomposition has been pursued by Bein (Bein, W. W., Kamburowski, J. E. and M. M. Stallmann, 1992) based on the concept of reduction. They adopt the following classification of project networks:

**A - Series-parallel networks**

These networks are directed acyclic ones and they don't contain any network of the type presented in Fig.3.9, called **interdictive network**.

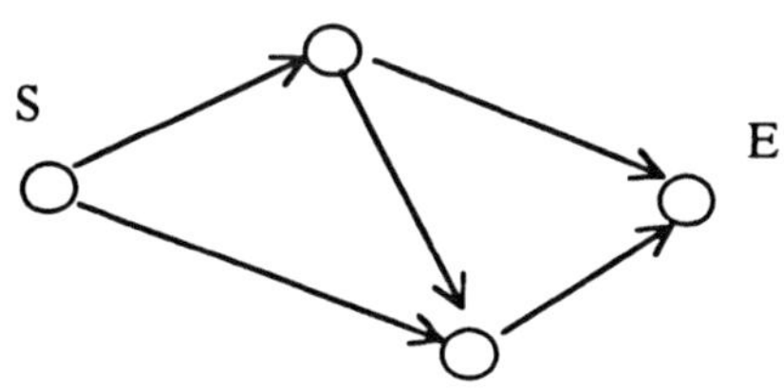


FIGURE 3.9
THE INTERDICTIVE NETWORK

This means that the network can be considered as a set of series and/or of parallel networks which were already discussed.

**B - Non series-parallel network**

All the other networks.

The construction of a series-parallel network can be iteratively obtained by starting with the most elementary network including just one arc and the S and E nodes. Obviously, this network is a series-parallel one. Then, two types of **composition** can be done:

a) **parallel composition**

Let the network $\Omega_1$, be a series-parallel one with ($S_1$, $E_1$) and $\Omega_2$ another series--parallel network with ($S_2$, $E_2$). Then, the network composing $\Omega_1$ and $\Omega_2$ by making $S_1 = S_2$ and $F_1 = F_2$ is also a series-parallel network.

b) **series composition**

Let the network $\Omega_1$, be a series-parallel one with ($S_1$, $E_1$) and $\Omega_2$ another series-parallel network with ($S_2$, $E_2$). Then, the new network composing $\Omega_1$ and $\Omega_2$ with $F_1 = S_2$ is also a series-parallel network.

Similarly, any given network $\Omega$ can be reduced to a series-parallel network by adopting three alternative types of **reduction**:

a) **parallel reduction**

Two or more arcs connecting two nodes (Fig. 3.10) can be substituted by one equivalent arc.

b) **series reduction**

If the node v has I (v) = O (v) = 1 and being u, w, the adjacent nodes (Fig.3.10), then (u, v, w) can be reduced to (u, w) where the arc (u, w) will be equivalent to the arcs between (u, v) and between (v, w).

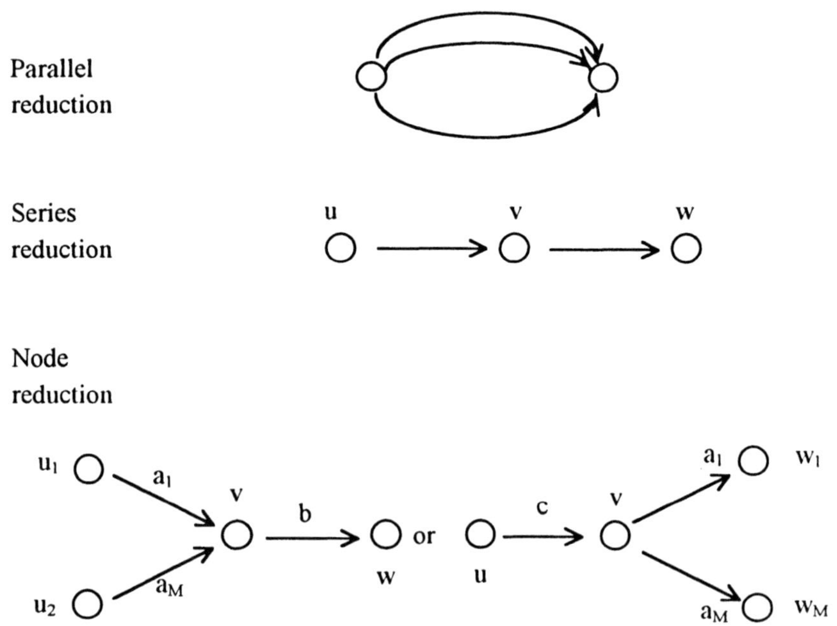


**FIGURE 3.10**
**THE THREE TYPES OF REDUCTION**

c) **node reduction**

This reduction is just feasible if O(v) or I(v) is equal to 1 (Fig.3.10). This node can be suppressed by creating a set of arcs $\{a'_1, ..., a'_M\}$ being $a'_m$ equivalent to $(a_m, b)$ or $(c, a_m)$ with $m = 1, ..., M$.

It should be noted that the reduction a) and b) use sub-networks which are **modules** of the studied network. However, this is not the case c).

Actually, these sub-networks are not necessarily modules due to several possible reasons:

- $w_m$ can also receive other arcs besides $\{a_1, ..., a_M\}$;
- other arcs besides $\{a_1, ..., a_M\}$ can also leave from $\{u_1, ..., u_M\}$;
- w can receive other arcs besides b;
- other arcs can besides c can start from u.

Furthermore, the new arcs $\{a'_1, ...a'_M\}$ are not independent as there is a common arc, (v, w) or (u, v).

The processes of reduction a) and b) are equivalent ways of describing the network but the process c) is really changing the network as it is substituting it by a simpler one.

Obviously, a series-parallel network can be aggregated until becoming just one arc between S and E without any node reductions. However, this is not the case of the other networks and therefore an indicator can be proposed to assess such complexity of reduction:

The **reduction complexity** of a network is the minimal number of **node reductions** (besides the series-parallel reductions) to reduce it to a single arc between S and E.

Obviously, this indicator is zero for a series-parallel network. This concept can be illustrated for two networks showing that this indicator is equal to 2 and 3 for $\Omega_1$ and $\Omega_2$, respectively (Fig. 3.11 a and b).

$\Omega_1$:

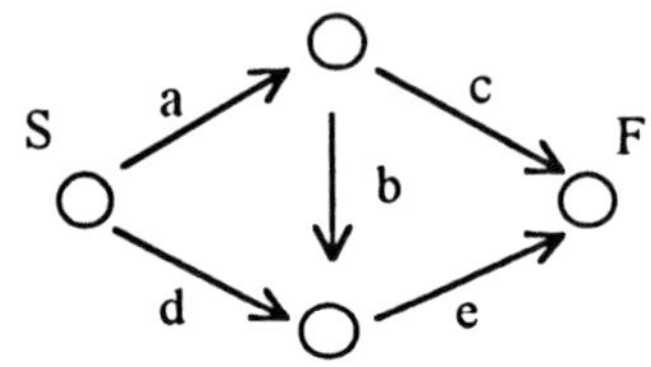


Node reduction 1:

1)

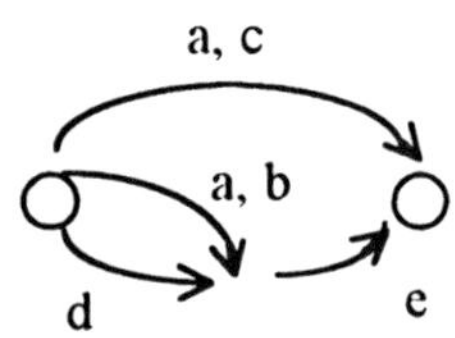


Node reduction 2:

2)

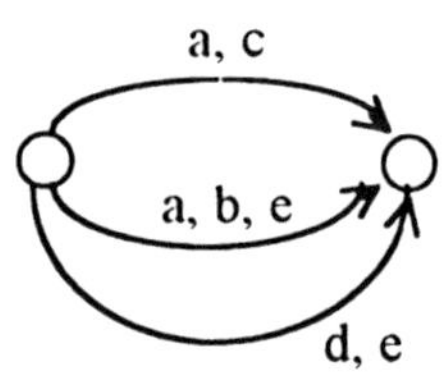


Parallel reduction:

3)

**Conclusion: reduction complexity = 2**

**FIG. 3.11 A**
**NETWORK REDUCTION OF NETWORK**

$\Omega_2$:

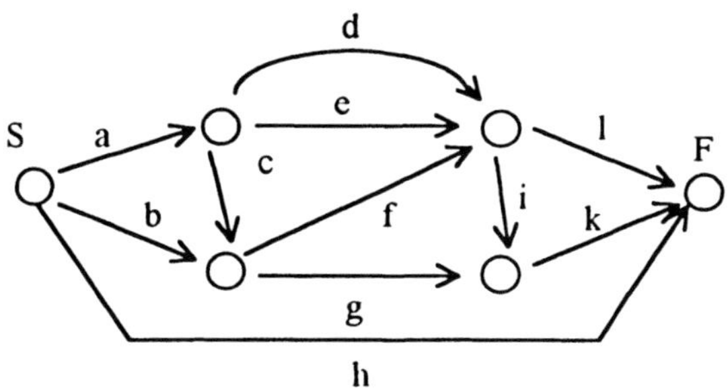


Parallel reduction 1:

1)

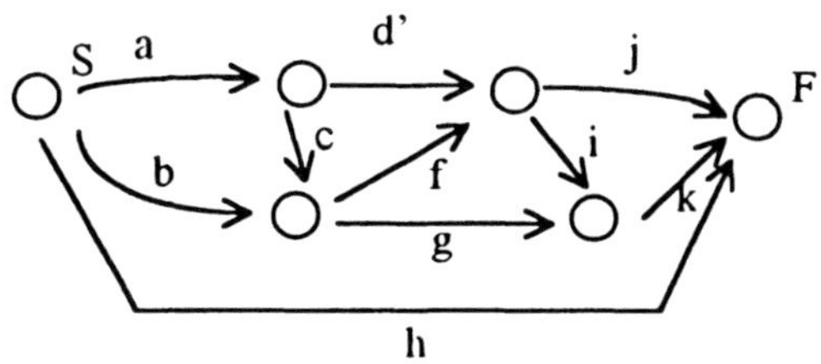


Node reduction 1:

2)

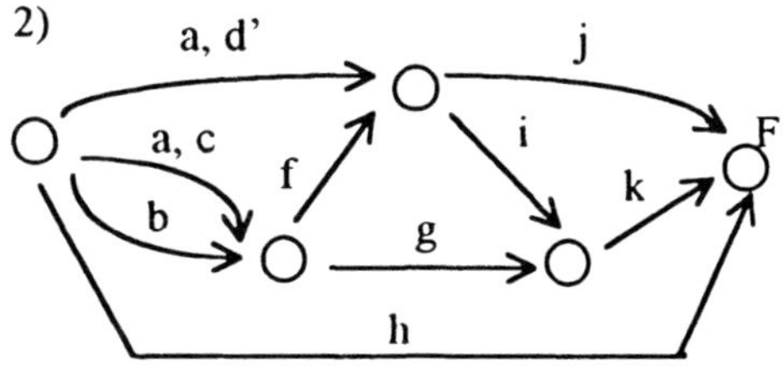


Parallel reduction:

3)

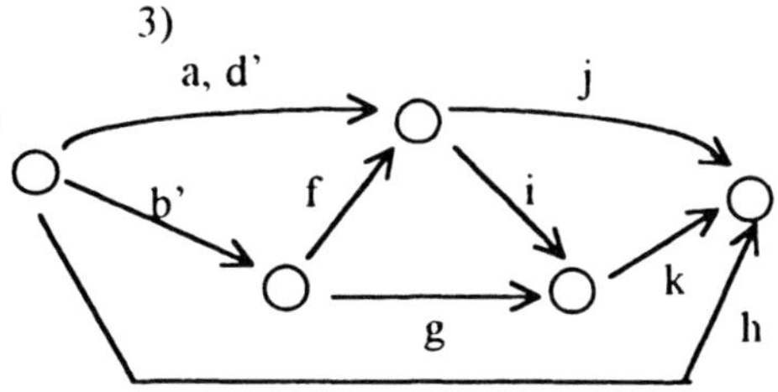


Node reduction 2:

4)

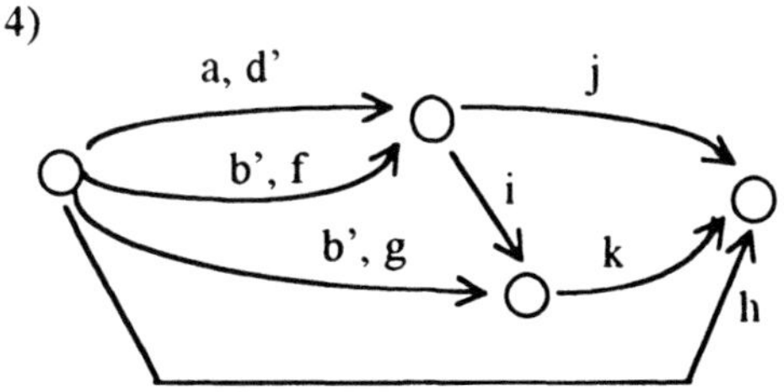


Parallel reduction

5)

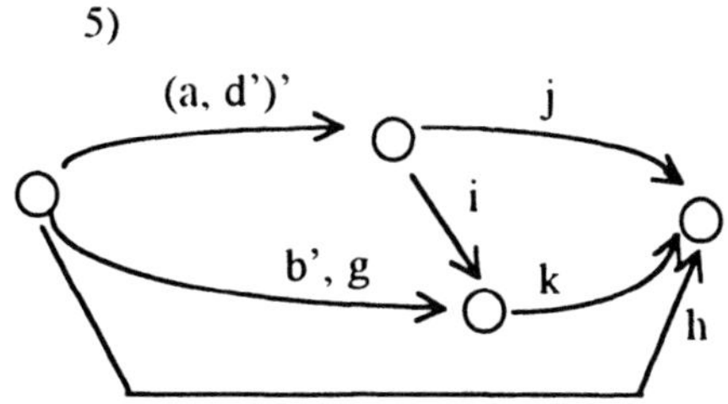

Node reduction 3:

6)

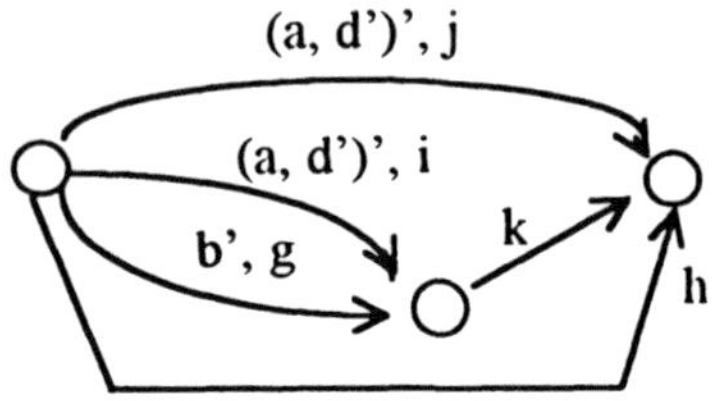


Parallel reduction

7)

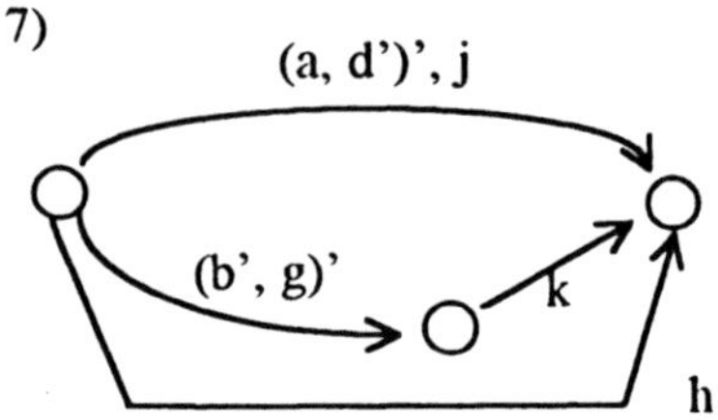


Series reduction

8)

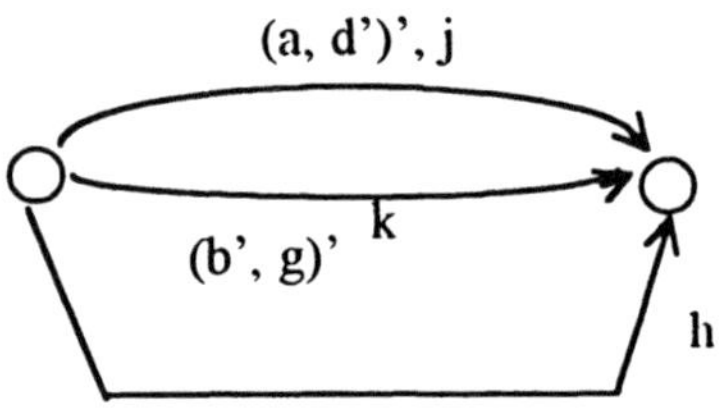


Parallel reduction

9)

**Conclusion: reduction complexity = 3**

**FIG. 3.11**
**NETWORK REDUCTION**

## 4. Hierarchical levels

The concept of topological length of a path can be very useful to analyze a network and to understand its structure.

The topological length $d[L(i,j)]$ of a path from node i to node j is defined by one plus the number of nodes, k, belonging to L (i, j) and satisfying $k \in J_j \cap k \notin J_i$.

Therefore, using an AoN representation, $d[L(i,j)]$ means the number of activities belonging to such path and between i and j plus one.

This concept is very useful to understand the relative position of any activity i belonging to a network as it allows the definition of hierarchical level. The relative position of i is important as the activities which can have an impact on the schedule and, eventually, on the delay of i are those preceding i.

Two major concepts now can be introduced to understand the relative position of any activity within the network:

- **Progressive hierarchical level** or, simply, **progressive level** of activity i, p(i): p(i) = Max d [L (s, i)] where L (s, i) denotes any path starting at the starting node and finishing at activity i.

  This means that [p (i)] is the maximal length of the topological paths from **s** to i and it can be easily computed by an iterative procedure similar to the one adopted for the earliest time of occurrence of any node:

  $p(i) = \max_{j \in J_i}[p(j)] + 1$ and p (i) = 0 if J(i) = Ø which corresponds to having i as the initial node. Therefore, p(i) also can be considered as the earliest time of starting i if assuming that the duration of each precedent activity always will be equal to 1 and starting the project (starting node) at time zero. The maximal progressive level corresponds to the final node, **e**, and p(**e**) will be denoted by P.

- **Regressive hierarchical level**, or simply **regressive level**, q (i)

  q(i) = p(**e**) - [Max d [L (i, **e**)]] where **e** denotes the final node of the network.

Again, q (i) can be considered as the latest finishing time of i if the final node occurs at time p(e). An iterative procedure similar to the one presented to compute the latest occurrence times of nodes can be adopted to determine q (i):

$q(i) = \underset{k \in K_i}{\mathrm{Min}}[q(k)] - 1$ and $q(i) = p(e)$ if $K_i = \varnothing$ which corresponds to having i as the final node.

The levels of the network presented in Fig. 2.2 are presented in Fig. 3.12.

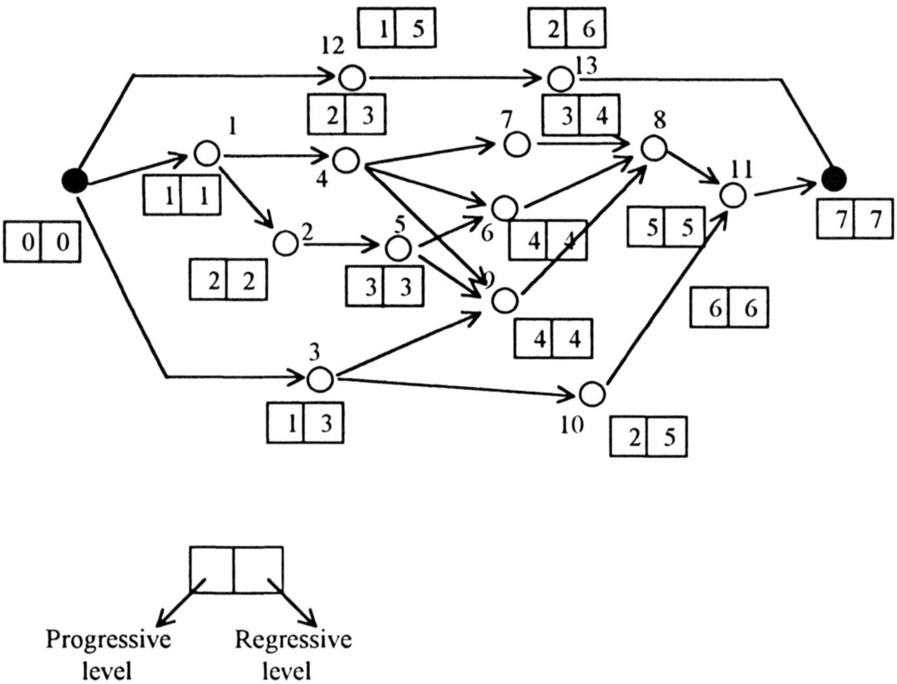


**FIGURE 3.12**
**PROGRESSIVE AND REGRESSIVE LEVELS OF THE NETWORK PRESENTED IN FIG. 2.2**

Obviously, the concept of **topological float** or **level float** now can be defined by:

$\Delta'(i) = q(i) - p(i)$ with $q(i) \geq p(i)$ and a **topological critical path** can be defined by the set of nodes with zero float, that is, with q(i)=p(i).

It can be proved easily that there exists at least one activity with p(i) = q(i) for i=0, ..., N by using AoN and making the duration of each activity equal to one.

The progressive (or regressive) level is a key concept also to describe the structure of a network because it can model the network as a sequence of **stages** (See the pioneering work of Ferreira, 1989, Tavares, 1989 and Tavares, 1990):

a) The **progressive stage s** occurs when the first activity with p(i) = s is starting;
b) The **regressive stage s'** occurs when the last activity to be finished with q (i) = s' is ending.

The sequences of the progressive and regressive stages are presented in Fig. 3.13 for the previous example of Figure 2.3.

**FIGURE 3.13**

**τ (S) AND τ' (S') FOR THE NETWORK STUDIED IN FIG. 2.2 USING THE EARLIEST STARTING TIMES**

**Calendar of the network of Fig. 2.2 adopting the earliest starting times**

| Activity | $t^s(l)$ | $t^f(l)$ | p(i) | q(i) | Δ(i) |
|---|---|---|---|---|---|
| 1 | 0 | 10 | 1 | 1 | 0 |
| 2 | 10 | 13 | 2 | 2 | 0 |
| 3 | 0 | 7 | 1 | 3 | 2 |
| 4 | 10 | 15 | 2 | 3 | 1 |
| 5 | 13 | 21 | 3 | 3 | 0 |
| 6 | 21 | 23 | 4 | 4 | 0 |
| 7 | 15 | 26 | 3 | 4 | 1 |
| 8 | 27 | 31 | 5 | 5 | 0 |
| 9 | 21 | 27 | 4 | 4 | 0 |
| 10 | 7 | 14 | 2 | 5 | 3 |
| 11 | 31 | 37 | 6 | 6 | 0 |
| 12 | 0 | 9 | 1 | 5 | 4 |
| 13 | 9 | 16 | 2 | 6 | 4 |

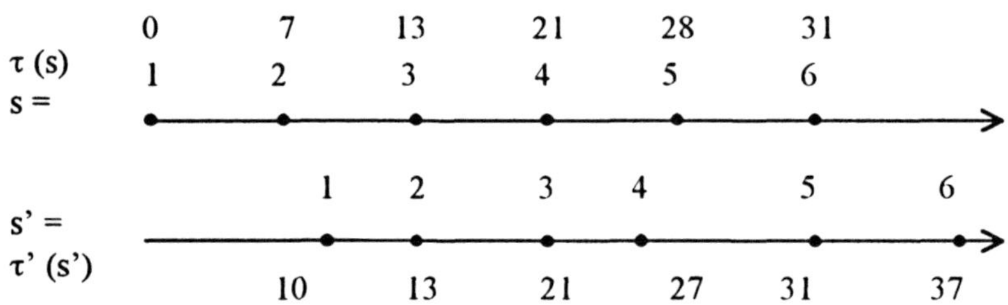


Obviously, the occurrence of the progressive stage (s+1) does not mean that all activities with p(i)=s have been completed but just that completion of one or more activities have allowed the start of the first one of the next stage with p(k)=s+1.

Similarly, the occurrence of the regressive stage s' does not mean that all activities with q(i)=s'+1 have not yet been completed but just that the last one with q(i)=s' is being finished at time τ' (s').

Unfortunately, most of the literature of Project Management has ignored these key concepts.

According to this approach, the scheduling of a project can be developed in terms of two types of decisions:

- Setting up $\{\tau(s)\}$ and $\{\tau'(s')\}$
- Optimizing t (i) under $\{\tau(s)\}$ and $\{\tau'(s')\}$.

The major advantage of this approach is allowing optimization methods like Dynamic Programming to study the first step and then searching for optimal schedules within more restricted sets of feasible solutions.

The deduction of conditions between the occurrence times of stages can be very useful and the following three theorems are presented with this objective:

**Theorem 1:**

$$\tau(s+1) \geq \tau(s) + \underset{p(j)=s}{\mathrm{Min}}\, D(j) \text{ with } s = 0,\dots,P-1$$

where P is the maximal progressive level of the network, and D(j) is the duration of the activity j.

Proof:

By definition, $\tau(s+1)$ corresponds to the start of an activity i* with p (i*)=s+1 and therefore denoting the starting time of i by $t^s(i)$ and its finishing time by $t^f(i)$, one should have:

$$t^s(i^*) \geq \underset{j \in J_{i^*}}{\mathrm{Max}}\left[t^f(j) = t^s(j) + D(j)\right] \geq \underset{j \in J_{i^*} \cap j:p(j)=s}{\mathrm{Max}}\left[t^f(j)\right]$$

The set $j \in J_i \cap j{:}p(j)=s$ is not empty because p(i*)=s+1. Then:

$$t^s(j) \geq \tau(s)$$

and

$$D(j) \geq \underset{p(j)=s}{\mathrm{Min}}\, D(j)$$

which means that

$$\tau(s+1) \geq \tau(s) + \underset{p(j)=s}{\mathrm{Min}}\, D(j)$$

**Theorem 2:**

$$\tau'(s+1) \geq \tau'(s) + \min_{q(k)=s+1} D(k)$$

with s=0, ..., P-1.

By definition, $\tau'(s)$ is determined by the finishing time of an activity, i*. Let be $k \in K(i^*)$ with q (k)=s+1.

This activity has to exist because otherwise q (i) ≠s.

Then

$$t^f(k) = \tau'(s) + D(k)$$

and by definition

$$\tau'(s) + D(k) \leq \tau'(s+1)$$

Finally,

$$\tau'(s+1) \geq \tau'(s) + \min_{q(k)=s+1} D(k)$$

**Theorem 3:**

$$\tau'(s) \geq \tau(s) + \text{Max}\left[\min_{p(i)=s} D(i);\ \min_{q(i)=s} D(i)\right]$$

By definition,

$$\tau'(s) \geq t^f(k) = t^s(k) + D(k)$$

for some activity k with q(k)=s and p(k)=s.

This activity has to exist because it was shown that the topological critical path includes at least one activity with p(i) = q(i) for i = 0,...,P .

Then

$$t^s(k) \geq \tau(s)$$

and

$$D(k) \geq \text{Max}\left[\min_{p(i)=s} D(i);\ \min_{q(i)=s} D(i)\right]$$

## *5. Level adjacency matrix*

The concept of **hierarchical level** is also very useful to describe networks generalizing the notion already presented of **Adjacency Matrix** {A(i, j)} with i, j = 1, ..., N.

Any network can be described by this matrix as is exemplified by the network in the figure below.

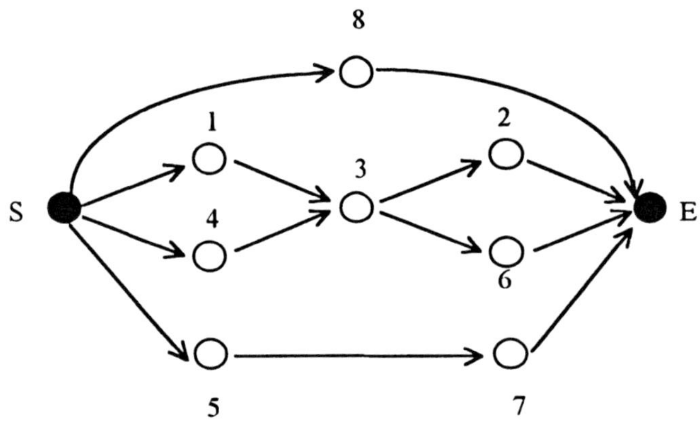


**FIGURE 3.14**
**AN EXAMPLE OF NETWORK**

The matrix A (i, j) will be given by:

Precedence link from i to j — Activity j → ; Activity i ↓

| | S | 1 | 2 | 3 | 4 | 5 | 6 | 7 | 8 | E |
|---|---|---|---|---|---|---|---|---|---|---|
| S | - | 1 | - | - | 1 | 1 | - | - | 1 | - |
| 1 | - | - | - | 1 | - | - | - | - | - | - |
| 2 | - | - | - | - | - | - | - | - | - | 1 |
| 3 | - | - | 1 | - | - | - | 1 | - | - | - |
| 4 | - | - | - | 1 | - | - | - | - | - | - |
| 5 | - | - | - | - | - | - | - | 1 | - | - |
| 6 | - | - | - | - | - | - | - | - | - | 1 |
| 7 | - | - | - | - | - | - | - | - | - | 1 |
| 8 | - | - | - | - | - | - | - | - | - | 1 |
| E | - | - | - | - | - | - | - | - | - | - |

This matrix is often rather sparse and its size is quite large for large networks becoming less useful for presenting a network with N>50 or 100.

However, a new matrix can be proposed using the concept of progressive (or regressive) hierarchical level: the **Level Adjacency Matrix**:
B (k, m) with k, m = 0, ..., M,

where M is the maximal progressive level and being B (k, m) defined by the number of precedence links between activities of order k and m.

For the given example, one has:

| m= | 0 | 1 | 2 | 3 | 4 |
|---|---|---|---|---|---|
| k=0 | | 4 | 0 | 0 | 0 |
| 1 | | | 3 | 0 | 1 |
| 2 | | | | 2 | 1 |
| 3 | | | | | 2 |
| 4 | | | | | |

and it is quite obvious that the maximal number of links in each element of the line adjacent to the diagonal, B (k, k+1), with k=0, ..., (M-1), is given by $N_k . N_{k+1}$ where N. is the number of activities of level . but the maximal number of links between k and (k+i) with i=2, 3, ... is lower than $N_k . N_{k+i}$ because this number includes the redundant links. Another example is presented in Fig. 3.15 and

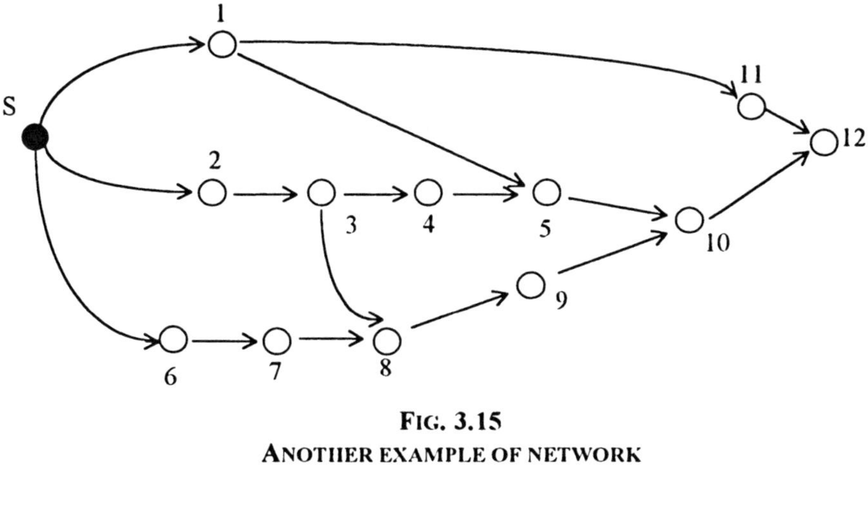


**FIG. 3.15**
**ANOTHER EXAMPLE OF NETWORK**

the correspondent level adjacency matrix is:

| | 0 | 1 | 2 | 3 | 4 | 5 | 6 |
|---|---|---|---|---|---|---|---|
| 0 | | 3 | 0 | 0 | 0 | 0 | 0 |
| 1 | | | 3 | 0 | 1 | 0 | 0 |
| 2 | | | | 3 | 0 | 0 | 1 |
| 3 | | | | | 2 | 0 | 0 |
| 4 | | | | | | 2 | 0 |
| 5 | | | | | | | 1 |
| 6 | | | | | | | |

Usually, the elements of the matrix far form the diagonal tend to be zero and therefore a more economical representation can be constructed substituting the columns j by k = j-i with $j \geq i$.

This means that each element of the matrix (i, k) with $k \neq i$ corresponds to the number of direct links between the progressive level i and j = k + i. For the case k = 0, the number of activities belonging to the level i is given (i, k = 0).

The previous matrix is now presented by:

k

i =

| | 0 | 1 | 2 | 3 | 4 |
|---|---|---|---|---|---|
| 0 | 1 | 3 | 0 | 0 | 0 |
| 1 | 3 | 3 | 0 | 1 | 0 |
| 2 | 3 | 3 | 0 | 0 | 1 |
| 3 | 2 | 2 | 0 | 0 | 0 |
| 4 | 2 | 2 | 0 | 0 | 0 |
| 5 | 1 | 1 | 0 | 0 | 0 |

The networks with most links of a lower "depth" have a sum of elements outside the line adjacent to the diagonal lower than those with "deeper" links. Therefore, the following two indicators can be proposed:

$$I_1 = \sum_{k=0}^{M-1} \frac{B(k, k+1)}{N_k \cdot N_{k+1}}$$

$$I_2 = \left[ \sum_{\substack{k=o \\ l>k+1}}^{M-1} B(k, l) \right] \Big/ \sum_{k=0}^{M-1} B(k, k+1)$$

where $I_1$ represents the existing number of links with length equal to 1 as a fraction of the maximal number of such links and where $I_2$ represents the importance of the number of links with length higher than 1 if compared with those with length equal to 1.

This matrix can be very useful to describe large networks (e.g., N=1000) because the matrix has just the dimension $M^2$.

The developed software - RISKNET - includes the Module 3 to produce the level adjacency matrix for the studied network (Annex 1).

# 4

# MORPHOLOGY AND SIMULATION OF PROJECT NETWORKS

## *1. Morphology*

So far, project networks have been studied in terms of:

- the definition of their elements (arcs and nodes);
- their analytical properties (sequentiality, criticality, etc.);
- alternative meaning to represent projects (AoA and AoN);
- structural concepts (hierarchical levels, concepts decomposition and reduction analysis)

However, most project managers rely heavily on their assessment of the so-called "type" of network which is really based on its "shape" or morphology given a representation systems.

As was mentioned before, the AoA may not produce an unique representation of each network and therefore the AoN representation will be used.

Each node is represented by a small circle of the same size excepting the nodes with a single out-link converging to the end which are represented by a small square. This notation has the advantage of avoiding the representation of the End node as well as all in-links of this node, making drawings much clearer. The start node is represented by a solid circle. The precedence links are represented by arrows, as usual.

Obviously, there are two very specific and extreme shapes of networks already mentioned and easily recognized:

a) Series - network

   In this case, $J(i)=i-1$ with $i=1, ..., N$ for a project network with N activities (the start node corresponds to $i=0$) numbered in sequential order. (Fig.4.1).

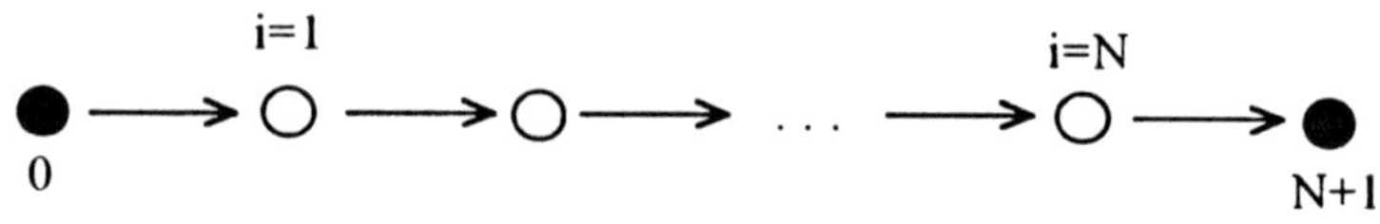


**FIGURE 4.1**
**THE SERIES-NETWORK**

b) Parallel - network

Now, J(i) = 0 for any i = 1, ..., N, which means that are no precedence restrictions, as it is shown in Figure 4.2.

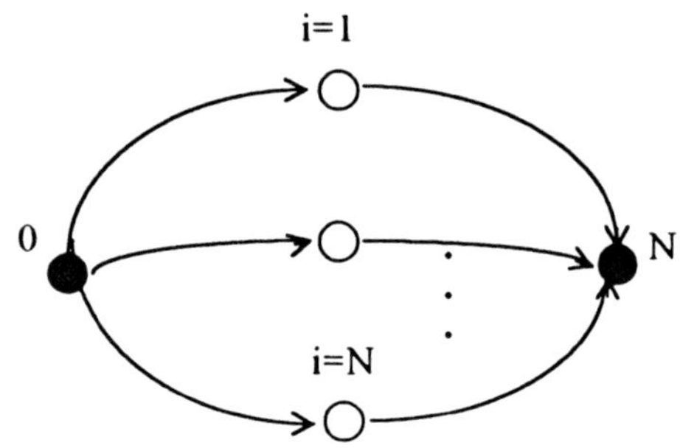


**FIGURE 4.2**
**THE PARALLEL-NETWORK**

However, most networks do not belong to any of these two groups and therefore their morphology should be studied.

This problem has been hardly studied in the literature and so three major perspectives are proposed to describe the morphology of the network:

A - graphical shape of the border containing the represented network ("**external shape**")

B - number of the non-redundant precedence links ("**density**")

C - level length of these links ("**depth**")

The study of these three perspectives will be carried through the following proposed indicators:

The **perspective A** can be synthetized by:

- **Size**

  The corresponding indicator $I_1$, is defined by the number of the set of activities; $\Omega$:

  $$I_1 = N$$

- **Level Length**

  The absolute level length is defined by the dimension of the longest path measured in terms of the hierarchical levels:

  $$M = \max_{i \in \Omega} a_i$$

  The corresponding indicator $I_2$, is the relative length and measures how serial is the shape of the network (for N>1):

  $$I_2 = \frac{M-1}{N-1}$$

  where $0 \le I_2 \le 1$ corresponding 0 to the parallel case and 1 to the serial one.

- **Width**

  This magnitude can be defined by the number of activities at each progressive order,
  W(a), with a = 1, ..., M being, W = 1, 2, ... .

  The maximal width is denoted by MW and several shapes can be adopted to describe the function W(a) (see Figure 4.3).

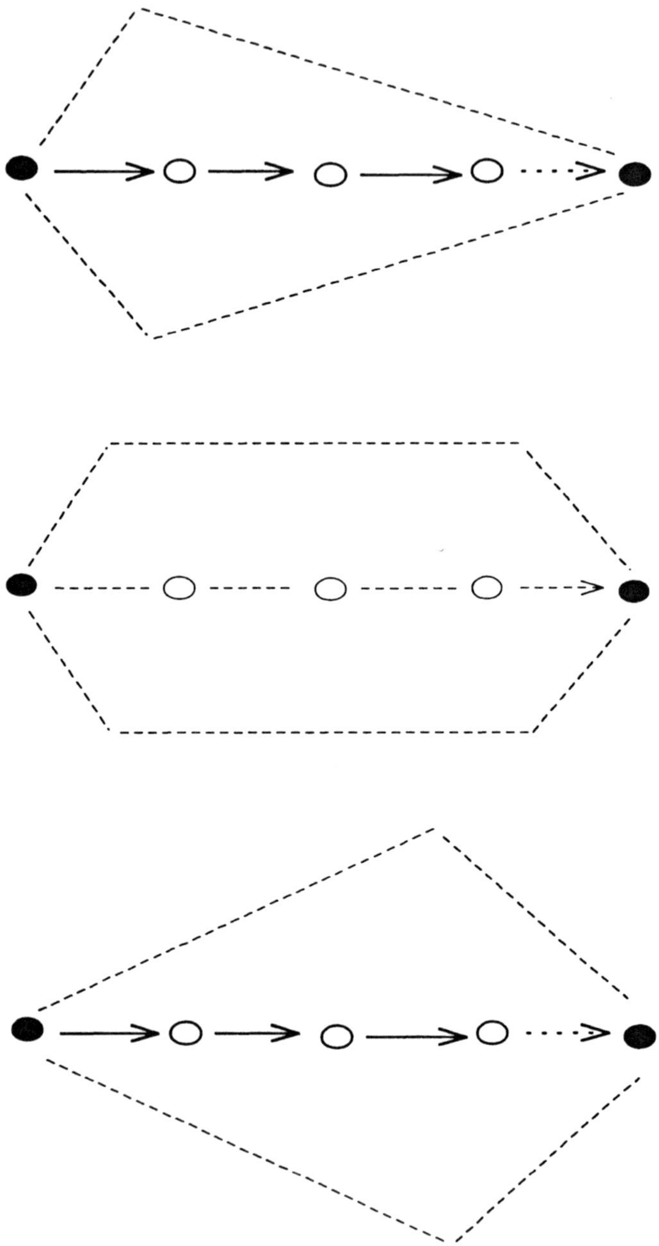

FIGURE 4.3
SEVERAL SHAPES FOR THE WIDTH OF THE NETWORK DESCRIBED BY W(a)

The proposed indicator, $I_3$, is only meaningful if $N > M$ and it is defined by the function $I_3(a) = \frac{W(a) - 1}{N - M}$ with $1 \leq a \leq M$. Obviously, one should have $\sum_{a=1}^{M} I_3(a) = 1$.

Often, a linear shape is assumed for W(a) having MW at level $\frac{a}{2}$ (or at the two nearest integer values) and W(1) = W(M) (symmetrical variation). In this case, a simplified indicator can be proposed: $I_3^* = \frac{W(1)}{MW}$

The **perspective B** can be described in terms of the number of non-redundant direct precedence links per activity.

The minimal number of these links for a network with N activities and M progressive levels is equal to N if the links between the initial node and the activities of level one also will be counted but ignore the connections between the activities of the last level and the final node. This is the convention adopted in this book.

Obviously, many other direct precedence links can be received by an activity i, with length equal to 1, 2, ..., $a_i$ where $a_i$ is the progressive order of i.

The maximal number of these links occurs when all activities of a progressive level are directly linked with all the activities of the level i-1.

The **maximal number, D, of the non-redundant precedence links with level length equal to one** is obtained by the connection of each activity, i, belonging to the level a (a = 1, ..., M-1) with all the activities belonging to the level (a+1). Therefore, D is given by:

$$D = W(1) + \sum_{a=1}^{M-1} W(a).W(a+1)$$

and the proposed indicator, $I_4$, expressing the relative complexity index for links with a length equal to 1, is defined by:

$$I_4 = \frac{n(1) - N}{D - N}$$

which will satisfy $O \leq I_4 \leq 1$ as $n(1) \geq N$.

The **perspective C** can be synthetized by the distribution of the level float of the activities or by the distribution of the level length of each direct precedence link. A project network with longer links has higher floats and these magnitudes are strongly correlated.

Usually, the number of links of level length L, n (L), decreases with their level length L and therefore the following assumption may often be adopted

$$n(L+1) = n(L) \cdot p$$

with L=1, 2, ..., V where V is the maximal length and where p is a factor between 0 and 1. It should be noted that V cannot be higher than (M-1).

The proposed indicators to express this perspective are

$$0 \leq I_5 = p \leq 1$$

and

$$0 \leq I_6 = \frac{V-1}{M-1} \leq 1.$$

The average level length, $\overline{L}$, of the direct precedence links can be easily determined by:

$$\overline{L} = \frac{V\,p^V(p-1) + \left(1-p^V\right)}{\left(1-p^V\right)(1-p)}$$

and the total number, T, of non-redundant direct precedence links is given by:

$$T = n(1). \frac{1-p^V}{1-p}$$

The study of the network's morphology requires a good ability to produce the automatic drawing of the network, for obvious reasons. Unfortunately, the ability of commercial packages to achieve this task is rather poor and therefore a specific software was developed: RISKNET - Module 4 -, Annex 1.

This software is based on the following procedures:

- the activities are represented by nodes and they are located in terms of their progressive order and are equally spaced along each order. They are located along a central strip (S) of the plane. Every **node** is represented by a **circle** except the **starting node** and those nodes not **preceding** any other activity which are represented by a **small square**. The final node is not represented.

- a specific indicator, $\pi_I$, measuring for each activity, i, the degree of difficulty for drawing its link is computed by: $\pi_i = \sum_{L=1}^{V} n_i(L).2^{2(L-1)}$ i where $n_i(L)$ is the number of links including i with length L=1,2,...V. (excepting those connecting the activity to the final node). This function is adopted because a longer link is more difficult to draw without confusion with other links than a shorter one.

- at each order, the nodes are ordered in terms of $\pi_i$ allocating an activity to each node from bottom up. This procedure guarantees that the most difficult nodes will be located near the upper border which is convenient to reduce the number of long links crossing most of the network.

- the links with length equal to one are represented by bold full straight lines;

- the links longer than one are represented by curves distributed on strips of space outside S. A link with length L'>L will go through a strip more peripheral than the curve representing the link with length equal to L. These curves are described by B - splines (Foley, J. D. and A. Van Dams, 1982) and the following conventions are used:

    - thick full line (L=1)
    - full line (L=2)
    - thin full line (L≥3)

- the links connecting each activity (without succeeding activities) with the final node are not represented as they are not required to understand the network.

Therefore, without loss of generality, these links are not considered to compute $I_4$, $I_5$ or $I_6$.

## 2. The generation method

The project network will be generated (Tavares, Ferreira and Coelho, 1997) in terms of the defined indicators, $I_1$, $I_2$, $I_3$, $I_4$, $I_5$ and $I_6$:

$I_1 = N > 1$

$$0 \le I_2 = \frac{M-1}{N-1} \le 1$$

$$0 \le I_3(a) = \frac{W(a)-1}{N-M} \le 1 \text{ with } 1 \le a \le M \text{ and } \sum_{a=1}^{M} I_3(a) = 1$$

$$0 \le I_4 = \frac{n(1)-N}{D-N} \le 1$$

$$0 \le I_5 = p \le 1$$

$$0 \le I_6 = \frac{V-1}{M-1} \le 1$$

N defines the size of the network and $I_2$ describes how much it has a serial morphology. Usual networks have $I_2 < 0.5$.

The selection of $I_3$ is easier to make in terms of $I_3^*$ and two typical cases are $I_3^* = 1$ (rectangular shape) and $I_3^* = 0.5$ (triangular shape).

The selection of $I_4$ defines the density of links with level length equal to one and, obviously, the number of possible non-redundant links with L>1 is very small if $I_4$ is too high.

Therefore, this difficulty is particularly severe if $I_5$ (=p) is high because then we should also have a large number of links with L>1. Usual networks often have $I_5<0.3$ and the numerical experiments carried out have shown that $I_4$ also should be $I_4<0.3$ to avoid a significant probability of generating an unfeasible network.

The importance of $I_6$ grows when $I_5$ >0.1 and its recommended range is between 0.1 and 0.3.

The user should specify a value for each indicator of the set $I_1$, ..., $I_6$ and then the following procedure to generate a network is adopted.

1- Generation of N and M given $I_1$, $I_2$.

2- Allocation of the activities to the levels, a = 1, ..., M, in terms of $I_3$ (a) and introduction of precedences with level length equal to one between arbitrarily chosen activities, one from each level.

3- Generation of the population of possible links with level length L, P (L), starting with L=1.
Generation of the population of activities of level a = 2, ..., M not yet receiving a precedence link from level (a - 1) (population, $\overline{P}$).

4- Random selection from P (L) of a link and eventual updating of $\overline{P}$. Repetition of this step until the number prescribed by $I_4$ is reached.

Every time a link is selected, a test of feasibility is performed to verify that the number of activities not yet receiving a link from the previous level is not higher than the number of links with level length equal to one that should be added to the network to obtain n(1).

If the answer to this test is negative, then the last link generated is deleted and the test is repeated.

If the answer remains negative, another deletion is introduced. After obtaining a positive answer, the process of introducing a new link goes on.

It should be noted that there is no risk of redundancy with links having L=1 and therefore the redundancy check is just carried out for L>1.

5- After completing the generation of links with level length equal to one, the links with level length equal to two, three, ..., V are generated using a similar procedure:

- generation of P (L)
- random selection of a link from P (L)
- checking of non-redundancy until the prescribed number of links for L=2,..., V is reached.

If no set of non-redundant links of level length L is found fulfilling n (L), a recursive search is developed starting at the previous level length (L-1) until a feasible solution is achieved or until a maximal time limit is reached.

A software program - RISKNET Module 5 - (Annex 1) - was developed in Microsoft Visual $C^{++}$ 2.2 and most examples with 50-100 activities are generated in less than one second using a Pentium 166 MHz.

As was mentioned before, the links between any activity without succeeding activity and the final node are not considered.

Two examples are given for

a) $I_1$ = 14; $I_2$ = 0.30; $I_3$ = 0.14; 0.22; 0.29; 0.22; 0.14; for a = 1; 2; 3; 4; 5, respectively; $I_4$=0.10; $I_5$ = 0.35; $I_6$ = 1.00.

b) $I_1$=75; $I_2$=0.20; $I_3$=0.03; 0.04; 0.05; 0.06; 0.07; 0.08; 0.08; 0.09; 0.09; 0.08; 0.08; 0.07; 0.06; 0.05; 0.04; 0.03; $I_4$= 0.00; $I_5$=0.10; $I_6$=0.25.

The generation time was less than one second for a) or b). The obtained results are presented in Tables 4.1 and 4.2 and in Figures 4.4 and 4.5.

TABLE 4.1

FEATURES OF THE GENERATED NETWORK WITH 14 ACTIVITIES

| Activity | Progressive Level | Regressive Level | Immediate Precedent Activities |
|---|---|---|---|
| 0 | 0 | 0 | |
| 1 | 1 | 1 | 0; |
| 2 | 1 | 2 | 0; |
| 3 | 2 | 2 | 1; |
| 4 | 2 | 3 | 1; |
| 5 | 2 | 3 | 2; |
| 6 | 3 | 3 | 2; 3; |
| 7 | 3 | 4 | 4; |
| 8 | 3 | 5 | 5; |
| 9 | 3 | 3 | 3; |
| 10 | 4 | 4 | 4; 5; 6; |
| 11 | 4 | 4 | 9; |
| 12 | 4 | 5 | 2; 7; |
| 13 | 5 | 5 | 7; 9; 10; |
| 14 | 5 | 5 | 2; 4; 11; |

TABLE 4.2
FEATURES OF THE GENERATED NETWORK WITH 75 ACTIVITIES

| Activity | Progressive Level | Regressive Level | Immediate Precedent Activities |
|---|---|---|---|
| 1 | 0 | 0 | |
| 2 | 1 | 1 | 1; |
| 3 | 1 | 16 | 1; |
| 4 | 1 | 14 | 1; |
| 5 | 2 | 2 | 2; |
| 6 | 2 | 15 | 2; |
| 7 | 2 | 15 | 4; |
| 8 | 3 | 3 | 5; |
| 9 | 3 | 16 | 7; |
| 10 | 3 | 7 | 5; |
| 11 | 3 | 16 | 6; |
| 12 | 4 | 4 | 8; |
| 13 | 4 | 8 | 10; |
| 14 | 4 | 12 | 8; |
| 15 | 4 | 9 | 8; |
| 16 | 5 | 5 | 12; |
| 17 | 5 | 10 | 15; |
| 18 | 5 | 9 | 13; |
| 19 | 5 | 16 | 15; |
| 20 | 5 | 13 | 14; |
| 21 | 6 | 6 | 16; |
| 22 | 6 | 15 | 18; |
| 23 | 6 | 14 | 20; |
| 24 | 6 | 11 | 17; |
| 25 | 6 | 10 | 18; |
| 26 | 6 | 8 | 16; |
| 27 | 7 | 7 | 21; |
| 28 | 7 | 9 | 26; |
| 29 | 7 | 16 | 22; |
| 30 | 7 | 15 | 23; |
| 31 | 7 | 16 | 25; |
| 32 | 7 | 11 | 25; |
| 33 | 8 | 8 | 27; |
| 34 | 8 | 16 | 28; |
| 35 | 8 | 16 | 22; 30; |
| 36 | 8 | 10 | 28; |
| 37 | 8 | 16 | 28; |

| Activity | Progressive Level | Regressive Level | Immediate Precedent Activities |
|---|---|---|---|
| 38 | 8 | 12 | 24; 32; |
| 39 | 8 | 16 | 24; 27; |
| 40 | 9 | 9 | 33; |
| 41 | 9 | 16 | 33; |
| 42 | 9 | 15 | 33; |
| 43 | 9 | 16 | 38; |
| 44 | 9 | 12 | 36; |
| 45 | 9 | 11 | 36; |
| 46 | 9 | 13 | 38; |
| 47 | 10 | 10 | 40; |
| 48 | 10 | 13 | 45; |
| 49 | 10 | 16 | 46; |
| 50 | 10 | 12 | 45; |
| 51 | 10 | 15 | 45; |
| 52 | 10 | 16 | 42; |
| 53 | 11 | 11 | 47; |
| 54 | 11 | 16 | 51; |
| 55 | 11 | 13 | 44; 50; |
| 56 | 11 | 14 | 46; 48; |
| 57 | 11 | 16 | 44; 51; |
| 58 | 12 | 12 | 53; |
| 59 | 12 | 15 | 50; 56; |
| 60 | 12 | 14 | 55; |
| 61 | 12 | 15 | 40; 55; |
| 62 | 12 | 16 | 53; |
| 63 | 13 | 13 | 58; |
| 64 | 13 | 16 | 61; |
| 65 | 13 | 15 | 60; |
| 66 | 13 | 16 | 59; |
| 67 | 14 | 14 | 63; |
| 68 | 14 | 15 | 63; |
| 69 | 14 | 14 | 63; |
| 70 | 14 | 16 | 63; |
| 71 | 15 | 15 | 67; |
| 72 | 15 | 16 | 65; 68; |
| 73 | 15 | 15 | 69; |
| 74 | 16 | 16 | 71; |
| 75 | 16 | 16 | 73; |
| 76 | 16 | 16 | 73; |

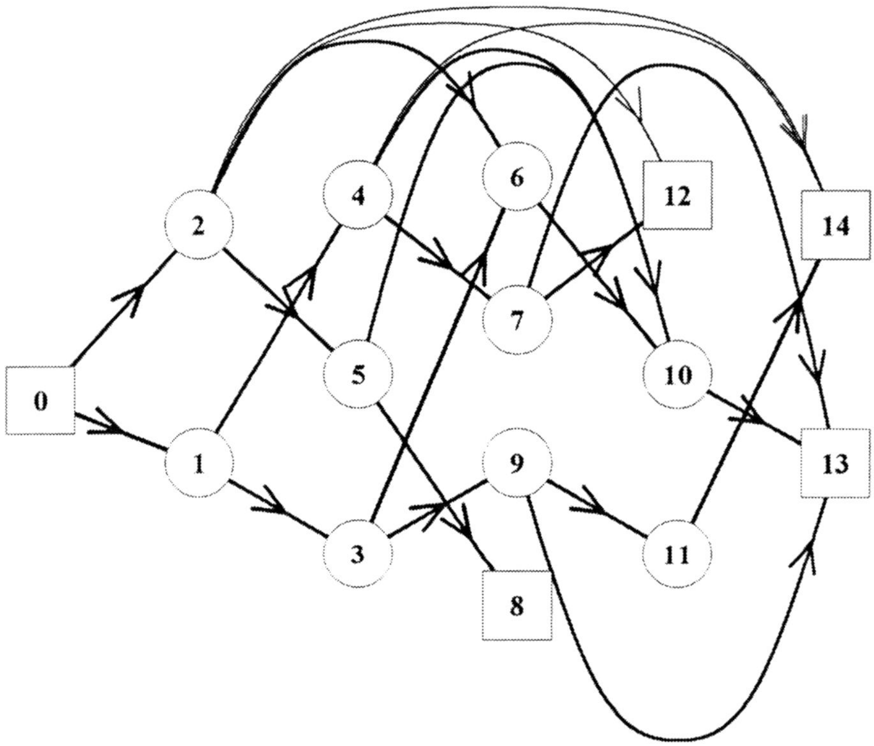


**FIGURE 4.4**
**THE GRAPHICAL REPRESENTATION OF THE NETWORK IN TABLE 4.1**

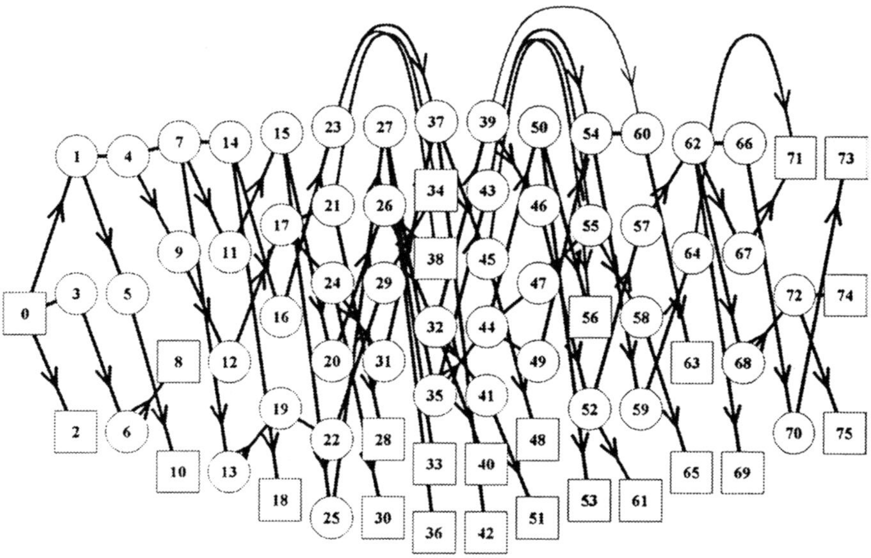


**FIGURE 4.5**
**THE GRAPHICAL REPRESENTATION OF THE NETWORK IN TABLE 4.2**

## 3. Comparative analysis

After presenting the proposed method, a critical comparison with other methods can be presented.

Actually, the major generators already published were proposed by Demeulemeester, Dodin and Herroelen 1993, by Kolish, Sprecher and Drexl 1995 and by (Agrawal, Elmaghraby and Herroelen 1996).

The analysis of any simulation method has to give special attention to the features which are supposed to be preserved by the generated sample of instances and therefore the features preserved by these three methods should be considered:

A - Demeulemeester, Dodin and Herroelen, 1993

This method just preserves the number activities and nodes. The AoA (Activity-on-Arc) representation has to be adopted as the problem of redundancy is not considered.

Actually, each generated arc is an activity and hence the problem of redundancy does not exist. However, there are networks requiring the introduction of dummy activities to be represented by AoA (see Fig. 4.6,) but these type of activities are not considered by the authors. Obviously, checking redundancy is required if dummy activities are included as was shown in the previous figure. Therefore, this method of generation implicitly excludes all networks requiring such types of activities.

B - Kolish, Sprecher and Drexl, 1995

This generator, Progen, adopts the AoN (Activity-on-Node) representation and attempts to preserve the number of non-redundant arcs per node (network complexity).

Unfortunately, there is no guarantee of producing an acyclical network.

C - Agrawal, Elmaghraby and Herroelen, 1996

This procedure, DAGEN, is based on the "**reduction complexity**" index proposed by Bein, Kamburowski and Stallman 1992, measuring the non-conformity of a network to the series-parallel case.

DAGEN receives from the user this index, the number of nodes and arcs and tries to get a feasible solution.

There is no checking of redundancy on the arcs which means that this generator has the same drawback of the first one.

Several remarks can be made about these three contributions:

- Methods A and C imply the AoA representation and dummy activities are not considered. However, as it is recognized by one of the authors of C (Elmaghraby, 1995, pages 386 and 388), "the AoA representation of the same project, commonly referred to as the graph D is **not** unique, and is cluttered with dummy activities and dummy events..." and "Undoubtedly, the AoN representation is the more direct, the more frugal, and is unique". Thus, this drawback which is common to A and C may be quite a serious one.

- The adoption of the AoA representation has another serious drawback if used for simulation: it requires receiving the number of the nodes from the user. Unfortunately, the number of nodes in AoA is more an output than an input as it is a function of the number of the activities and of the adopted precedence structure than an input. Thus, this is not an easy input to be given.

- The first method just preserves the number of activities and nodes and the two others just introduce another indicator. All the variety of shapes and additional features of precedence structures are simply ignored.

Therefore, these methods cannot guarantee the reproduction of different types and shapes of project networks. Furthermore, the size of the generated networks is below 100: the results obtained in A are discussed with no more than 60 activities, the application in B is presented with 15 activities and the program of DAGEN does not accept more than 100 nodes (AoA).

This limitation is confirmed by recent experiments using generated networks with no more than 50 activities to compare alternative heuristics (Ahn and Erengue, 1998 and Sprecher and Drexl, 1998).

The model already presented adopts a radically different approach which is based on:

- the use of AoN representation to avoid the drawback of requiring from the user the input of the number of nodes (events, using AoA) which is unknown, most often;

- the adoption of the hierarchical level as the basic concept to study the morphology of the network;

- the construction of a framework of six morphological indicators (instead of just two as was proposed before) to synthetize the morphologic type of the network;

- the construction of a generator based on these indicators;

Obviously, the generation of the specific features of each activity (duration, cost, etc.) easily can be carried out using standard procedures once the network has been generated.

Example I

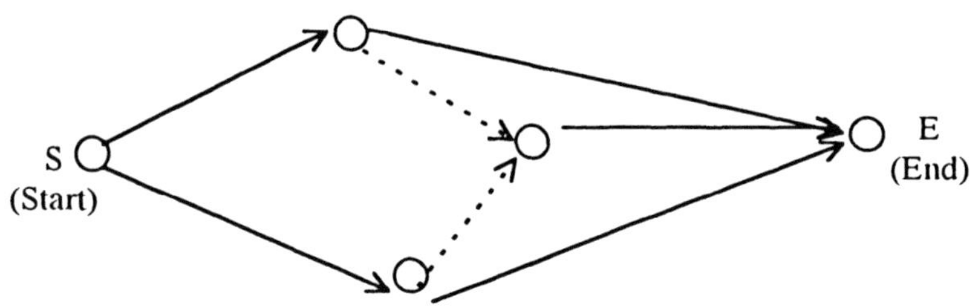


Example II

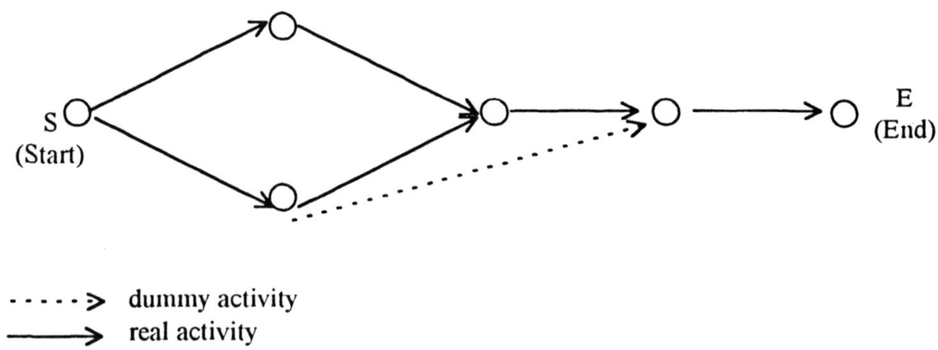


**Example of a network (I) requiring two dummy activities to be represented by AoA and an example with a redundant dummy activity (II)**
**FIGURE 4.6**

## 4. A test set of generated networks

A collection of 216 test problems was generated with

$I_1$ = 10; 25; 50; 100; 250; 500; 750; 1000
$I_2$ = 0.1; 0.3
$I_3^*$ = 0.5; 1
$I_4$ = 0.1; 0.3
$I_5$ =0.1; 0.3
$I_6$ =0.1; 0.3

The largest networks with N=1000 are generated in less than 15 seconds.

For $N \geq 100$, the cases corresponding to $I_4 = 0.3$ and $I_5 = 0.3$ are not considered as they are usually unfeasible because too many links for L=1 and for L>1 are being required.

This set of 216 test problems was compacted so it could be saved on a 1.4 Mb diskette. This diskette with this collection of 216 test problems is available from the author and at the Kluwer web site.

A few examples of the generated networks can be presented using $I_3^* = 0.5$

Example 1 → N=25 (Adjacency matrix, level adjacency matrix and network in Figs. 4.7, 4.8 and 4.9).

Example 2 → N=50 with M=7 (Adjacency matrix, level adjacency matrix and network in Figs. 4.10, 4.11 and 4.12).

Example 3 → N=50 with M=7 (Adjacency matrix, level adjacency matrix and network in Figs. 4.13, 4.14 and 4.15).

Example 4 → N=100 with M=12 (Level adjacency matrix, Fig. 4.16).

Example 5 → N=1000 with M= 22 (Level adjacency matrix, Fig. 4.17).

The network and the adjacency matrix are not presented for the examples 4 and 5 because of their size.

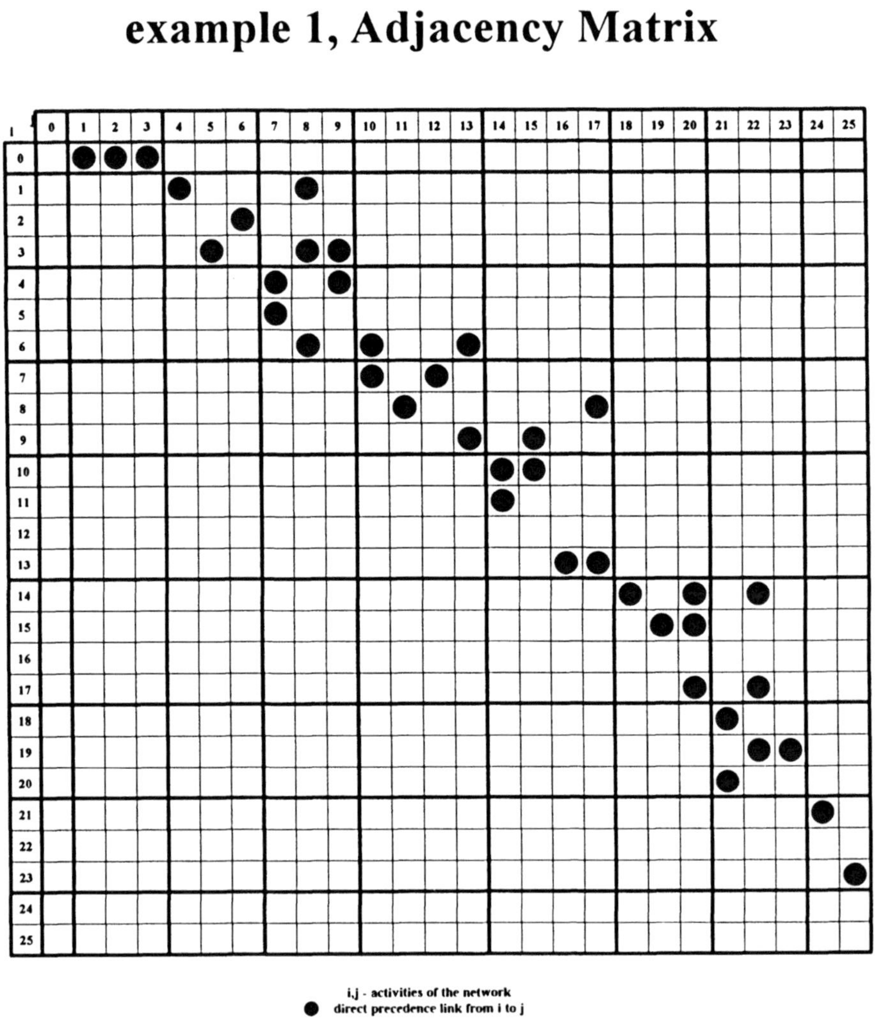


**FIG. 4.7**
**ADJACENCY MATRIX OF EXAMPLE 1**

| i \ k | 0 | 1 | 2 |
|---|---|---|---|
| 0 | 1 | 3 | |
| 1 | 3 | 3 | 3 |
| 2 | 3 | 4 | 2 |
| 3 | 3 | 4 | 2 |
| 4 | 4 | 5 | |
| 5 | 4 | 5 | 2 |
| 6 | 3 | 4 | |
| 7 | 3 | 2 | |
| 8 | 2 | | |

**Number of precedence links between each progressive level,**
**i, and the level given by (i+k) with k not equal to 0.**
**The element (i, k=0) represents the number of activities belonging to the level i.**

Note:

For k ≠ 0, the ratio between each element (i, k) and its corresponding maximal value is computed and a darker shade indicates a higher ratio (Black corresponds to 100%and white to 0%).

FIG. 4.8
LEVEL ADJACENCY MATRIX FOR EXAMPLE 1

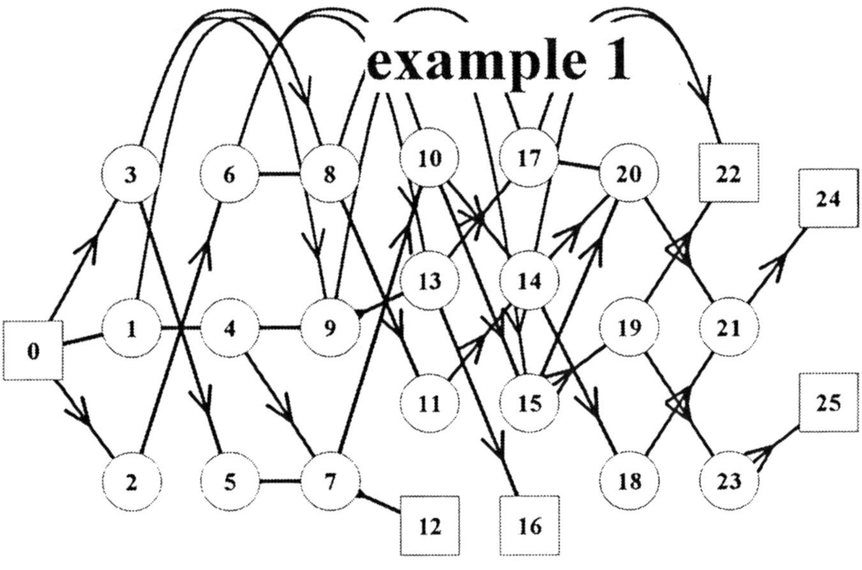


FIG. 4.9
NETWORK OF EXAMPLE 1

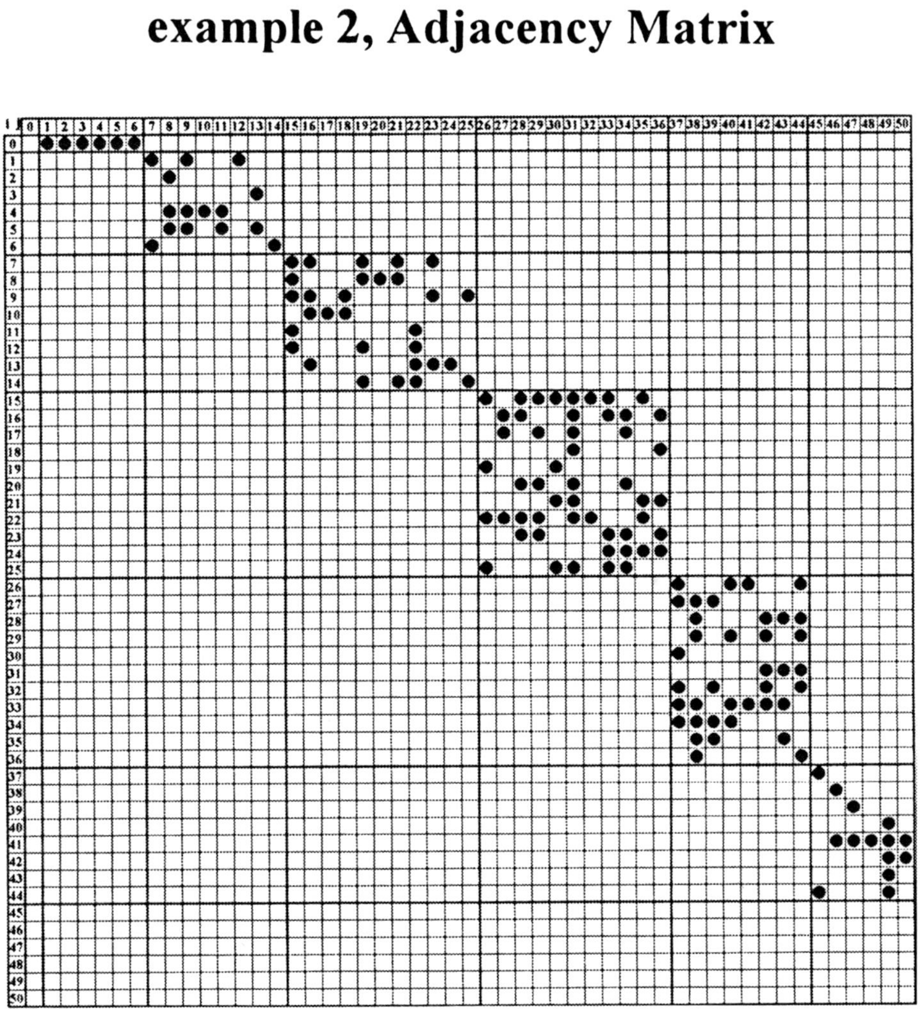


**FIG. 4.10**
**ADJACENCY MATRIX OF EXAMPLE 2**

| i \ k | 0 | 1 |
|---|---|---|
| 0 | 1 | 6 |
| 1 | 6 | 15 |
| 2 | 8 | 30 |
| 3 | 11 | 51 |
| 4 | 11 | 38 |
| 5 | 8 | 14 |
| 6 | 6 | |

FIG. 4.11
LEVEL ADJACENCY MATRIX OF EXAMPLE 2

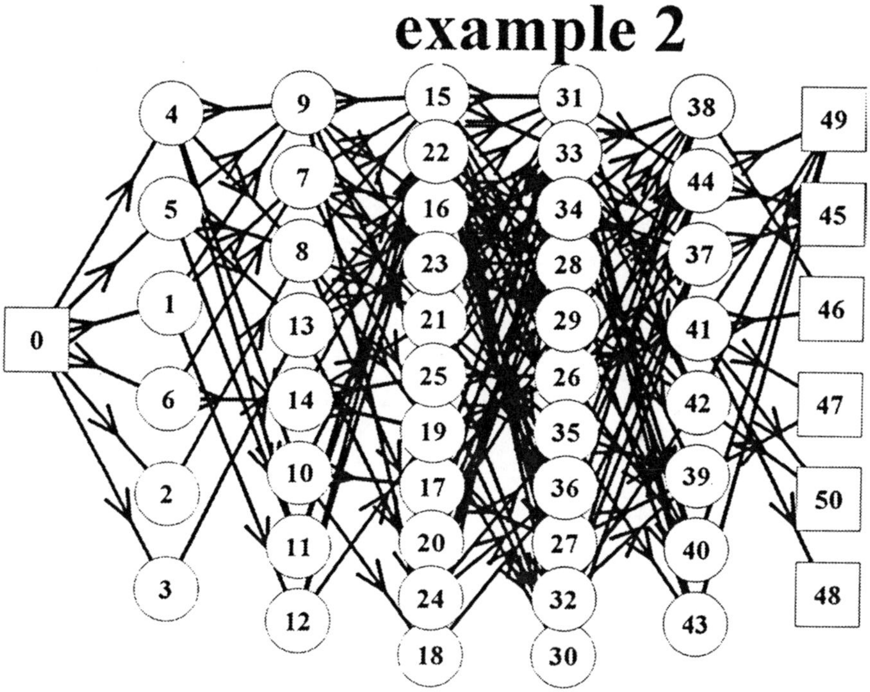


FIG. 4.12
NETWORK OF EXAMPLE 2

# example 3, Adjacency Matrix

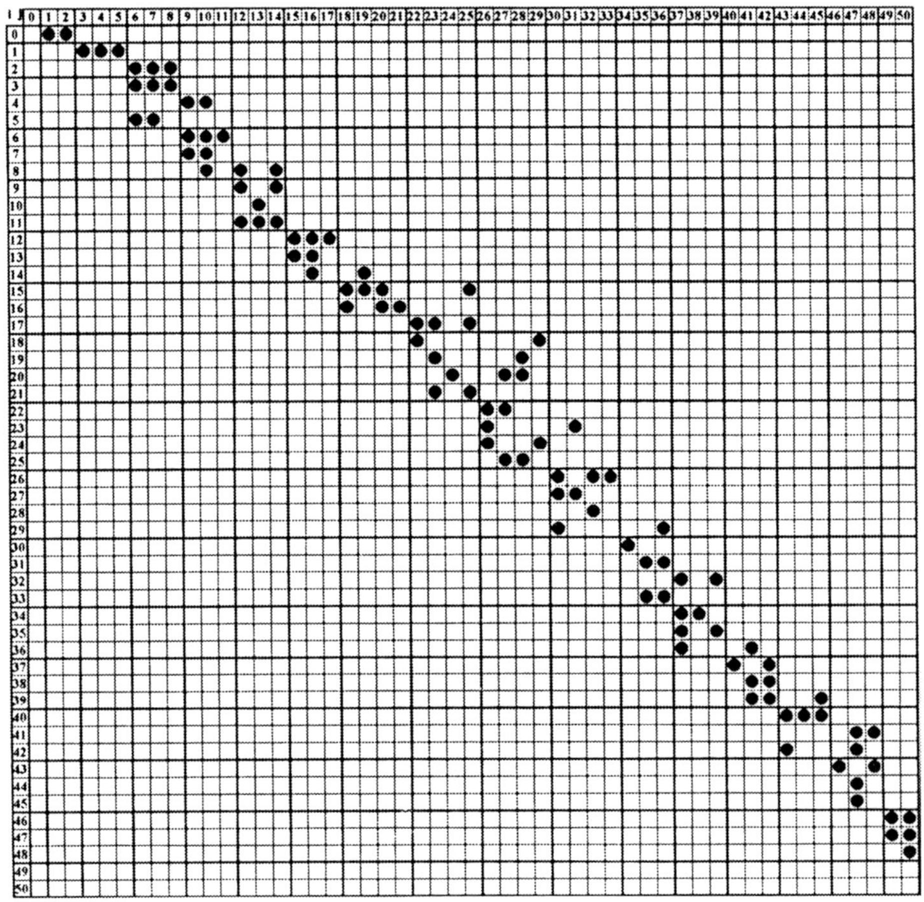

FIG. 4.13
ADJACENCY MATRIX OF EXAMPLE 3

| i \ k | 0 | 1 | 2 |
|---|---|---|---|
| 0 | 1 | 2 | |
| 1 | 2 | 3 | 3 |
| 2 | 3 | 5 | 2 |
| 3 | 3 | 6 | 2 |
| 4 | 3 | 6 | |
| 5 | 3 | 6 | 1 |
| 6 | 3 | 6 | 4 |
| 7 | 4 | 5 | 4 |
| 8 | 4 | 7 | 1 |
| 9 | 4 | 7 | 1 |
| 10 | 4 | 5 | 2 |
| 11 | 3 | 5 | 1 |
| 12 | 3 | 6 | 1 |
| 13 | 3 | 4 | 3 |
| 14 | 3 | 4 | |
| 15 | 3 | 5 | |
| 16 | 2 | | |

**FIG. 4.14**
**LEVEL ADJACENCY MATRIX OF EXAMPLE 3**

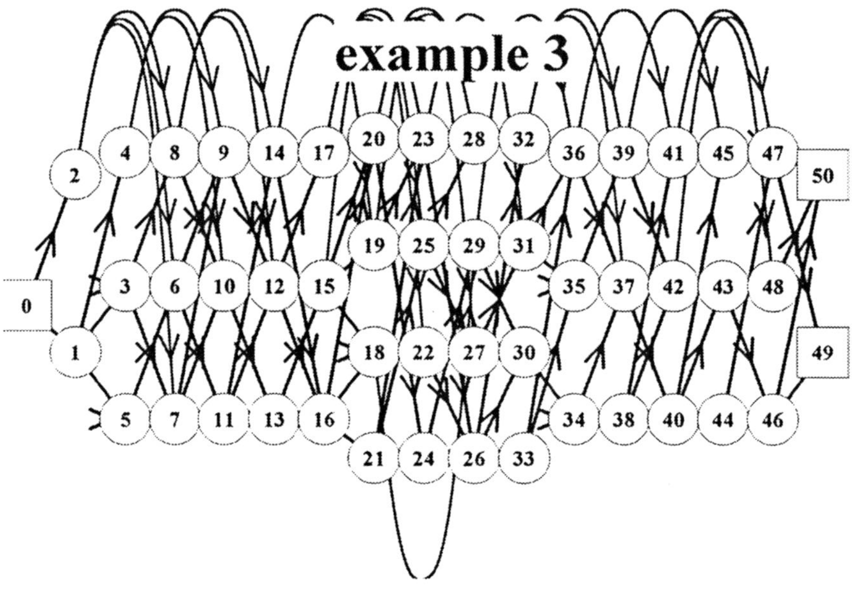


FIG. 4.15
NETWORK OF EXAMPLE 3

| i \ k | 0 | 1 | 2 |
|---|---|---|---|
| 0 | 1 | 6 | |
| 1 | 6 | 13 | 2 |
| 2 | 8 | 16 | 2 |
| 3 | 9 | 16 | 3 |
| 4 | 10 | 21 | 3 |
| 5 | 11 | 22 | |
| 6 | 12 | 25 | 3 |
| 7 | 11 | 20 | 3 |
| 8 | 10 | 21 | 1 |
| 9 | 9 | 13 | 1 |
| 10 | 8 | 8 | |
| 11 | 6 | | |

FIG. 4.16
LEVEL ADJACENCY MATRIX OF EXAMPLE 4

| i \ k | 0 | 1 | 2 | 3 |
|---|---|---|---|---|
| 0 | 1 | 33 | | |
| 1 | 33 | 57 | 6 | |
| 2 | 36 | 58 | 9 | 1 |
| 3 | 39 | 72 | 11 | |
| 4 | 42 | 75 | 5 | 1 |
| 5 | 45 | 78 | 11 | |
| 6 | 48 | 109 | 11 | 3 |
| 7 | 51 | 116 | 6 | 4 |
| 8 | 55 | 122 | 14 | |
| 9 | 58 | 132 | 14 | 1 |
| 10 | 61 | 145 | 20 | 2 |
| 11 | 64 | 160 | 19 | 2 |
| 12 | 61 | 131 | 8 | 4 |
| 13 | 58 | 111 | 10 | |
| 14 | 55 | 124 | 11 | |
| 15 | 51 | 87 | 13 | |
| 16 | 48 | 90 | 8 | |
| 17 | 45 | 77 | 12 | |
| 18 | 42 | 63 | 3 | 1 |
| 19 | 39 | 59 | 4 | |
| 20 | 36 | 48 | | |
| 21 | 33 | | | |

FIG. 4.17
LEVEL ADJACENCY MATRIX OF EXAMPLE 5

# 5

# DURATION OF PROJECTS

## *1. Modelling*

The study of the duration of a project under uncertainty conditions is no doubt responsible for a large fraction of the developments in Project Management.

Two major problems can be considered:

A - Modeling the duration of each activity.

B - Studying the statistical distribution of the total duration under a fixed static scheduling rule, for instance, starting each activity as early as possible.

Problem A

The initial work proposing PERT (see Kelley et al, 1959 and Malcolm et al, 1958) has adopted three estimates for the distribution of each duration:

a) the optimistic estimate → o
b) the most likely duration → m
c) the pessimist estimate → p

Clark suggested the use of the **Beta law** with

$$\mu = \frac{1}{6}(o + 4m + p)$$

$$\sigma^2 = \left(\frac{p - o}{6}\right)^2$$

Many other alternative continuous distributions have been proposed and the **Normal or Lognormal** law is often adopted.

The former may be preferred when the duration can be considered as a sum of partial durations independently and equally distributed as happens in the case of some projects of Marketing (opinion surveys, etc).

The latter is more appropriate to describe the distribution of activities with a finite lower bound due to physical or contractual reasons. For instance, many types of construction activities cannot be carried out in less than a minimal time because of physical constraints (consolidation of concrete, etc.). Also, there is no motivation to carry out an activity in less than a lower limit if it is sub-contrated to another institution and if there is no reward for finishing it before such limit.

The adoption of the lognormal law seems very appropriate to all cases with a lower bound (which can be zero) and with a positive skewness which represents the possibility of occurring unbounded long delays with a small probability due to unforeseen reasons or factors. Dean, Burton, S. Mertel Jr., L. Roepke, 1969 have shown that this distribution fits quite well to several cases.

The Negative Exponential Law also can be used (Maggot and Skudlarski, 1993) and it may be considered as an extreme and limit case of the previous distribution when the mode and the lower bound have the same value.

Another type of situation can be considered if two main outcomes occur:

**a** - no delay: then the activity has a fixed duration according to the corresponding contract.

**b** - delay: in this case the activity has the contractual duration plus a random delay.

This type of duration can be modelled by a mixed distribution with a finite probability allocated to **a** and a continuous law describing **b** (see Fig. 5.1):

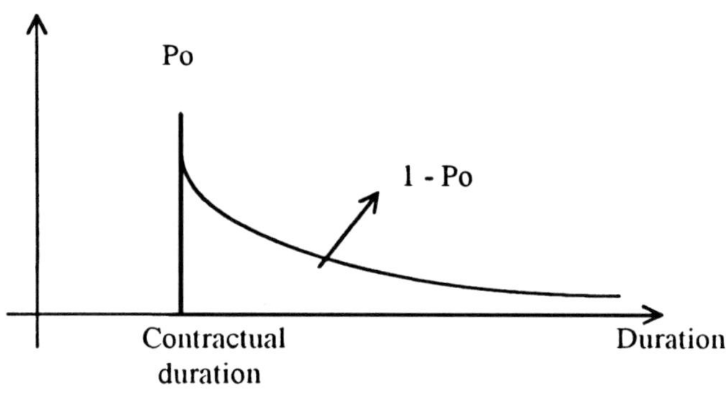


**FIG. 5.1**
**EXAMPLE OF A MIXED DISTRIBUTION**

The continuous part of the distribution can be modelled by different laws but the Negative Exponential Law may be a very convenient choice.

This model was proposed by Tavares 1986 and it seems to be very appropriate to many examples where the activity is sub-contracted to another organization.

Obviously, it can be generalized to the case of offering a prize if the contract is achieved in a smaller duration and then we may have (Fig. 5.2):

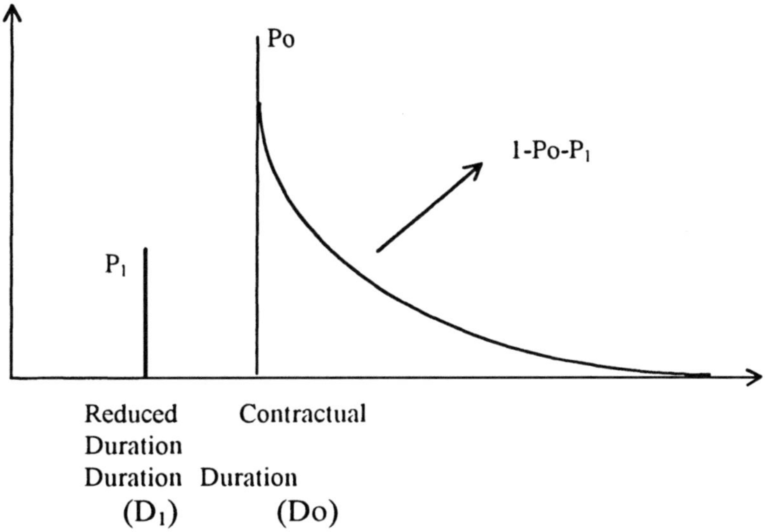


**FIG. 5.2**
**EXAMPLE OF A MIXED DISTRIBUTION WITH A REDUCED DURATION**

The assignment of no probability to the duration between $D_o$ and $D_1$ is due to the lack of incentives to the organization for having $D_1<D<D_D$.

The importance of the exponential law also stems from results deduced by Kamburowski, 1985.

A few definitions and deductions have to be presented to understand the final conclusions.

1. Mean of the random duration, X (with $X \geq 0$)

The usual definition is:

$\mu_X = \int_0^\infty f \cdot X dX$ where O is the lower bound of its domain but it can be easily shown that

$$\mu_X = \int_O^{\infty}\left[1 - F_x(t)\right]dt$$

because

$$\frac{d}{dt}\left[(1 - F_X(t).t\right] = (1 - F_X(t)) - f_X(t) \cdot t$$

and so

$$\int_O^{\infty}\frac{d}{dt}\left[(1 - F_X(t)) \cdot t\right]dt = \int_O^{\infty}(1 - F_X(t))dt - \int_O^{\infty} f(x) \cdot t\,dt$$

with $\left\{\left[1 - F_X(t) \cdot t\right]\right\}_O^{\infty} = O$

whenever $\lim\limits_{t \to \infty} \dfrac{1 - F_X(t)}{1/t} = \dfrac{f_X(t)}{{}^1/_{t^2}} = 0$

which means that $f_x$ tends to zero with an order higher than 2 if $t \to \infty$.

2. Given two random variables, X, Y $\geq$ O, X is said stochastically smaller than Y, $X \prec Y$, if, for any L $\geq$ O, one has $\int_L^{\infty}(1 - F_X(t))dt \leq \int_L^{\infty}(1 - F_y(t))dt$.

Therefore, using the previous result, one has, with L=0, $\mu_x \leq \mu_y$.

3. If $X_i \prec Y_i$ with i = 1, ..., N where these are independent random variables, then

$E\left[g(X_1 \ldots X_N)\right] \leq E\left[g(Y_1 \ldots Y_N\right]$ if g is a non-decreasing convex function.

The sum and the maximum are examples of this type of function.

4. A distribution is called NBUE if $\int_L^{\infty}(1 - F_X(t))dt \leq \mu_X \cdot (1 - F_X(L))$ with L $\geq$ 0, and equality holds if the distribution follows a negative exponential law.

Most of the adopted distributions like Normal, Beta or Lognormal are NBUE.

5. If X is NBUE and Y is negative exponential with $\mu_x = \mu_y$, then $X \prec Y$.

6. Being $X_1, \ldots, X_N$ independent random variables, and $Z = \max(X_1, \ldots, X_N)$; $W = \sum_{i=1}^{N} X_i$ one has that Z, W are NBUE. Therefore, $Z \prec u$; $W \prec v$,

where u, v, are negative exponential variables with $\mu_u = \mu_z$ and $\mu_v = \mu_w$.

This conclusion is quite useful to estimate an upper bound for the expected total duration because the distribution of the duration of the activities can be substituted by a negative exponential law with the guarantee of obtaining an upper bound.

Problem B

The study of the total duration cannot be pursued by exact methods because the analytical determination of its distribution is usually too complex as was mentioned before.

Three major approaches have been developed to overcome this problem: approximate estimation of the distribution of the total duration; computing lower of upper bounds for this distribution; estimation of bounds for the expected project total duration.

Most of the methods are not easily applied and have been rarely used in real applications.

Therefore, simulation seems to be the most appropriate approach and this now can be more easily carried out due to the simulation model presented in Chapter 4.

For each set of values of the indicators $I_1, \ldots, I_6$, a certain number, S, of networks is generated.

The realization of each network implies the generation of the duration of each activity.

Then, the statistical distribution of the duration of each activity i, with $i=1, \ldots, N$ has to be defined. For comparative purposes, the following assumptions are adopted:

- the distribution follows either the normal (Case A) the lognormal law (Case B) or the negative exponential law (Case C);

- the distribution is specified by the mean, $\mu_i$, and by the coefficient of variation ($C_v$) for Case A or by $C_V$ and the lower bound ($B_i$) for Case B or by its mean, $\mu_i$ for Case C. $C_v$ is constant for all activities and $B_i = \mu_i/2$.

- for each activity, i, the mean $\mu_i$ is generated using a uniform generator, $a_i$, within $[a-\Delta, a+\Delta]$ and scaled in order that the total duration , TD, of the critical path, CP, assuming that the duration of each activity, i, will be equal to its mean, $\mu_i$, will be made equal to 1000:

$$\mu_i = a_i \text{ x } \frac{1000}{\sum\limits_{i \in CP} a_i}$$

Under these assumptions, the realization of each network is generated Z times. Therefore, the simulation of each case requires the generation of S.Z instances.

For each case, the statistical distribution of the total duration, $D_T$, divided by the duration TD = 1000, t = TD/1000 can be estimated in terms of the morphological features described by $I_1,...,I_6$.

These results allow the direct estimation of any measure of the risk of delay.

The estimated parameters of the distribution of t are the mean $\mu$, the standard deviation $\sigma$ and the 95% - 5% quantiles, $Q_{0.95}$, $Q_{0.05}$.

This methodology for simulating project networks can be used to estimate the relationship between the total duration of the project, $D_T$, and the morphologic features of its network.

## 2. *Experimental results*

The application of this methodology (Tavares, 1985 and Tavares, Ferreira and Coelho, 1997) was done using the following data:

$S = 100$; $Z = 100$
$C_V = 0.5$; $a = 1000$ and $\Delta = 10$
$I_1 = 10 \rightarrow 100$
$I_2 = 0.1 \rightarrow 1.0$
$I_3$ corresponding to a width for the first or last hierarchical level equal to half of the value for the central level and a linear variation between them.

$I_4 = 0 \rightarrow 0.4$
$I_5 = 0 \rightarrow 0.3$
$I_6 = 0 \rightarrow 0.3$

The upper bounds of $I_4$, $I_5$, $I_6$ have been chosen lower than 1 because the networks generated with these indicators above the selected bounds have precedence links much longer than is usually found in real project networks. The statistical study of t is done through the estimated average, $\mu_t$ and standard deviation, $\sigma_t$, and of the estimated 5% and 95% quantiles, $Q_{0.05}$ and $Q_{0.95}$.

Several types of analysis can be carried out for the studied cases using the generated results:

A - t is studied in terms of $I_1$ ($I_1 = 10 \rightarrow I_1 = 100$) making the other indicators constant and equal to $I_2 = 0.20$; $I_4 = 0.10$; $I_5 = 0.20$; $I_6 = 0.20$.

The software RISKNET - module 6 - presented in Annex 1 allows the estimation of all these features for any set of given or generated networks.

The obtained results are presented in Figures 5.3, 5.4 and 5.5 showing that $\mu$ ,$\sigma$, and the upper quantiles decrease when $I_1$ increases as could be expected because a larger number of activities is responsible for a decrease of the variance of the total duration along the relevant paths. On the contrary, $Q_{0.05}$ increases with $I_1$ but the gap between $Q_{0.95}$ and $\mu$ in Case B or C is much higher than between $\mu$ and $Q_{0.05}$ due to the positive skewness of the distribution of $D_T$. $Q_{0.095}$ and $\mu$ is higher for Case C than for the other cases as it should be expected.

B - t as a function of the number of levels

In this case, t is studied in terms of $I_2$ ($I_2$ = 0.1→1.0) measuring how serial is network. The other indicators are made constant and equal to $I_1$ = 50; $I_4$ = 0.10; $I_5$ = 0.20; $I_6$ = 0.20.

The obtained results are presented in Figure 5.6, 5.7 and 5.8 and the following comments can be made:

- there is an exponential decrease of $D_T$ if $I_2$ is increased which is due to the increase of the number of activities belonging to the critical path.
- the gap between $Q_{0.95}$ or $Q_{0.05}$ decreases rapidly until $I_2$ is around 0.5.
- $Q_{0.95}$ and $\mu$ for Case C exceeds the same estimates for the other cases, as it should be.

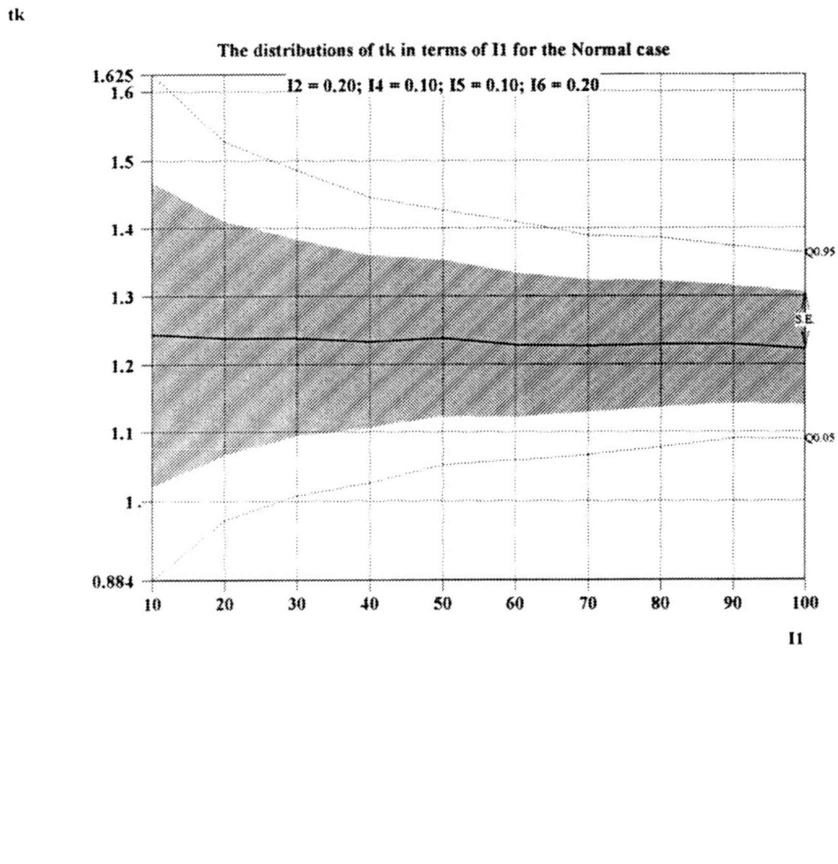


FIGURE 5.3
THE DISTRIBUTION OF $t_K$ IN TERMS OF I1 FOR THE NORMAL CASE

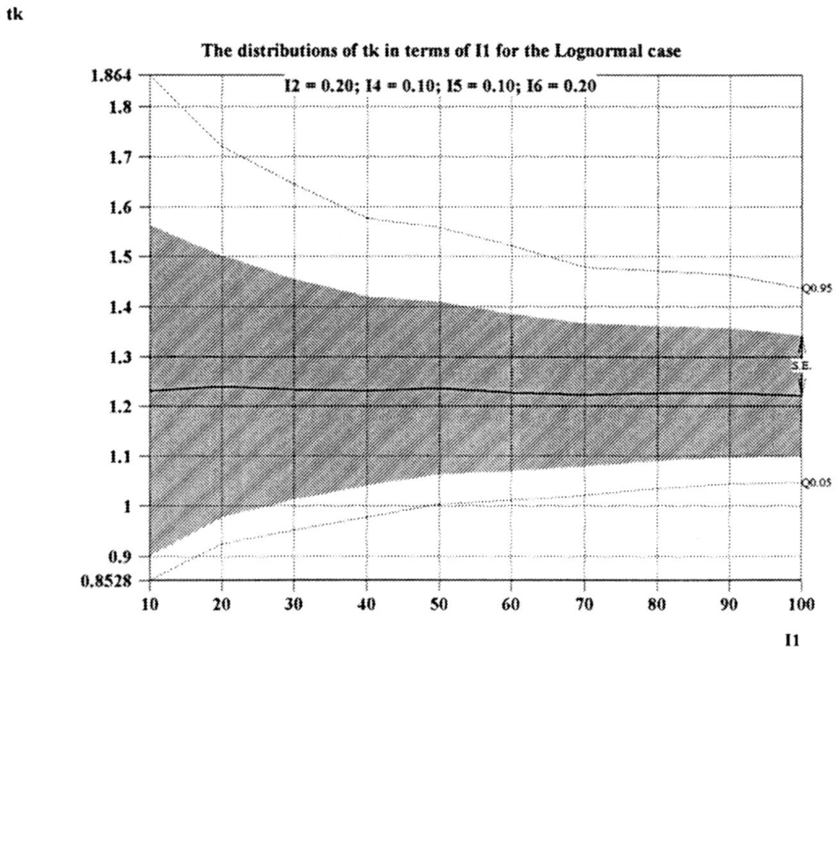


FIGURE 5.4
THE DISTRIBUTION OF $t_K$ IN TERMS OF I1 FOR THE LOGNORMAL CASE

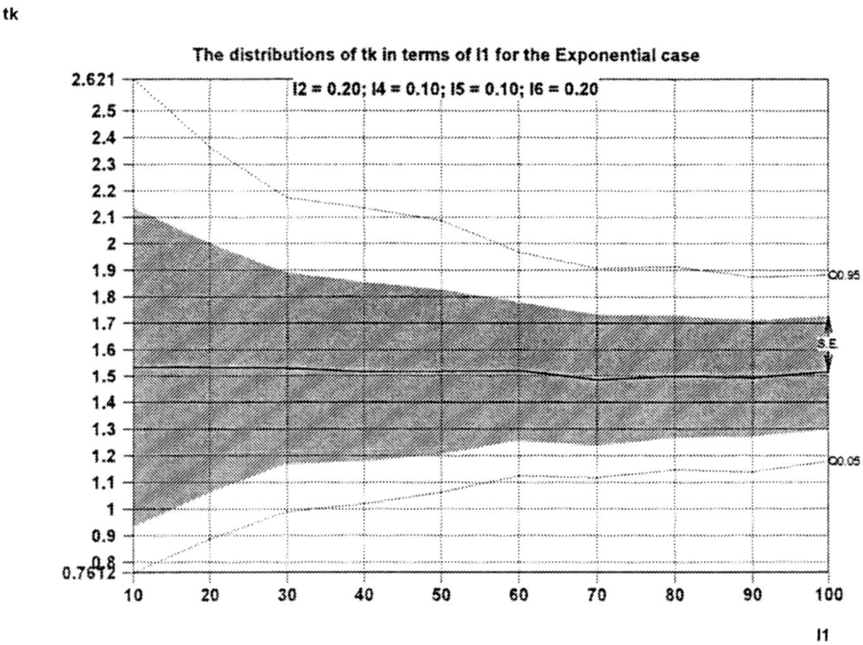


FIGURE 5.5
THE DISTRIBUTION OF $t_K$ IN TERMS OF I1 FOR THE EXPONENTIAL CASE

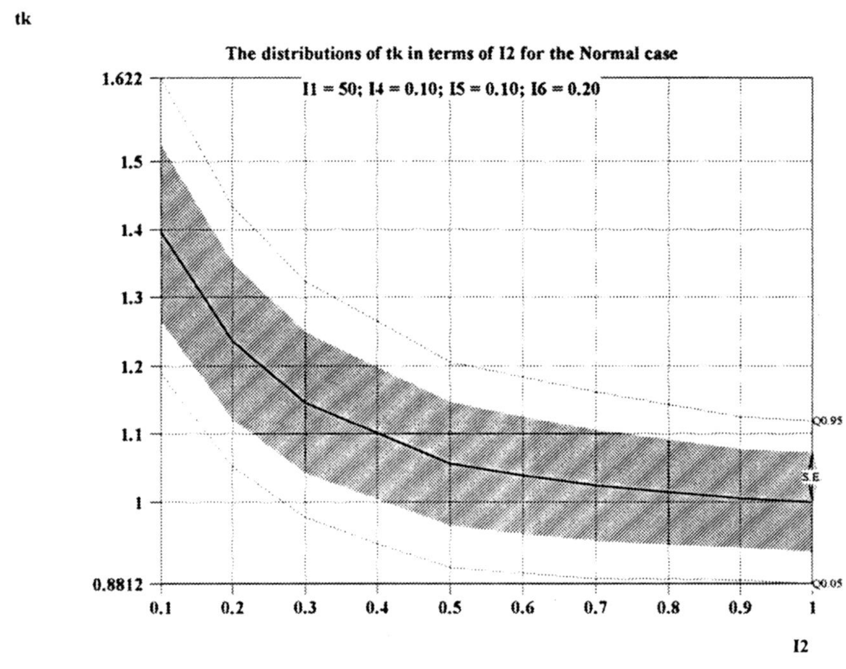


FIGURE 5.6
THE DISTRIBUTION OF $t_K$ IN TERMS OF I2 FOR THE NORMAL CASE

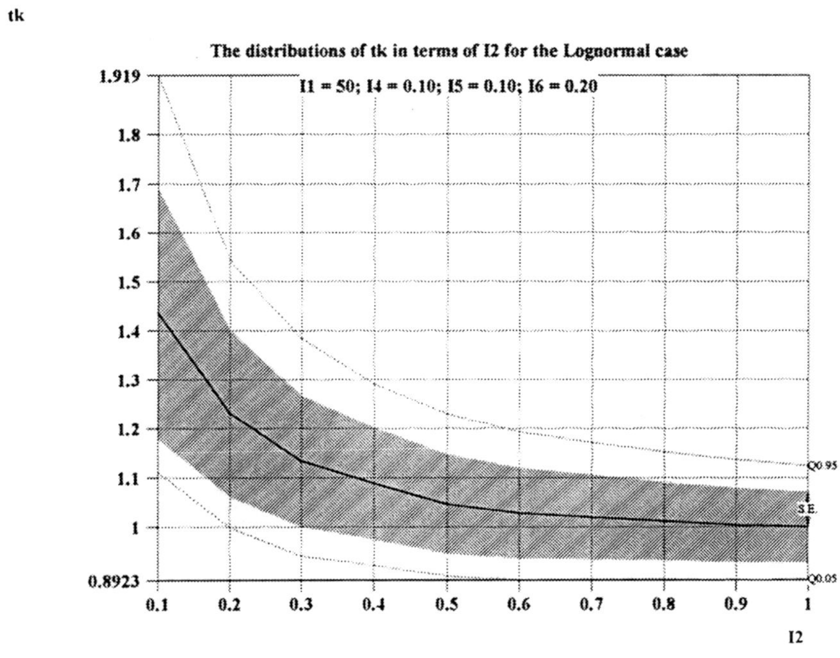


FIGURE 5.7
THE DISTRIBUTION OF $t_K$ IN TERMS OF I2 FOR THE LOGNORMAL CASE

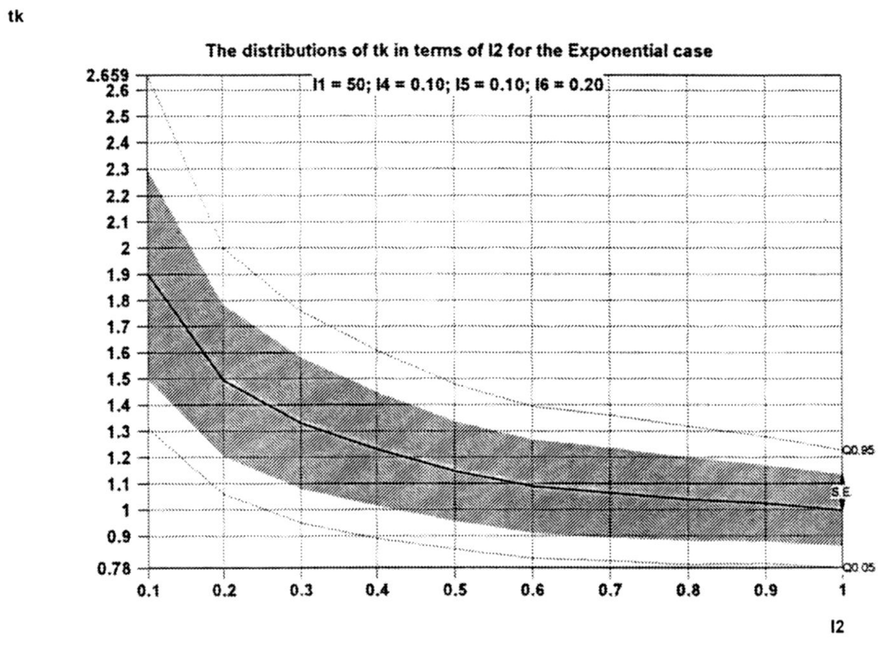


**FIGURE 5.8**

**THE DISTRIBUTION OF $t_K$ IN TERMS OF I2 FOR THE EXPONENTIAL CASE**

C - t as a function of the number of precedence links

t is studied in terms of $I_4$ (0→0.5) making the other indicators constant and equal to $I_1 = 50$; $I_2 = 0.20$; $I_5 = 0.20$; $I_6 = 0.20$.

The results are presented in Figures 5.9, 5.10 and 5.11 showing that there is a steady increase of t with $I_4$. This means that a variation of the number of links easily can increase the average duration from 1.1 (Cases A and B) or 1.25 (Case C) to more than 1.4 (Cases A and B) or 1.9 (Case C).

D - $D_T$ as a function of the depth of the precedence links

$D_T$ is studied in terms of $I_5$ or $I_6$ ($I_5 = I_6 = 0 \rightarrow 0.40$) making constant the other indicators and equal to $I_1 = 0.50$; $I_2 = 0.20$; $I_4 = 0.10$.

The obtained results show that the statistical parameters of $D_T$ are not sensitive to these indicators. As an example, the Figures 5.12, 5.13 and 5.14 show the variation in terms of $I_6$.

Therefore, the statistical study of $\mu_t$, $\sigma_t$, $Q_{0.05}$ and $Q_{0.95}$ should be carried out just in terms of $I_1$, $I_2$ and $I_4$ because $I_6$ (or $I_5$) have no significant effects.

Tables 5.1 to 5.12 summarize the results obtained which can be described by a regression model as is shown in Table 5.13 and in Figures 5.15 to 5.26.

The proposed model allows the estimation of the parameters of the distribution of $D_T$ in terms of the morphology of the network.

Obviously, the risk of a delay above a specific threshold can be estimated then. The upper quantile, $Q_{0.95}$, is also an interesting measure of an extreme delay (just exceeded with a probability of 5%) and it is directly given by the model.

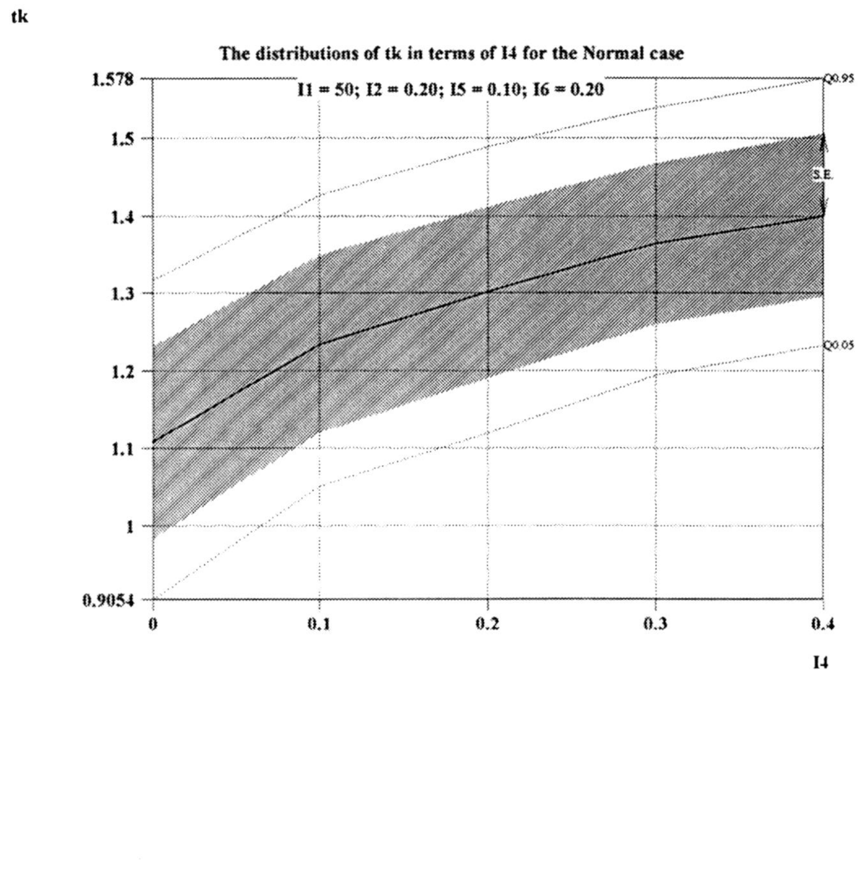


FIGURE 5.9

THE DISTRIBUTION OF $t_K$ IN TERMS OF I4 FOR THE NORMAL CASE

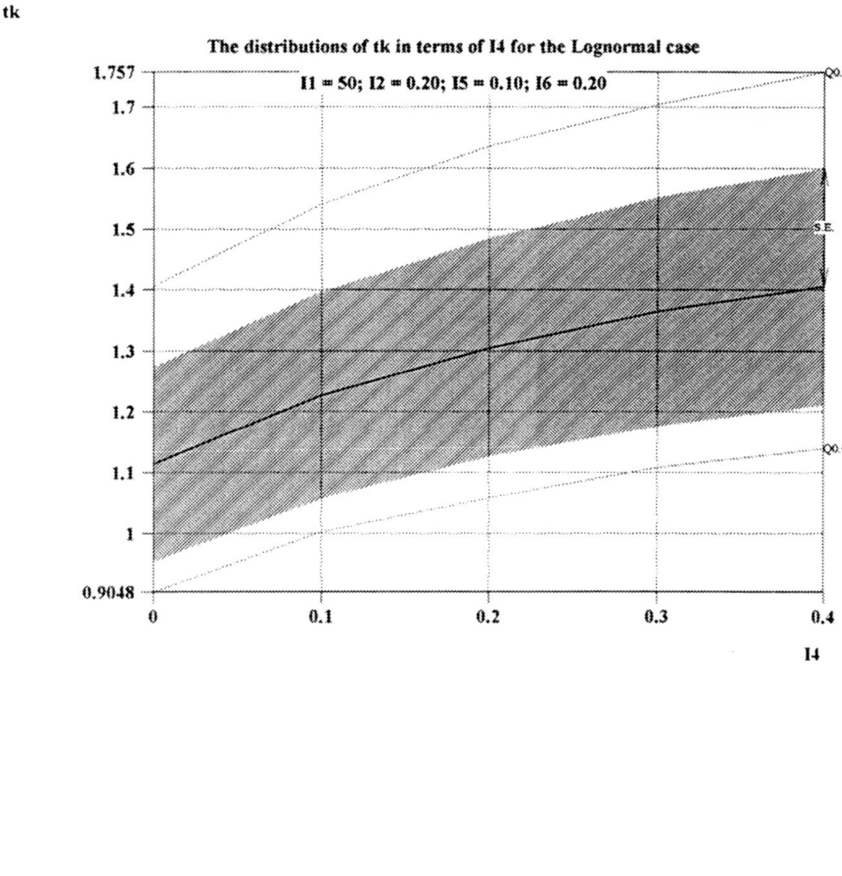


**FIGURE 5.10**
**THE DISTRIBUTION OF $t_K$ IN TERMS OF I4 FOR THE LOGNORMAL CASE**

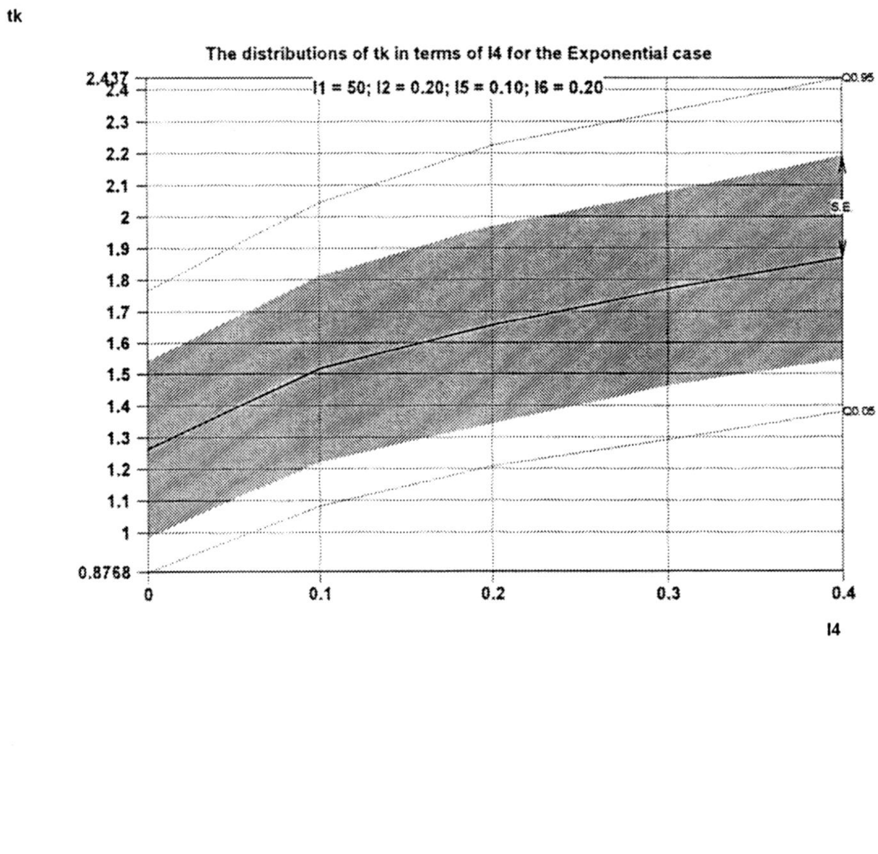


**FIGURE 5.11**
**THE DISTRIBUTION OF $t_K$ IN TERMS OF I4 FOR THE EXPONENTIAL CASE**

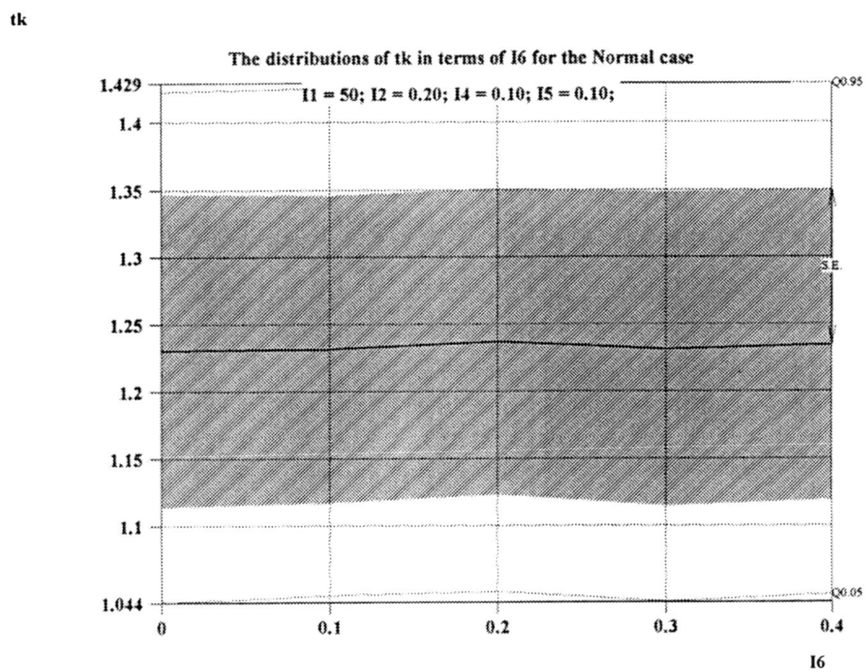
tk
The distributions of tk in terms of I6 for the Normal case
I1 = 50; I2 = 0.20; I4 = 0.10; I5 = 0.10;
1.429
1.4
1.35
1.3
1.25
1.2
1.15
1.1
1.044
Q0.95
S.E.
Q0.05
0
0.1
0.2
0.3
0.4
I6

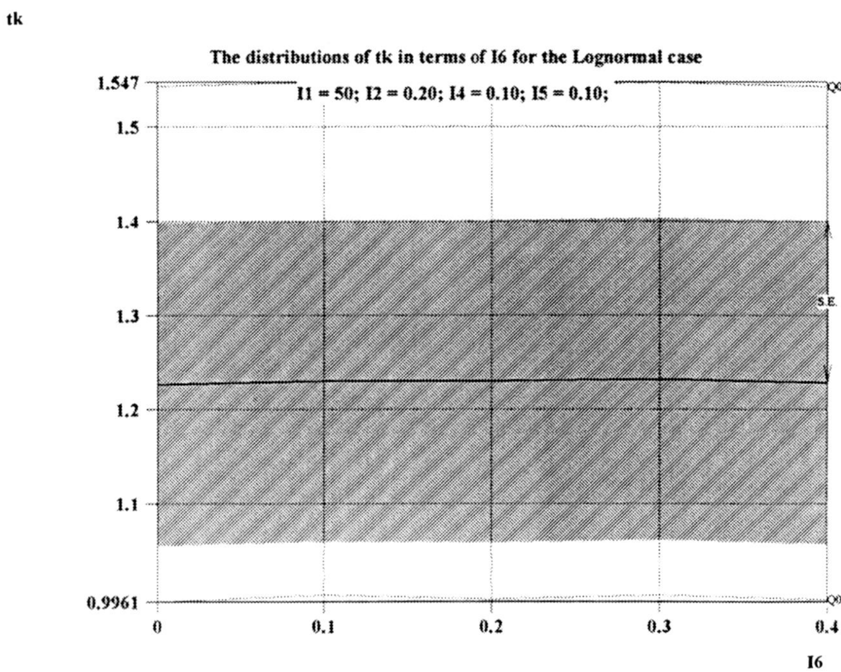


FIGURE 5.13

THE DISTRIBUTION OF $t_K$ IN TERMS OF I6 FOR THE LOGNORMAL CASE

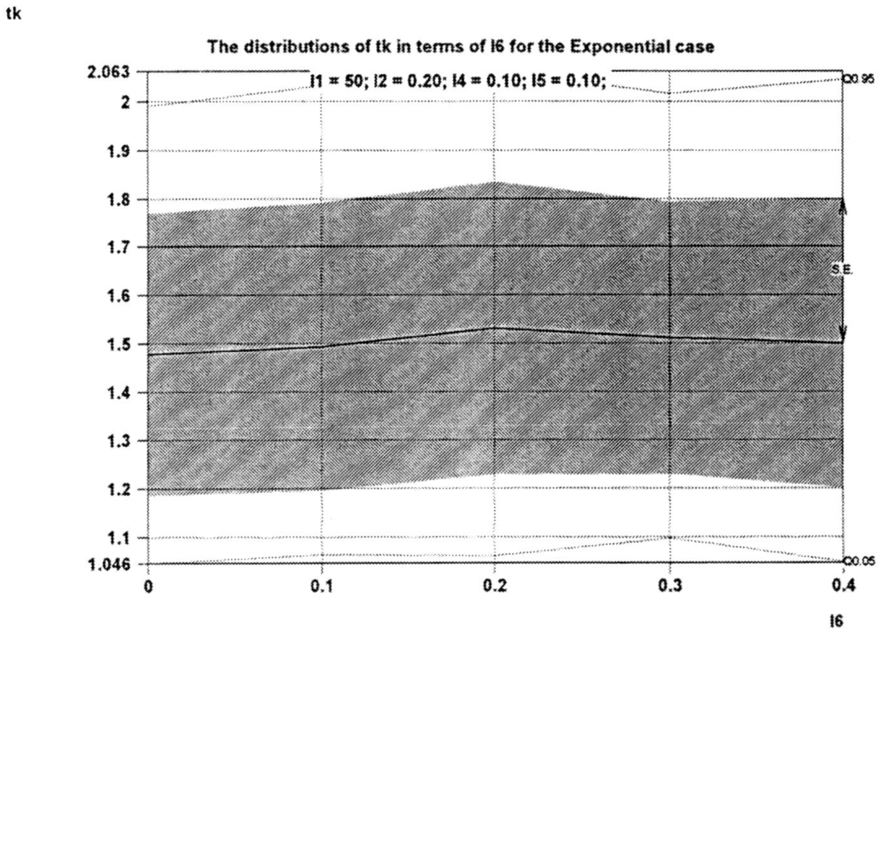


FIGURE 5.14
THE DISTRIBUTION OF $t_K$ IN TERMS OF I6 FOR THE EXPONENTIAL CASE

| | | I2 | | | | | | | | | |
|---|---|---|---|---|---|---|---|---|---|---|---|
| **I1** | **I4** | **0.1** | **0.2** | **0.3** | **0.4** | **0.5** | **0.6** | **0.7** | **0.8** | **0.9** | **1** |
| **10** | **0** | 1.380 | 1.218 | 1.116 | 1.084 | 1.023 | 1.020 | 1.004 | 1.002 | 1.000 | 1.001 |
| | **0.1** | 1.424 | 1.249 | 1.158 | 1.085 | 1.020 | 1.020 | 1.004 | 1.000 | 1.001 | 1.002 |
| | **0.2** | 1.457 | 1.290 | 1.185 | 1.117 | 1.056 | 1.061 | 1.037 | 0.999 | 0.998 | 1.002 |
| | **0.3** | 1.482 | 1.333 | 1.210 | 1.150 | 1.091 | 1.089 | 1.039 | 1.001 | 0.997 | 1.001 |
| | **0.4** | 1.502 | 1.341 | 1.245 | 1.183 | 1.088 | 1.088 | 1.066 | 1.032 | 0.999 | 1.003 |
| | **0.5** | 1.521 | 1.368 | 1.257 | 1.202 | 1.108 | 1.113 | 1.067 | 1.033 | 1.002 | 1.001 |
| **20** | **0** | 1.311 | 1.159 | 1.077 | 1.048 | 1.014 | 1.002 | 1.001 | 1.002 | 1.000 | 1.000 |
| | **0.1** | 1.411 | 1.240 | 1.147 | 1.095 | 1.038 | 1.030 | 1.024 | 1.001 | 1.002 | 1.001 |
| | **0.2** | 1.471 | 1.302 | 1.197 | 1.133 | 1.072 | 1.065 | 1.036 | 1.015 | 1.002 | 0.998 |
| | **0.3** | 1.510 | 1.339 | 1.231 | 1.155 | 1.107 | 1.080 | 1.051 | 1.032 | 1.000 | 1.002 |
| | **0.4** | 1.540 | 1.377 | 1.267 | 1.192 | 1.119 | 1.101 | 1.060 | 1.031 | 1.017 | 1.001 |
| | **0.5** | 1.563 | 1.403 | 1.296 | 1.215 | 1.144 | 1.115 | 1.071 | 1.040 | 1.016 | 1.002 |
| **30** | **0** | 1.285 | 1.123 | 1.059 | 1.033 | 1.008 | 1.003 | 1.001 | 1.000 | 1.000 | 1.001 |
| | **0.1** | 1.412 | 1.233 | 1.151 | 1.093 | 1.042 | 1.031 | 1.015 | 1.014 | 1.002 | 1.001 |
| | **0.2** | 1.486 | 1.296 | 1.203 | 1.132 | 1.080 | 1.057 | 1.033 | 1.022 | 1.010 | 1.000 |
| | **0.3** | 1.535 | 1.350 | 1.243 | 1.165 | 1.112 | 1.075 | 1.054 | 1.030 | 1.011 | 0.999 |
| | **0.4** | 1.568 | 1.388 | 1.278 | 1.197 | 1.131 | 1.101 | 1.061 | 1.034 | 1.018 | 1.000 |
| | **0.5** | 1.598 | 1.419 | 1.307 | 1.224 | 1.156 | 1.114 | 1.071 | 1.041 | 1.017 | 1.000 |
| **40** | **0** | 1.251 | 1.101 | 1.047 | 1.025 | 1.007 | 1.001 | 1.001 | 1.001 | 1.001 | 1.000 |
| | **0.1** | 1.403 | 1.227 | 1.145 | 1.089 | 1.047 | 1.037 | 1.021 | 1.010 | 0.999 | 1.002 |
| | **0.2** | 1.487 | 1.301 | 1.200 | 1.140 | 1.085 | 1.060 | 1.035 | 1.023 | 1.007 | 1.002 |
| | **0.3** | 1.537 | 1.353 | 1.243 | 1.171 | 1.112 | 1.082 | 1.049 | 1.028 | 1.013 | 0.999 |
| | **0.4** | 1.576 | 1.394 | 1.280 | 1.201 | 1.136 | 1.102 | 1.066 | 1.038 | 1.013 | 1.000 |
| | **0.5** | 1.608 | 1.428 | 1.314 | 1.228 | 1.160 | 1.114 | 1.073 | 1.043 | 1.017 | 1.000 |
| **50** | **0** | 1.224 | 1.098 | 1.037 | 1.021 | 1.004 | 1.000 | 1.000 | 1.000 | 1.000 | 1.000 |
| | **0.1** | 1.391 | 1.228 | 1.143 | 1.097 | 1.048 | 1.030 | 1.017 | 1.010 | 1.000 | 0.999 |
| | **0.2** | 1.480 | 1.307 | 1.204 | 1.139 | 1.086 | 1.059 | 1.035 | 1.020 | 1.005 | 1.000 |
| | **0.3** | 1.540 | 1.359 | 1.244 | 1.174 | 1.118 | 1.079 | 1.049 | 1.027 | 1.010 | 1.001 |
| | **0.4** | 1.580 | 1.402 | 1.283 | 1.201 | 1.136 | 1.099 | 1.064 | 1.036 | 1.013 | 1.000 |
| | **0.5** | 1.611 | 1.439 | 1.317 | 1.230 | 1.162 | 1.117 | 1.074 | 1.043 | 1.020 | 1.000 |
| **60** | **0** | 1.202 | 1.070 | 1.029 | 1.016 | 1.004 | 1.001 | 0.999 | 0.999 | 1.000 | 1.000 |
| | **0.1** | 1.387 | 1.225 | 1.146 | 1.090 | 1.050 | 1.032 | 1.019 | 1.013 | 1.005 | 0.998 |
| | **0.2** | 1.483 | 1.300 | 1.200 | 1.135 | 1.090 | 1.062 | 1.040 | 1.023 | 1.010 | 0.999 |
| | **0.3** | 1.545 | 1.356 | 1.248 | 1.176 | 1.118 | 1.086 | 1.051 | 1.029 | 1.013 | 1.000 |
| | **0.4** | 1.583 | 1.399 | 1.285 | 1.206 | 1.141 | 1.097 | 1.066 | 1.036 | 1.016 | 0.998 |
| | **0.5** | 1.618 | 1.433 | 1.318 | 1.231 | 1.166 | 1.117 | 1.074 | 1.044 | 1.018 | 0.999 |
| **70** | **0** | 1.183 | 1.062 | 1.027 | 1.013 | 1.005 | 1.001 | 1.000 | 1.001 | 1.001 | 1.000 |
| | **0.1** | 1.390 | 1.231 | 1.138 | 1.090 | 1.050 | 1.035 | 1.021 | 1.009 | 1.005 | 0.999 |
| | **0.2** | 1.486 | 1.304 | 1.202 | 1.137 | 1.088 | 1.063 | 1.038 | 1.021 | 1.009 | 0.999 |
| | **0.3** | 1.547 | 1.360 | 1.251 | 1.173 | 1.119 | 1.084 | 1.053 | 1.030 | 1.010 | 0.999 |
| | **0.4** | 1.588 | 1.398 | 1.284 | 1.206 | 1.141 | 1.103 | 1.064 | 1.036 | 1.016 | 1.000 |
| | **0.5** | 1.623 | 1.435 | 1.320 | 1.232 | 1.168 | 1.118 | 1.077 | 1.043 | 1.019 | 1.000 |
| **80** | **0** | 1.168 | 1.062 | 1.021 | 1.012 | 1.003 | 1.001 | 0.999 | 1.000 | 1.000 | 0.999 |
| | **0.1** | 1.385 | 1.225 | 1.141 | 1.089 | 1.052 | 1.037 | 1.021 | 1.014 | 1.005 | 0.999 |
| | **0.2** | 1.483 | 1.305 | 1.203 | 1.138 | 1.088 | 1.059 | 1.039 | 1.024 | 1.010 | 0.999 |
| | **0.3** | 1.547 | 1.362 | 1.248 | 1.174 | 1.119 | 1.083 | 1.053 | 1.028 | 1.012 | 1.000 |
| | **0.4** | 1.592 | 1.406 | 1.288 | 1.204 | 1.144 | 1.101 | 1.063 | 1.037 | 1.017 | 1.000 |
| | **0.5** | 1.624 | 1.443 | 1.322 | 1.233 | 1.169 | 1.117 | 1.075 | 1.043 | 1.019 | 1.000 |
| **90** | **0** | 1.166 | 1.054 | 1.021 | 1.012 | 1.003 | 1.001 | 1.000 | 1.001 | 0.999 | 0.999 |
| | **0.1** | 1.389 | 1.223 | 1.141 | 1.090 | 1.050 | 1.037 | 1.021 | 1.011 | 1.004 | 0.999 |
| | **0.2** | 1.490 | 1.305 | 1.199 | 1.139 | 1.092 | 1.062 | 1.038 | 1.023 | 1.009 | 0.999 |
| | **0.3** | 1.554 | 1.361 | 1.250 | 1.173 | 1.125 | 1.082 | 1.052 | 1.030 | 1.012 | 1.000 |
| | **0.4** | 1.598 | 1.408 | 1.290 | 1.207 | 1.143 | 1.101 | 1.067 | 1.039 | 1.015 | 0.998 |
| | **0.5** | 1.632 | 1.445 | 1.321 | 1.233 | 1.171 | 1.120 | 1.074 | 1.044 | 1.019 | 0.999 |
| **100** | **0** | 1.148 | 1.044 | 1.016 | 1.010 | 1.003 | 1.000 | 0.998 | 0.999 | 1.000 | 1.000 |
| | **0.1** | 1.383 | 1.217 | 1.137 | 1.093 | 1.051 | 1.033 | 1.021 | 1.012 | 1.004 | 0.999 |
| | **0.2** | 1.486 | 1.302 | 1.200 | 1.138 | 1.090 | 1.060 | 1.039 | 1.021 | 1.008 | 1.000 |
| | **0.3** | 1.550 | 1.360 | 1.248 | 1.174 | 1.120 | 1.082 | 1.053 | 1.030 | 1.012 | 0.999 |
| | **0.4** | 1.596 | 1.406 | 1.290 | 1.208 | 1.146 | 1.102 | 1.066 | 1.037 | 1.016 | 0.998 |
| | **0.5** | 1.630 | 1.441 | 1.324 | 1.236 | 1.170 | 1.118 | 1.074 | 1.044 | 1.017 | 1.000 |

**TABLE 5.1**

**ESTIMATION OF THE MEAN OF $t_K$ IN TERMS OF $I_1$, $I_2$ AND $I_4$ (CASE A)**

| | | I2 | | | | | | | | | |
|---|---|---|---|---|---|---|---|---|---|---|---|
| **I1** | **I4** | **0.1** | **0.2** | **0.3** | **0.4** | **0.5** | **0.6** | **0.7** | **0.8** | **0.9** | **1** |
| **10** | **0** | 1.399 | 1.212 | 1.105 | 1.074 | 1.021 | 1.015 | 1.004 | 1.002 | 0.998 | 1.000 |
| | **0.1** | 1.441 | 1.238 | 1.142 | 1.074 | 1.018 | 1.019 | 1.003 | 1.000 | 0.999 | 1.001 |
| | **0.2** | 1.464 | 1.277 | 1.164 | 1.100 | 1.048 | 1.050 | 1.029 | 0.999 | 0.997 | 1.000 |
| | **0.3** | 1.494 | 1.313 | 1.186 | 1.127 | 1.076 | 1.074 | 1.031 | 1.001 | 0.996 | 1.000 |
| | **0.4** | 1.513 | 1.320 | 1.215 | 1.152 | 1.070 | 1.072 | 1.054 | 1.025 | 0.998 | 1.000 |
| | **0.5** | 1.534 | 1.340 | 1.221 | 1.167 | 1.088 | 1.093 | 1.054 | 1.025 | 1.001 | 1.001 |
| **20** | **0** | 1.351 | 1.163 | 1.075 | 1.042 | 1.012 | 1.003 | 1.001 | 1.001 | 0.999 | 0.999 |
| | **0.1** | 1.448 | 1.236 | 1.135 | 1.081 | 1.031 | 1.025 | 1.019 | 1.001 | 1.001 | 1.000 |
| | **0.2** | 1.508 | 1.292 | 1.179 | 1.112 | 1.060 | 1.054 | 1.030 | 1.011 | 1.000 | 0.998 |
| | **0.3** | 1.558 | 1.328 | 1.206 | 1.133 | 1.090 | 1.067 | 1.042 | 1.026 | 1.000 | 1.000 |
| | **0.4** | 1.584 | 1.364 | 1.239 | 1.165 | 1.099 | 1.083 | 1.049 | 1.024 | 1.011 | 0.999 |
| | **0.5** | 1.610 | 1.388 | 1.265 | 1.182 | 1.118 | 1.095 | 1.058 | 1.031 | 1.012 | 1.001 |
| **30** | **0** | 1.324 | 1.125 | 1.058 | 1.030 | 1.006 | 1.002 | 1.000 | 1.000 | 0.999 | 1.000 |
| | **0.1** | 1.454 | 1.230 | 1.137 | 1.081 | 1.035 | 1.026 | 1.013 | 1.011 | 1.002 | 1.000 |
| | **0.2** | 1.538 | 1.292 | 1.184 | 1.115 | 1.067 | 1.046 | 1.027 | 1.018 | 1.008 | 1.000 |
| | **0.3** | 1.595 | 1.344 | 1.220 | 1.143 | 1.094 | 1.062 | 1.044 | 1.025 | 1.009 | 0.999 |
| | **0.4** | 1.637 | 1.381 | 1.253 | 1.170 | 1.109 | 1.083 | 1.050 | 1.027 | 1.014 | 1.000 |
| | **0.5** | 1.669 | 1.412 | 1.281 | 1.191 | 1.129 | 1.095 | 1.057 | 1.034 | 1.013 | 0.999 |
| **40** | **0** | 1.279 | 1.101 | 1.041 | 1.022 | 1.006 | 1.000 | 1.000 | 1.000 | 1.000 | 1.000 |
| | **0.1** | 1.447 | 1.223 | 1.128 | 1.081 | 1.036 | 1.030 | 1.017 | 1.008 | 1.000 | 0.998 |
| | **0.2** | 1.541 | 1.295 | 1.179 | 1.123 | 1.070 | 1.052 | 1.030 | 1.019 | 1.005 | 0.999 |
| | **0.3** | 1.602 | 1.349 | 1.223 | 1.148 | 1.094 | 1.067 | 1.039 | 1.025 | 1.010 | 1.000 |
| | **0.4** | 1.659 | 1.389 | 1.254 | 1.174 | 1.110 | 1.082 | 1.051 | 1.031 | 1.011 | 1.000 |
| | **0.5** | 1.687 | 1.422 | 1.284 | 1.195 | 1.130 | 1.098 | 1.062 | 1.035 | 1.015 | 1.000 |
| **50** | **0** | 1.250 | 1.093 | 1.036 | 1.018 | 1.004 | 0.999 | 1.000 | 1.001 | 0.999 | 0.998 |
| | **0.1** | 1.438 | 1.223 | 1.125 | 1.083 | 1.043 | 1.026 | 1.015 | 1.006 | 0.999 | 1.000 |
| | **0.2** | 1.544 | 1.300 | 1.184 | 1.121 | 1.072 | 1.049 | 1.030 | 1.015 | 1.003 | 1.000 |
| | **0.3** | 1.615 | 1.355 | 1.227 | 1.152 | 1.099 | 1.065 | 1.042 | 1.023 | 1.007 | 1.000 |
| | **0.4** | 1.660 | 1.401 | 1.262 | 1.176 | 1.117 | 1.083 | 1.051 | 1.031 | 1.011 | 1.000 |
| | **0.5** | 1.698 | 1.435 | 1.290 | 1.199 | 1.134 | 1.097 | 1.063 | 1.036 | 1.016 | 0.998 |
| **60** | **0** | 1.225 | 1.074 | 1.029 | 1.015 | 1.005 | 1.000 | 0.999 | 0.999 | 1.000 | 1.000 |
| | **0.1** | 1.435 | 1.218 | 1.131 | 1.077 | 1.041 | 1.028 | 1.016 | 1.009 | 1.002 | 1.000 |
| | **0.2** | 1.546 | 1.297 | 1.186 | 1.125 | 1.073 | 1.050 | 1.035 | 1.017 | 1.007 | 1.001 |
| | **0.3** | 1.616 | 1.358 | 1.229 | 1.153 | 1.100 | 1.070 | 1.044 | 1.023 | 1.011 | 0.999 |
| | **0.4** | 1.666 | 1.399 | 1.264 | 1.176 | 1.116 | 1.082 | 1.055 | 1.031 | 1.013 | 0.999 |
| | **0.5** | 1.707 | 1.434 | 1.290 | 1.200 | 1.138 | 1.098 | 1.062 | 1.035 | 1.014 | 0.999 |
| **70** | **0** | 1.198 | 1.066 | 1.026 | 1.012 | 1.003 | 0.999 | 1.000 | 1.000 | 0.999 | 0.998 |
| | **0.1** | 1.428 | 1.220 | 1.123 | 1.079 | 1.042 | 1.031 | 1.018 | 1.009 | 1.003 | 0.999 |
| | **0.2** | 1.548 | 1.299 | 1.190 | 1.122 | 1.074 | 1.052 | 1.033 | 1.017 | 1.006 | 0.998 |
| | **0.3** | 1.617 | 1.359 | 1.228 | 1.150 | 1.100 | 1.066 | 1.044 | 1.025 | 1.008 | 0.999 |
| | **0.4** | 1.677 | 1.399 | 1.264 | 1.180 | 1.118 | 1.085 | 1.052 | 1.031 | 1.014 | 0.999 |
| | **0.5** | 1.715 | 1.439 | 1.295 | 1.202 | 1.136 | 1.097 | 1.062 | 1.035 | 1.015 | 0.998 |
| **80** | **0** | 1.183 | 1.060 | 1.021 | 1.010 | 1.002 | 0.998 | 0.999 | 0.998 | 1.000 | 0.998 |
| | **0.1** | 1.426 | 1.217 | 1.127 | 1.077 | 1.045 | 1.029 | 1.016 | 1.010 | 1.003 | 1.000 |
| | **0.2** | 1.544 | 1.300 | 1.186 | 1.119 | 1.074 | 1.050 | 1.032 | 1.017 | 1.009 | 0.998 |
| | **0.3** | 1.627 | 1.361 | 1.231 | 1.152 | 1.101 | 1.067 | 1.044 | 1.025 | 1.009 | 1.000 |
| | **0.4** | 1.681 | 1.404 | 1.265 | 1.179 | 1.120 | 1.085 | 1.052 | 1.029 | 1.013 | 0.998 |
| | **0.5** | 1.719 | 1.445 | 1.297 | 1.203 | 1.139 | 1.097 | 1.062 | 1.035 | 1.015 | 0.998 |
| **90** | **0** | 1.177 | 1.056 | 1.018 | 1.010 | 1.001 | 0.999 | 0.998 | 0.999 | 0.998 | 0.999 |
| | **0.1** | 1.426 | 1.221 | 1.130 | 1.078 | 1.042 | 1.029 | 1.020 | 1.009 | 1.002 | 0.999 |
| | **0.2** | 1.553 | 1.300 | 1.185 | 1.123 | 1.076 | 1.054 | 1.031 | 1.019 | 1.007 | 0.999 |
| | **0.3** | 1.636 | 1.360 | 1.233 | 1.155 | 1.100 | 1.068 | 1.042 | 1.024 | 1.010 | 0.998 |
| | **0.4** | 1.693 | 1.408 | 1.267 | 1.182 | 1.120 | 1.082 | 1.054 | 1.031 | 1.011 | 0.998 |
| | **0.5** | 1.737 | 1.443 | 1.296 | 1.203 | 1.140 | 1.096 | 1.062 | 1.035 | 1.016 | 0.999 |
| **100** | **0** | 1.159 | 1.047 | 1.018 | 1.010 | 1.001 | 1.000 | 0.999 | 0.999 | 0.999 | 0.998 |
| | **0.1** | 1.427 | 1.212 | 1.127 | 1.081 | 1.041 | 1.029 | 1.017 | 1.008 | 1.002 | 0.999 |
| | **0.2** | 1.554 | 1.298 | 1.185 | 1.121 | 1.077 | 1.050 | 1.031 | 1.017 | 1.006 | 0.999 |
| | **0.3** | 1.632 | 1.361 | 1.230 | 1.154 | 1.100 | 1.070 | 1.044 | 1.024 | 1.010 | 0.999 |
| | **0.4** | 1.688 | 1.406 | 1.267 | 1.181 | 1.123 | 1.083 | 1.054 | 1.030 | 1.013 | 1.000 |
| | **0.5** | 1.733 | 1.443 | 1.300 | 1.205 | 1.139 | 1.096 | 1.062 | 1.036 | 1.014 | 0.998 |

**TABLE 5.2**

**ESTIMATION OF THE MEAN OF $t_K$ IN TERMS OF $I_1$, $I_2$ AND $I_4$ (CASE B)**

| | | I2 | | | | | | | | | |
|---|---|---|---|---|---|---|---|---|---|---|---|
| **I1** | **I4** | **0.1** | **0.2** | **0.3** | **0.4** | **0.5** | **0.6** | **0.7** | **0.8** | **0.9** | **1** |
| **10** | **0** | 1.871 | 1.482 | 1.286 | 1.184 | 1.088 | 1.087 | 1.038 | 1.013 | 1.002 | 1.002 |
| | **0.1** | 1.932 | 1.541 | 1.341 | 1.181 | 1.089 | 1.086 | 1.035 | 1.013 | 1.005 | 0.999 |
| | **0.2** | 2.002 | 1.616 | 1.373 | 1.247 | 1.150 | 1.137 | 1.099 | 1.013 | 1.002 | 0.997 |
| | **0.3** | 2.039 | 1.667 | 1.412 | 1.303 | 1.183 | 1.184 | 1.089 | 1.021 | 1.002 | 1.002 |
| | **0.4** | 2.097 | 1.701 | 1.485 | 1.330 | 1.187 | 1.184 | 1.129 | 1.064 | 1.006 | 0.994 |
| | **0.5** | 2.122 | 1.765 | 1.508 | 1.377 | 1.213 | 1.214 | 1.132 | 1.064 | 1.002 | 1.003 |
| **20** | **0** | 1.777 | 1.378 | 1.207 | 1.123 | 1.056 | 1.032 | 1.008 | 0.999 | 0.999 | 1.001 |
| | **0.1** | 1.936 | 1.521 | 1.313 | 1.205 | 1.093 | 1.075 | 1.047 | 1.000 | 0.998 | 0.998 |
| | **0.2** | 2.066 | 1.643 | 1.400 | 1.256 | 1.163 | 1.129 | 1.072 | 1.036 | 0.997 | 1.002 |
| | **0.3** | 2.151 | 1.722 | 1.460 | 1.312 | 1.210 | 1.147 | 1.089 | 1.054 | 1.001 | 0.998 |
| | **0.4** | 2.222 | 1.786 | 1.531 | 1.375 | 1.227 | 1.194 | 1.116 | 1.055 | 1.025 | 0.997 |
| | **0.5** | 2.278 | 1.845 | 1.578 | 1.409 | 1.266 | 1.215 | 1.134 | 1.072 | 1.020 | 0.996 |
| **30** | **0** | 1.684 | 1.312 | 1.161 | 1.089 | 1.038 | 1.014 | 1.001 | 0.999 | 1.003 | 0.996 |
| | **0.1** | 1.926 | 1.502 | 1.321 | 1.184 | 1.090 | 1.073 | 1.030 | 1.020 | 1.000 | 1.001 |
| | **0.2** | 2.093 | 1.626 | 1.411 | 1.258 | 1.167 | 1.113 | 1.066 | 1.039 | 1.014 | 0.999 |
| | **0.3** | 2.211 | 1.728 | 1.490 | 1.325 | 1.224 | 1.146 | 1.095 | 1.050 | 1.013 | 0.999 |
| | **0.4** | 2.318 | 1.812 | 1.557 | 1.373 | 1.248 | 1.186 | 1.113 | 1.063 | 1.028 | 1.003 |
| | **0.5** | 2.376 | 1.894 | 1.612 | 1.425 | 1.286 | 1.213 | 1.134 | 1.079 | 1.027 | 1.000 |
| **40** | **0** | 1.602 | 1.259 | 1.119 | 1.065 | 1.029 | 1.005 | 1.000 | 0.998 | 0.999 | 1.001 |
| | **0.1** | 1.910 | 1.481 | 1.297 | 1.191 | 1.104 | 1.072 | 1.040 | 1.017 | 0.998 | 0.998 |
| | **0.2** | 2.115 | 1.631 | 1.397 | 1.274 | 1.163 | 1.117 | 1.065 | 1.042 | 1.015 | 0.998 |
| | **0.3** | 2.240 | 1.743 | 1.498 | 1.330 | 1.218 | 1.156 | 1.095 | 1.054 | 1.024 | 0.996 |
| | **0.4** | 2.350 | 1.829 | 1.563 | 1.383 | 1.253 | 1.185 | 1.120 | 1.073 | 1.023 | 1.001 |
| | **0.5** | 2.421 | 1.914 | 1.624 | 1.432 | 1.297 | 1.217 | 1.133 | 1.082 | 1.031 | 0.999 |
| **50** | **0** | 1.536 | 1.224 | 1.098 | 1.049 | 1.028 | 1.002 | 0.996 | 0.998 | 0.998 | 0.998 |
| | **0.1** | 1.894 | 1.489 | 1.291 | 1.194 | 1.103 | 1.056 | 1.031 | 1.012 | 0.998 | 1.000 |
| | **0.2** | 2.117 | 1.642 | 1.404 | 1.268 | 1.174 | 1.110 | 1.071 | 1.035 | 1.010 | 1.000 |
| | **0.3** | 2.255 | 1.760 | 1.494 | 1.339 | 1.223 | 1.155 | 1.092 | 1.052 | 1.021 | 1.000 |
| | **0.4** | 2.359 | 1.858 | 1.570 | 1.388 | 1.256 | 1.188 | 1.116 | 1.066 | 1.028 | 1.000 |
| | **0.5** | 2.449 | 1.926 | 1.630 | 1.439 | 1.303 | 1.222 | 1.135 | 1.076 | 1.033 | 0.999 |
| **60** | **0** | 1.485 | 1.189 | 1.080 | 1.044 | 1.019 | 1.003 | 0.999 | 1.002 | 1.000 | 1.001 |
| | **0.1** | 1.883 | 1.479 | 1.293 | 1.181 | 1.106 | 1.062 | 1.038 | 1.025 | 1.011 | 1.000 |
| | **0.2** | 2.113 | 1.635 | 1.404 | 1.269 | 1.175 | 1.111 | 1.075 | 1.038 | 1.018 | 1.002 |
| | **0.3** | 2.264 | 1.753 | 1.490 | 1.333 | 1.224 | 1.150 | 1.097 | 1.051 | 1.024 | 1.001 |
| | **0.4** | 2.368 | 1.847 | 1.575 | 1.387 | 1.264 | 1.180 | 1.123 | 1.069 | 1.030 | 0.998 |
| | **0.5** | 2.462 | 1.927 | 1.633 | 1.440 | 1.307 | 1.217 | 1.139 | 1.081 | 1.034 | 1.002 |
| **70** | **0** | 1.448 | 1.166 | 1.074 | 1.035 | 1.020 | 1.001 | 1.000 | 1.001 | 1.001 | 0.999 |
| | **0.1** | 1.879 | 1.470 | 1.289 | 1.179 | 1.101 | 1.064 | 1.048 | 1.021 | 1.007 | 0.999 |
| | **0.2** | 2.118 | 1.643 | 1.416 | 1.263 | 1.170 | 1.118 | 1.076 | 1.045 | 1.016 | 1.002 |
| | **0.3** | 2.278 | 1.760 | 1.496 | 1.333 | 1.229 | 1.155 | 1.097 | 1.058 | 1.020 | 1.001 |
| | **0.4** | 2.390 | 1.862 | 1.572 | 1.397 | 1.269 | 1.187 | 1.122 | 1.069 | 1.030 | 1.000 |
| | **0.5** | 2.478 | 1.937 | 1.644 | 1.453 | 1.314 | 1.216 | 1.142 | 1.082 | 1.035 | 1.000 |
| **80** | **0** | 1.414 | 1.155 | 1.062 | 1.031 | 1.016 | 1.001 | 1.001 | 0.999 | 0.999 | 0.998 |
| | **0.1** | 1.870 | 1.469 | 1.291 | 1.176 | 1.101 | 1.063 | 1.040 | 1.026 | 1.006 | 1.001 |
| | **0.2** | 2.114 | 1.637 | 1.409 | 1.265 | 1.170 | 1.111 | 1.075 | 1.041 | 1.019 | 1.000 |
| | **0.3** | 2.273 | 1.762 | 1.506 | 1.340 | 1.223 | 1.149 | 1.100 | 1.055 | 1.024 | 1.001 |
| | **0.4** | 2.390 | 1.867 | 1.582 | 1.397 | 1.269 | 1.187 | 1.118 | 1.071 | 1.032 | 1.000 |
| | **0.5** | 2.483 | 1.950 | 1.645 | 1.450 | 1.313 | 1.216 | 1.141 | 1.081 | 1.033 | 1.000 |
| **90** | **0** | 1.387 | 1.137 | 1.054 | 1.027 | 1.016 | 1.001 | 0.998 | 0.999 | 1.000 | 1.000 |
| | **0.1** | 1.873 | 1.469 | 1.280 | 1.176 | 1.104 | 1.066 | 1.042 | 1.022 | 1.007 | 1.000 |
| | **0.2** | 2.129 | 1.642 | 1.404 | 1.274 | 1.171 | 1.115 | 1.073 | 1.044 | 1.019 | 1.001 |
| | **0.3** | 2.288 | 1.766 | 1.507 | 1.337 | 1.226 | 1.149 | 1.096 | 1.055 | 1.026 | 0.999 |
| | **0.4** | 2.413 | 1.867 | 1.574 | 1.397 | 1.268 | 1.189 | 1.120 | 1.071 | 1.029 | 1.001 |
| | **0.5** | 2.515 | 1.954 | 1.649 | 1.449 | 1.312 | 1.217 | 1.139 | 1.079 | 1.034 | 1.000 |
| **100** | **0** | 1.358 | 1.115 | 1.048 | 1.024 | 1.010 | 1.001 | 1.000 | 1.000 | 1.000 | 1.000 |
| | **0.1** | 1.869 | 1.461 | 1.285 | 1.184 | 1.101 | 1.062 | 1.039 | 1.020 | 1.005 | 1.000 |
| | **0.2** | 2.119 | 1.638 | 1.403 | 1.268 | 1.170 | 1.116 | 1.072 | 1.040 | 1.017 | 0.999 |
| | **0.3** | 2.290 | 1.765 | 1.506 | 1.337 | 1.231 | 1.152 | 1.094 | 1.057 | 1.022 | 0.999 |
| | **0.4** | 2.415 | 1.865 | 1.582 | 1.400 | 1.274 | 1.186 | 1.119 | 1.069 | 1.029 | 0.999 |
| | **0.5** | 2.509 | 1.949 | 1.654 | 1.447 | 1.316 | 1.214 | 1.137 | 1.080 | 1.036 | 0.998 |

**TABLE 5.3**

**ESTIMATION OF THE MEAN OF $t_K$ IN TERMS OF $I_1$, $I_2$ AND $I_4$ (CASE C)**

| | | I2 | | | | | | | | | |
|---|---|---|---|---|---|---|---|---|---|---|---|
| **I1** | **I4** | **0.1** | **0.2** | **0.3** | **0.4** | **0.5** | **0.6** | **0.7** | **0.8** | **0.9** | **1** |
| **10** | **0** | 0.248 | 0.223 | 0.211 | 0.201 | 0.190 | 0.190 | 0.183 | 0.175 | 0.165 | 0.157 |
| | **0.1** | 0.245 | 0.225 | 0.208 | 0.200 | 0.189 | 0.191 | 0.184 | 0.172 | 0.167 | 0.157 |
| | **0.2** | 0.240 | 0.220 | 0.209 | 0.202 | 0.190 | 0.187 | 0.182 | 0.176 | 0.165 | 0.159 |
| | **0.3** | 0.240 | 0.213 | 0.205 | 0.197 | 0.186 | 0.185 | 0.184 | 0.174 | 0.166 | 0.156 |
| | **0.4** | 0.234 | 0.215 | 0.202 | 0.191 | 0.184 | 0.185 | 0.180 | 0.173 | 0.168 | 0.156 |
| | **0.5** | 0.236 | 0.212 | 0.202 | 0.190 | 0.184 | 0.189 | 0.181 | 0.172 | 0.167 | 0.159 |
| **20** | **0** | 0.199 | 0.177 | 0.165 | 0.157 | 0.143 | 0.141 | 0.134 | 0.126 | 0.118 | 0.111 |
| | **0.1** | 0.191 | 0.171 | 0.162 | 0.154 | 0.144 | 0.139 | 0.130 | 0.125 | 0.117 | 0.113 |
| | **0.2** | 0.186 | 0.169 | 0.158 | 0.146 | 0.140 | 0.137 | 0.129 | 0.124 | 0.118 | 0.112 |
| | **0.3** | 0.185 | 0.165 | 0.151 | 0.146 | 0.139 | 0.135 | 0.128 | 0.121 | 0.117 | 0.112 |
| | **0.4** | 0.184 | 0.161 | 0.149 | 0.142 | 0.137 | 0.134 | 0.130 | 0.123 | 0.117 | 0.111 |
| | **0.5** | 0.182 | 0.159 | 0.147 | 0.139 | 0.132 | 0.133 | 0.127 | 0.120 | 0.117 | 0.112 |
| **30** | **0** | 0.171 | 0.154 | 0.143 | 0.133 | 0.122 | 0.117 | 0.109 | 0.101 | 0.095 | 0.092 |
| | **0.1** | 0.163 | 0.146 | 0.136 | 0.126 | 0.118 | 0.116 | 0.108 | 0.101 | 0.096 | 0.091 |
| | **0.2** | 0.156 | 0.139 | 0.129 | 0.123 | 0.117 | 0.111 | 0.106 | 0.100 | 0.096 | 0.091 |
| | **0.3** | 0.155 | 0.135 | 0.126 | 0.121 | 0.114 | 0.111 | 0.106 | 0.100 | 0.096 | 0.091 |
| | **0.4** | 0.151 | 0.136 | 0.123 | 0.118 | 0.112 | 0.109 | 0.104 | 0.099 | 0.095 | 0.092 |
| | **0.5** | 0.153 | 0.134 | 0.122 | 0.115 | 0.111 | 0.110 | 0.104 | 0.100 | 0.095 | 0.092 |
| **40** | **0** | 0.156 | 0.141 | 0.127 | 0.120 | 0.106 | 0.101 | 0.094 | 0.089 | 0.084 | 0.080 |
| | **0.1** | 0.145 | 0.129 | 0.119 | 0.112 | 0.104 | 0.100 | 0.093 | 0.088 | 0.084 | 0.080 |
| | **0.2** | 0.140 | 0.123 | 0.113 | 0.107 | 0.101 | 0.099 | 0.093 | 0.087 | 0.083 | 0.080 |
| | **0.3** | 0.138 | 0.120 | 0.109 | 0.105 | 0.100 | 0.095 | 0.091 | 0.087 | 0.083 | 0.080 |
| | **0.4** | 0.134 | 0.118 | 0.109 | 0.102 | 0.097 | 0.095 | 0.090 | 0.087 | 0.084 | 0.079 |
| | **0.5** | 0.135 | 0.115 | 0.107 | 0.101 | 0.096 | 0.094 | 0.090 | 0.086 | 0.082 | 0.078 |
| **50** | **0** | 0.144 | 0.130 | 0.116 | 0.106 | 0.097 | 0.090 | 0.085 | 0.079 | 0.076 | 0.072 |
| | **0.1** | 0.132 | 0.116 | 0.108 | 0.100 | 0.094 | 0.090 | 0.083 | 0.079 | 0.076 | 0.072 |
| | **0.2** | 0.124 | 0.109 | 0.103 | 0.097 | 0.092 | 0.087 | 0.082 | 0.078 | 0.075 | 0.072 |
| | **0.3** | 0.123 | 0.107 | 0.100 | 0.094 | 0.089 | 0.087 | 0.082 | 0.078 | 0.074 | 0.071 |
| | **0.4** | 0.123 | 0.105 | 0.097 | 0.092 | 0.088 | 0.085 | 0.081 | 0.077 | 0.074 | 0.071 |
| | **0.5** | 0.124 | 0.104 | 0.096 | 0.090 | 0.085 | 0.083 | 0.081 | 0.077 | 0.074 | 0.072 |
| **60** | **0** | 0.136 | 0.123 | 0.108 | 0.098 | 0.089 | 0.083 | 0.078 | 0.072 | 0.070 | 0.066 |
| | **0.1** | 0.122 | 0.107 | 0.098 | 0.091 | 0.087 | 0.081 | 0.076 | 0.072 | 0.069 | 0.066 |
| | **0.2** | 0.115 | 0.102 | 0.094 | 0.089 | 0.083 | 0.080 | 0.075 | 0.072 | 0.068 | 0.065 |
| | **0.3** | 0.113 | 0.098 | 0.091 | 0.086 | 0.083 | 0.077 | 0.075 | 0.071 | 0.069 | 0.065 |
| | **0.4** | 0.113 | 0.098 | 0.089 | 0.085 | 0.080 | 0.078 | 0.074 | 0.070 | 0.069 | 0.065 |
| | **0.5** | 0.113 | 0.096 | 0.087 | 0.083 | 0.078 | 0.076 | 0.073 | 0.071 | 0.068 | 0.066 |
| **70** | **0** | 0.128 | 0.115 | 0.103 | 0.092 | 0.082 | 0.077 | 0.073 | 0.068 | 0.064 | 0.060 |
| | **0.1** | 0.111 | 0.099 | 0.093 | 0.086 | 0.079 | 0.076 | 0.072 | 0.068 | 0.063 | 0.060 |
| | **0.2** | 0.106 | 0.095 | 0.088 | 0.082 | 0.077 | 0.073 | 0.070 | 0.067 | 0.063 | 0.062 |
| | **0.3** | 0.105 | 0.091 | 0.084 | 0.080 | 0.077 | 0.073 | 0.069 | 0.067 | 0.063 | 0.061 |
| | **0.4** | 0.105 | 0.089 | 0.083 | 0.078 | 0.075 | 0.071 | 0.068 | 0.066 | 0.063 | 0.061 |
| | **0.5** | 0.104 | 0.088 | 0.082 | 0.078 | 0.073 | 0.070 | 0.068 | 0.066 | 0.063 | 0.061 |
| **80** | **0** | 0.125 | 0.110 | 0.095 | 0.086 | 0.078 | 0.072 | 0.067 | 0.063 | 0.061 | 0.058 |
| | **0.1** | 0.106 | 0.096 | 0.087 | 0.081 | 0.075 | 0.070 | 0.066 | 0.063 | 0.059 | 0.057 |
| | **0.2** | 0.102 | 0.088 | 0.082 | 0.078 | 0.074 | 0.070 | 0.065 | 0.062 | 0.060 | 0.059 |
| | **0.3** | 0.099 | 0.085 | 0.079 | 0.076 | 0.072 | 0.067 | 0.064 | 0.062 | 0.060 | 0.056 |
| | **0.4** | 0.098 | 0.084 | 0.077 | 0.073 | 0.069 | 0.067 | 0.065 | 0.061 | 0.059 | 0.057 |
| | **0.5** | 0.098 | 0.083 | 0.075 | 0.072 | 0.068 | 0.066 | 0.064 | 0.061 | 0.060 | 0.057 |
| **90** | **0** | 0.117 | 0.105 | 0.091 | 0.082 | 0.073 | 0.069 | 0.064 | 0.060 | 0.058 | 0.054 |
| | **0.1** | 0.100 | 0.090 | 0.082 | 0.077 | 0.070 | 0.066 | 0.062 | 0.061 | 0.058 | 0.054 |
| | **0.2** | 0.095 | 0.085 | 0.078 | 0.073 | 0.068 | 0.066 | 0.062 | 0.059 | 0.056 | 0.055 |
| | **0.3** | 0.092 | 0.081 | 0.075 | 0.072 | 0.067 | 0.064 | 0.061 | 0.058 | 0.057 | 0.055 |
| | **0.4** | 0.092 | 0.079 | 0.072 | 0.069 | 0.066 | 0.063 | 0.060 | 0.058 | 0.057 | 0.055 |
| | **0.5** | 0.092 | 0.078 | 0.072 | 0.068 | 0.064 | 0.063 | 0.060 | 0.058 | 0.056 | 0.055 |
| **100** | **0** | 0.113 | 0.100 | 0.088 | 0.077 | 0.070 | 0.065 | 0.062 | 0.057 | 0.054 | 0.053 |
| | **0.1** | 0.095 | 0.084 | 0.079 | 0.073 | 0.068 | 0.064 | 0.060 | 0.057 | 0.054 | 0.053 |
| | **0.2** | 0.091 | 0.079 | 0.075 | 0.069 | 0.066 | 0.062 | 0.059 | 0.057 | 0.055 | 0.052 |
| | **0.3** | 0.089 | 0.078 | 0.072 | 0.068 | 0.063 | 0.060 | 0.058 | 0.056 | 0.054 | 0.053 |
| | **0.4** | 0.088 | 0.075 | 0.070 | 0.066 | 0.063 | 0.060 | 0.057 | 0.055 | 0.053 | 0.052 |
| | **0.5** | 0.089 | 0.075 | 0.069 | 0.064 | 0.061 | 0.060 | 0.057 | 0.055 | 0.054 | 0.052 |

**TABLE 5.4**

**ESTIMATION OF THE STANDARD DEVIATION OF $t_K$ IN TERMS OF $I_1$, $I_2$ AND $I_4$ (CASE A)**

| | | I2 | | | | | | | | | |
|---|---|---|---|---|---|---|---|---|---|---|---|
| **I1** | **I4** | **0.1** | **0.2** | **0.3** | **0.4** | **0.5** | **0.6** | **0.7** | **0.8** | **0.9** | **1** |
| **10** | **0** | 0.461 | 0.332 | 0.269 | 0.234 | 0.210 | 0.206 | 0.191 | 0.177 | 0.164 | 0.156 |
| | **0.1** | 0.473 | 0.334 | 0.273 | 0.237 | 0.205 | 0.210 | 0.190 | 0.175 | 0.166 | 0.156 |
| | **0.2** | 0.466 | 0.352 | 0.287 | 0.243 | 0.215 | 0.211 | 0.192 | 0.174 | 0.167 | 0.157 |
| | **0.3** | 0.489 | 0.357 | 0.289 | 0.247 | 0.216 | 0.218 | 0.193 | 0.177 | 0.166 | 0.156 |
| | **0.4** | 0.489 | 0.365 | 0.293 | 0.251 | 0.212 | 0.219 | 0.199 | 0.183 | 0.166 | 0.155 |
| | **0.5** | 0.494 | 0.362 | 0.290 | 0.253 | 0.220 | 0.227 | 0.197 | 0.179 | 0.168 | 0.159 |
| **20** | **0** | 0.362 | 0.250 | 0.195 | 0.172 | 0.150 | 0.144 | 0.134 | 0.125 | 0.117 | 0.110 |
| | **0.1** | 0.382 | 0.260 | 0.208 | 0.179 | 0.154 | 0.147 | 0.134 | 0.126 | 0.118 | 0.113 |
| | **0.2** | 0.389 | 0.269 | 0.218 | 0.179 | 0.157 | 0.150 | 0.136 | 0.126 | 0.118 | 0.112 |
| | **0.3** | 0.411 | 0.277 | 0.215 | 0.185 | 0.165 | 0.154 | 0.138 | 0.126 | 0.117 | 0.111 |
| | **0.4** | 0.407 | 0.283 | 0.220 | 0.189 | 0.164 | 0.156 | 0.141 | 0.129 | 0.117 | 0.111 |
| | **0.5** | 0.406 | 0.285 | 0.227 | 0.190 | 0.164 | 0.157 | 0.140 | 0.127 | 0.119 | 0.114 |
| **30** | **0** | 0.306 | 0.204 | 0.163 | 0.142 | 0.126 | 0.118 | 0.108 | 0.101 | 0.096 | 0.092 |
| | **0.1** | 0.328 | 0.219 | 0.174 | 0.147 | 0.129 | 0.121 | 0.111 | 0.103 | 0.096 | 0.091 |
| | **0.2** | 0.338 | 0.228 | 0.178 | 0.151 | 0.132 | 0.121 | 0.110 | 0.103 | 0.098 | 0.091 |
| | **0.3** | 0.355 | 0.234 | 0.183 | 0.155 | 0.136 | 0.124 | 0.114 | 0.106 | 0.098 | 0.091 |
| | **0.4** | 0.364 | 0.244 | 0.187 | 0.156 | 0.135 | 0.125 | 0.114 | 0.104 | 0.097 | 0.092 |
| | **0.5** | 0.371 | 0.248 | 0.194 | 0.157 | 0.139 | 0.129 | 0.114 | 0.107 | 0.097 | 0.091 |
| **40** | **0** | 0.261 | 0.171 | 0.141 | 0.123 | 0.110 | 0.103 | 0.095 | 0.089 | 0.084 | 0.080 |
| | **0.1** | 0.285 | 0.193 | 0.149 | 0.129 | 0.113 | 0.105 | 0.096 | 0.090 | 0.084 | 0.079 |
| | **0.2** | 0.306 | 0.200 | 0.156 | 0.131 | 0.116 | 0.108 | 0.097 | 0.089 | 0.083 | 0.079 |
| | **0.3** | 0.315 | 0.209 | 0.163 | 0.135 | 0.118 | 0.108 | 0.097 | 0.091 | 0.084 | 0.080 |
| | **0.4** | 0.328 | 0.210 | 0.165 | 0.139 | 0.122 | 0.109 | 0.100 | 0.091 | 0.085 | 0.079 |
| | **0.5** | 0.332 | 0.220 | 0.167 | 0.139 | 0.121 | 0.110 | 0.098 | 0.093 | 0.085 | 0.079 |
| **50** | **0** | 0.229 | 0.155 | 0.128 | 0.109 | 0.098 | 0.091 | 0.084 | 0.079 | 0.075 | 0.072 |
| | **0.1** | 0.255 | 0.172 | 0.135 | 0.116 | 0.102 | 0.094 | 0.086 | 0.080 | 0.073 | 0.071 |
| | **0.2** | 0.274 | 0.182 | 0.141 | 0.117 | 0.103 | 0.095 | 0.088 | 0.080 | 0.074 | 0.071 |
| | **0.3** | 0.291 | 0.189 | 0.144 | 0.122 | 0.107 | 0.096 | 0.088 | 0.080 | 0.077 | 0.072 |
| | **0.4** | 0.295 | 0.192 | 0.147 | 0.125 | 0.108 | 0.098 | 0.088 | 0.081 | 0.075 | 0.072 |
| | **0.5** | 0.299 | 0.200 | 0.154 | 0.127 | 0.109 | 0.099 | 0.089 | 0.081 | 0.075 | 0.070 |
| **60** | **0** | 0.209 | 0.144 | 0.116 | 0.102 | 0.092 | 0.084 | 0.077 | 0.073 | 0.068 | 0.066 |
| | **0.1** | 0.237 | 0.155 | 0.124 | 0.107 | 0.094 | 0.087 | 0.079 | 0.072 | 0.068 | 0.065 |
| | **0.2** | 0.255 | 0.166 | 0.129 | 0.109 | 0.096 | 0.086 | 0.080 | 0.074 | 0.069 | 0.066 |
| | **0.3** | 0.265 | 0.173 | 0.136 | 0.111 | 0.098 | 0.087 | 0.079 | 0.074 | 0.069 | 0.065 |
| | **0.4** | 0.267 | 0.179 | 0.138 | 0.114 | 0.098 | 0.088 | 0.080 | 0.074 | 0.070 | 0.066 |
| | **0.5** | 0.283 | 0.184 | 0.138 | 0.116 | 0.102 | 0.090 | 0.081 | 0.075 | 0.070 | 0.065 |
| **70** | **0** | 0.190 | 0.133 | 0.108 | 0.093 | 0.084 | 0.078 | 0.072 | 0.067 | 0.063 | 0.060 |
| | **0.1** | 0.220 | 0.147 | 0.115 | 0.099 | 0.086 | 0.080 | 0.073 | 0.067 | 0.064 | 0.062 |
| | **0.2** | 0.235 | 0.154 | 0.119 | 0.103 | 0.090 | 0.081 | 0.075 | 0.068 | 0.065 | 0.061 |
| | **0.3** | 0.251 | 0.159 | 0.123 | 0.104 | 0.092 | 0.081 | 0.075 | 0.069 | 0.064 | 0.061 |
| | **0.4** | 0.262 | 0.163 | 0.127 | 0.105 | 0.092 | 0.083 | 0.074 | 0.069 | 0.065 | 0.061 |
| | **0.5** | 0.260 | 0.171 | 0.130 | 0.107 | 0.092 | 0.083 | 0.076 | 0.069 | 0.064 | 0.060 |
| **80** | **0** | 0.173 | 0.123 | 0.100 | 0.088 | 0.078 | 0.071 | 0.067 | 0.063 | 0.060 | 0.058 |
| | **0.1** | 0.206 | 0.137 | 0.109 | 0.094 | 0.082 | 0.076 | 0.068 | 0.064 | 0.060 | 0.057 |
| | **0.2** | 0.222 | 0.142 | 0.113 | 0.095 | 0.084 | 0.076 | 0.070 | 0.064 | 0.060 | 0.058 |
| | **0.3** | 0.236 | 0.150 | 0.117 | 0.097 | 0.084 | 0.076 | 0.069 | 0.064 | 0.061 | 0.057 |
| | **0.4** | 0.239 | 0.153 | 0.119 | 0.099 | 0.086 | 0.077 | 0.071 | 0.064 | 0.061 | 0.056 |
| | **0.5** | 0.246 | 0.162 | 0.123 | 0.102 | 0.087 | 0.079 | 0.071 | 0.065 | 0.061 | 0.057 |
| **90** | **0** | 0.163 | 0.118 | 0.096 | 0.084 | 0.074 | 0.069 | 0.065 | 0.059 | 0.057 | 0.054 |
| | **0.1** | 0.195 | 0.132 | 0.103 | 0.087 | 0.076 | 0.070 | 0.065 | 0.061 | 0.058 | 0.056 |
| | **0.2** | 0.213 | 0.138 | 0.107 | 0.090 | 0.078 | 0.071 | 0.065 | 0.061 | 0.057 | 0.055 |
| | **0.3** | 0.220 | 0.141 | 0.110 | 0.094 | 0.081 | 0.073 | 0.065 | 0.060 | 0.057 | 0.055 |
| | **0.4** | 0.233 | 0.148 | 0.113 | 0.094 | 0.080 | 0.073 | 0.066 | 0.061 | 0.058 | 0.056 |
| | **0.5** | 0.234 | 0.150 | 0.116 | 0.096 | 0.083 | 0.074 | 0.066 | 0.061 | 0.058 | 0.055 |
| **100** | **0** | 0.157 | 0.109 | 0.092 | 0.079 | 0.070 | 0.065 | 0.062 | 0.058 | 0.056 | 0.053 |
| | **0.1** | 0.186 | 0.121 | 0.098 | 0.084 | 0.073 | 0.067 | 0.063 | 0.058 | 0.055 | 0.054 |
| | **0.2** | 0.204 | 0.131 | 0.102 | 0.086 | 0.075 | 0.068 | 0.063 | 0.058 | 0.055 | 0.052 |
| | **0.3** | 0.215 | 0.138 | 0.105 | 0.087 | 0.074 | 0.069 | 0.063 | 0.058 | 0.054 | 0.052 |
| | **0.4** | 0.221 | 0.140 | 0.107 | 0.089 | 0.076 | 0.069 | 0.062 | 0.058 | 0.054 | 0.053 |
| | **0.5** | 0.225 | 0.145 | 0.110 | 0.090 | 0.078 | 0.069 | 0.063 | 0.057 | 0.055 | 0.052 |

**TABLE 5.5**

**ESTIMATION OF THE STANDARD DEVIATION OF $t_K$ IN TERMS OF $I_1$, $I_2$ AND $I_4$ (CASE B)**

| | | I2 | | | | | | | | | |
|---|---|---|---|---|---|---|---|---|---|---|---|
| **I1** | **I4** | **0.1** | **0.2** | **0.3** | **0.4** | **0.5** | **0.6** | **0.7** | **0.8** | **0.9** | **1** |
| **10** | **0** | 0.741 | 0.573 | 0.484 | 0.442 | 0.395 | 0.393 | 0.371 | 0.351 | 0.328 | 0.315 |
| | **0.1** | 0.767 | 0.584 | 0.504 | 0.437 | 0.392 | 0.391 | 0.364 | 0.345 | 0.337 | 0.315 |
| | **0.2** | 0.791 | 0.608 | 0.507 | 0.454 | 0.410 | 0.405 | 0.378 | 0.348 | 0.333 | 0.311 |
| | **0.3** | 0.781 | 0.615 | 0.512 | 0.459 | 0.417 | 0.409 | 0.377 | 0.349 | 0.335 | 0.320 |
| | **0.4** | 0.801 | 0.641 | 0.531 | 0.469 | 0.414 | 0.410 | 0.383 | 0.354 | 0.334 | 0.311 |
| | **0.5** | 0.803 | 0.642 | 0.529 | 0.473 | 0.418 | 0.420 | 0.379 | 0.357 | 0.329 | 0.320 |
| **20** | **0** | 0.557 | 0.416 | 0.359 | 0.326 | 0.293 | 0.280 | 0.263 | 0.250 | 0.234 | 0.222 |
| | **0.1** | 0.597 | 0.446 | 0.381 | 0.339 | 0.294 | 0.285 | 0.265 | 0.246 | 0.235 | 0.223 |
| | **0.2** | 0.624 | 0.460 | 0.378 | 0.343 | 0.305 | 0.288 | 0.268 | 0.247 | 0.235 | 0.226 |
| | **0.3** | 0.636 | 0.481 | 0.391 | 0.338 | 0.308 | 0.291 | 0.268 | 0.253 | 0.235 | 0.222 |
| | **0.4** | 0.642 | 0.484 | 0.399 | 0.353 | 0.310 | 0.298 | 0.274 | 0.252 | 0.235 | 0.224 |
| | **0.5** | 0.652 | 0.488 | 0.405 | 0.355 | 0.311 | 0.303 | 0.271 | 0.252 | 0.237 | 0.223 |
| **30** | **0** | 0.470 | 0.350 | 0.295 | 0.273 | 0.241 | 0.230 | 0.216 | 0.203 | 0.191 | 0.181 |
| | **0.1** | 0.501 | 0.380 | 0.317 | 0.278 | 0.249 | 0.236 | 0.218 | 0.205 | 0.191 | 0.185 |
| | **0.2** | 0.520 | 0.385 | 0.326 | 0.281 | 0.254 | 0.237 | 0.221 | 0.207 | 0.192 | 0.181 |
| | **0.3** | 0.547 | 0.397 | 0.328 | 0.286 | 0.258 | 0.240 | 0.220 | 0.203 | 0.191 | 0.183 |
| | **0.4** | 0.562 | 0.409 | 0.334 | 0.291 | 0.261 | 0.241 | 0.223 | 0.205 | 0.192 | 0.184 |
| | **0.5** | 0.556 | 0.418 | 0.342 | 0.297 | 0.259 | 0.244 | 0.223 | 0.208 | 0.194 | 0.183 |
| **40** | **0** | 0.409 | 0.306 | 0.263 | 0.241 | 0.217 | 0.202 | 0.188 | 0.176 | 0.165 | 0.158 |
| | **0.1** | 0.446 | 0.326 | 0.279 | 0.245 | 0.221 | 0.205 | 0.188 | 0.180 | 0.164 | 0.157 |
| | **0.2** | 0.469 | 0.339 | 0.279 | 0.247 | 0.223 | 0.205 | 0.191 | 0.180 | 0.167 | 0.159 |
| | **0.3** | 0.486 | 0.350 | 0.288 | 0.250 | 0.221 | 0.210 | 0.190 | 0.178 | 0.167 | 0.158 |
| | **0.4** | 0.497 | 0.361 | 0.298 | 0.254 | 0.225 | 0.210 | 0.194 | 0.180 | 0.167 | 0.157 |
| | **0.5** | 0.501 | 0.368 | 0.299 | 0.260 | 0.228 | 0.213 | 0.189 | 0.180 | 0.166 | 0.156 |
| **50** | **0** | 0.363 | 0.279 | 0.238 | 0.212 | 0.194 | 0.182 | 0.168 | 0.157 | 0.149 | 0.140 |
| | **0.1** | 0.405 | 0.300 | 0.252 | 0.225 | 0.196 | 0.184 | 0.170 | 0.156 | 0.149 | 0.142 |
| | **0.2** | 0.427 | 0.312 | 0.253 | 0.223 | 0.201 | 0.184 | 0.171 | 0.157 | 0.146 | 0.142 |
| | **0.3** | 0.439 | 0.317 | 0.262 | 0.227 | 0.200 | 0.185 | 0.170 | 0.162 | 0.149 | 0.141 |
| | **0.4** | 0.450 | 0.326 | 0.265 | 0.230 | 0.201 | 0.187 | 0.173 | 0.160 | 0.152 | 0.142 |
| | **0.5** | 0.457 | 0.332 | 0.270 | 0.236 | 0.206 | 0.191 | 0.173 | 0.157 | 0.152 | 0.141 |
| **60** | **0** | 0.334 | 0.258 | 0.218 | 0.196 | 0.179 | 0.166 | 0.155 | 0.145 | 0.135 | 0.129 |
| | **0.1** | 0.366 | 0.273 | 0.231 | 0.202 | 0.185 | 0.167 | 0.156 | 0.144 | 0.137 | 0.132 |
| | **0.2** | 0.391 | 0.282 | 0.236 | 0.205 | 0.182 | 0.170 | 0.156 | 0.144 | 0.135 | 0.130 |
| | **0.3** | 0.407 | 0.290 | 0.237 | 0.208 | 0.181 | 0.169 | 0.160 | 0.144 | 0.137 | 0.129 |
| | **0.4** | 0.415 | 0.300 | 0.246 | 0.210 | 0.184 | 0.169 | 0.159 | 0.145 | 0.137 | 0.129 |
| | **0.5** | 0.422 | 0.302 | 0.249 | 0.213 | 0.189 | 0.175 | 0.158 | 0.145 | 0.137 | 0.129 |
| **70** | **0** | 0.309 | 0.235 | 0.205 | 0.184 | 0.165 | 0.157 | 0.142 | 0.134 | 0.127 | 0.118 |
| | **0.1** | 0.344 | 0.254 | 0.215 | 0.186 | 0.167 | 0.156 | 0.144 | 0.133 | 0.128 | 0.120 |
| | **0.2** | 0.365 | 0.263 | 0.219 | 0.188 | 0.170 | 0.157 | 0.146 | 0.135 | 0.125 | 0.119 |
| | **0.3** | 0.376 | 0.271 | 0.220 | 0.194 | 0.172 | 0.159 | 0.142 | 0.136 | 0.126 | 0.119 |
| | **0.4** | 0.386 | 0.275 | 0.226 | 0.196 | 0.176 | 0.158 | 0.146 | 0.136 | 0.127 | 0.119 |
| | **0.5** | 0.392 | 0.282 | 0.230 | 0.196 | 0.175 | 0.162 | 0.147 | 0.138 | 0.129 | 0.119 |
| **80** | **0** | 0.293 | 0.227 | 0.194 | 0.170 | 0.155 | 0.144 | 0.134 | 0.125 | 0.117 | 0.112 |
| | **0.1** | 0.328 | 0.241 | 0.201 | 0.177 | 0.157 | 0.146 | 0.134 | 0.125 | 0.119 | 0.110 |
| | **0.2** | 0.342 | 0.245 | 0.200 | 0.177 | 0.161 | 0.147 | 0.135 | 0.127 | 0.119 | 0.111 |
| | **0.3** | 0.358 | 0.254 | 0.208 | 0.182 | 0.162 | 0.150 | 0.134 | 0.126 | 0.120 | 0.112 |
| | **0.4** | 0.365 | 0.258 | 0.208 | 0.183 | 0.162 | 0.146 | 0.136 | 0.128 | 0.118 | 0.112 |
| | **0.5** | 0.375 | 0.270 | 0.214 | 0.186 | 0.162 | 0.151 | 0.138 | 0.127 | 0.118 | 0.112 |
| **90** | **0** | 0.273 | 0.218 | 0.185 | 0.162 | 0.146 | 0.135 | 0.128 | 0.117 | 0.111 | 0.106 |
| | **0.1** | 0.308 | 0.227 | 0.190 | 0.166 | 0.149 | 0.135 | 0.128 | 0.119 | 0.111 | 0.105 |
| | **0.2** | 0.324 | 0.232 | 0.195 | 0.169 | 0.150 | 0.139 | 0.128 | 0.119 | 0.111 | 0.106 |
| | **0.3** | 0.342 | 0.236 | 0.195 | 0.173 | 0.150 | 0.138 | 0.128 | 0.120 | 0.111 | 0.103 |
| | **0.4** | 0.346 | 0.244 | 0.202 | 0.172 | 0.154 | 0.140 | 0.129 | 0.118 | 0.112 | 0.104 |
| | **0.5** | 0.358 | 0.255 | 0.203 | 0.175 | 0.154 | 0.141 | 0.127 | 0.118 | 0.111 | 0.105 |
| **100** | **0** | 0.262 | 0.206 | 0.178 | 0.153 | 0.140 | 0.129 | 0.120 | 0.113 | 0.105 | 0.099 |
| | **0.1** | 0.296 | 0.216 | 0.181 | 0.158 | 0.143 | 0.130 | 0.121 | 0.111 | 0.106 | 0.101 |
| | **0.2** | 0.310 | 0.220 | 0.182 | 0.158 | 0.144 | 0.131 | 0.122 | 0.114 | 0.106 | 0.101 |
| | **0.3** | 0.318 | 0.228 | 0.186 | 0.162 | 0.145 | 0.133 | 0.121 | 0.113 | 0.105 | 0.099 |
| | **0.4** | 0.330 | 0.234 | 0.192 | 0.165 | 0.146 | 0.132 | 0.122 | 0.113 | 0.107 | 0.100 |
| | **0.5** | 0.340 | 0.242 | 0.194 | 0.167 | 0.148 | 0.134 | 0.124 | 0.114 | 0.106 | 0.099 |

**TABLE 5.6**

**ESTIMATION OF THE STANDARD DEVIATION OF $t_K$ IN TERMS OF $I_1$, $I_2$ AND $I_4$ (CASE C)**

| | | I2 | | | | | | | | | |
|---|---|---|---|---|---|---|---|---|---|---|---|
| **I1** | **I4** | **0.1** | **0.2** | **0.3** | **0.4** | **0.5** | **0.6** | **0.7** | **0.8** | **0.9** | **1** |
| **10** | **0** | 0.983 | 0.862 | 0.778 | 0.756 | 0.719 | 0.712 | 0.710 | 0.713 | 0.726 | 0.744 |
| | **0.1** | 1.037 | 0.893 | 0.817 | 0.761 | 0.713 | 0.709 | 0.709 | 0.721 | 0.724 | 0.744 |
| | **0.2** | 1.073 | 0.934 | 0.856 | 0.789 | 0.748 | 0.761 | 0.738 | 0.710 | 0.733 | 0.743 |
| | **0.3** | 1.106 | 0.991 | 0.882 | 0.830 | 0.789 | 0.790 | 0.736 | 0.711 | 0.721 | 0.741 |
| | **0.4** | 1.129 | 0.996 | 0.920 | 0.874 | 0.790 | 0.787 | 0.770 | 0.745 | 0.722 | 0.750 |
| | **0.5** | 1.152 | 1.031 | 0.934 | 0.894 | 0.810 | 0.808 | 0.771 | 0.750 | 0.730 | 0.739 |
| **20** | **0** | 0.999 | 0.878 | 0.811 | 0.792 | 0.780 | 0.772 | 0.778 | 0.795 | 0.810 | 0.815 |
| | **0.1** | 1.116 | 0.967 | 0.887 | 0.841 | 0.806 | 0.802 | 0.813 | 0.794 | 0.809 | 0.818 |
| | **0.2** | 1.179 | 1.032 | 0.938 | 0.893 | 0.847 | 0.845 | 0.825 | 0.813 | 0.808 | 0.813 |
| | **0.3** | 1.218 | 1.074 | 0.985 | 0.913 | 0.878 | 0.859 | 0.841 | 0.832 | 0.806 | 0.819 |
| | **0.4** | 1.245 | 1.120 | 1.031 | 0.961 | 0.895 | 0.881 | 0.846 | 0.831 | 0.823 | 0.817 |
| | **0.5** | 1.276 | 1.151 | 1.058 | 0.988 | 0.931 | 0.897 | 0.862 | 0.843 | 0.827 | 0.818 |
| **30** | **0** | 1.017 | 0.873 | 0.828 | 0.815 | 0.808 | 0.811 | 0.822 | 0.833 | 0.846 | 0.850 |
| | **0.1** | 1.155 | 0.998 | 0.930 | 0.886 | 0.848 | 0.840 | 0.839 | 0.849 | 0.846 | 0.851 |
| | **0.2** | 1.242 | 1.070 | 0.994 | 0.928 | 0.890 | 0.875 | 0.860 | 0.859 | 0.855 | 0.850 |
| | **0.3** | 1.291 | 1.133 | 1.043 | 0.967 | 0.926 | 0.893 | 0.881 | 0.868 | 0.855 | 0.850 |
| | **0.4** | 1.334 | 1.171 | 1.076 | 1.003 | 0.950 | 0.919 | 0.892 | 0.872 | 0.863 | 0.851 |
| | **0.5** | 1.352 | 1.203 | 1.108 | 1.035 | 0.972 | 0.933 | 0.900 | 0.878 | 0.863 | 0.849 |
| **40** | **0** | 1.007 | 0.875 | 0.838 | 0.829 | 0.831 | 0.835 | 0.850 | 0.855 | 0.865 | 0.873 |
| | **0.1** | 1.176 | 1.018 | 0.949 | 0.906 | 0.875 | 0.874 | 0.867 | 0.865 | 0.862 | 0.870 |
| | **0.2** | 1.267 | 1.104 | 1.015 | 0.966 | 0.918 | 0.898 | 0.881 | 0.880 | 0.872 | 0.871 |
| | **0.3** | 1.317 | 1.161 | 1.064 | 0.997 | 0.946 | 0.925 | 0.901 | 0.885 | 0.877 | 0.870 |
| | **0.4** | 1.365 | 1.207 | 1.102 | 1.034 | 0.978 | 0.946 | 0.920 | 0.896 | 0.877 | 0.871 |
| | **0.5** | 1.394 | 1.240 | 1.137 | 1.062 | 1.004 | 0.961 | 0.926 | 0.899 | 0.882 | 0.874 |
| **50** | **0** | 1.000 | 0.889 | 0.846 | 0.847 | 0.847 | 0.852 | 0.860 | 0.871 | 0.875 | 0.884 |
| | **0.1** | 1.184 | 1.042 | 0.969 | 0.932 | 0.893 | 0.881 | 0.880 | 0.881 | 0.876 | 0.882 |
| | **0.2** | 1.283 | 1.132 | 1.034 | 0.979 | 0.935 | 0.916 | 0.902 | 0.891 | 0.883 | 0.882 |
| | **0.3** | 1.347 | 1.184 | 1.081 | 1.020 | 0.970 | 0.935 | 0.915 | 0.897 | 0.888 | 0.884 |
| | **0.4** | 1.384 | 1.230 | 1.127 | 1.051 | 0.993 | 0.961 | 0.931 | 0.911 | 0.892 | 0.885 |
| | **0.5** | 1.416 | 1.272 | 1.161 | 1.082 | 1.020 | 0.980 | 0.943 | 0.917 | 0.900 | 0.884 |
| **60** | **0** | 0.992 | 0.867 | 0.851 | 0.856 | 0.860 | 0.865 | 0.871 | 0.880 | 0.887 | 0.892 |
| | **0.1** | 1.194 | 1.054 | 0.985 | 0.939 | 0.910 | 0.899 | 0.896 | 0.894 | 0.893 | 0.892 |
| | **0.2** | 1.301 | 1.136 | 1.048 | 0.989 | 0.954 | 0.930 | 0.918 | 0.905 | 0.896 | 0.892 |
| | **0.3** | 1.364 | 1.199 | 1.100 | 1.037 | 0.982 | 0.960 | 0.929 | 0.911 | 0.900 | 0.896 |
| | **0.4** | 1.404 | 1.241 | 1.142 | 1.066 | 1.011 | 0.970 | 0.944 | 0.918 | 0.905 | 0.894 |
| | **0.5** | 1.434 | 1.277 | 1.177 | 1.095 | 1.038 | 0.991 | 0.954 | 0.929 | 0.907 | 0.892 |
| **70** | **0** | 0.982 | 0.877 | 0.859 | 0.864 | 0.871 | 0.875 | 0.881 | 0.891 | 0.898 | 0.902 |
| | **0.1** | 1.215 | 1.071 | 0.987 | 0.951 | 0.922 | 0.911 | 0.904 | 0.899 | 0.903 | 0.902 |
| | **0.2** | 1.318 | 1.151 | 1.058 | 1.005 | 0.960 | 0.942 | 0.925 | 0.911 | 0.908 | 0.901 |
| | **0.3** | 1.380 | 1.212 | 1.114 | 1.042 | 0.994 | 0.965 | 0.941 | 0.921 | 0.907 | 0.903 |
| | **0.4** | 1.421 | 1.255 | 1.149 | 1.079 | 1.020 | 0.986 | 0.952 | 0.929 | 0.914 | 0.902 |
| | **0.5** | 1.457 | 1.292 | 1.186 | 1.105 | 1.049 | 1.004 | 0.966 | 0.936 | 0.915 | 0.901 |
| **80** | **0** | 0.970 | 0.882 | 0.864 | 0.871 | 0.874 | 0.883 | 0.890 | 0.899 | 0.903 | 0.907 |
| | **0.1** | 1.216 | 1.070 | 0.999 | 0.957 | 0.928 | 0.924 | 0.912 | 0.913 | 0.908 | 0.908 |
| | **0.2** | 1.319 | 1.162 | 1.067 | 1.011 | 0.966 | 0.946 | 0.934 | 0.923 | 0.913 | 0.907 |
| | **0.3** | 1.387 | 1.225 | 1.121 | 1.049 | 1.002 | 0.972 | 0.949 | 0.927 | 0.913 | 0.910 |
| | **0.4** | 1.433 | 1.269 | 1.160 | 1.084 | 1.030 | 0.993 | 0.957 | 0.939 | 0.922 | 0.908 |
| | **0.5** | 1.466 | 1.310 | 1.200 | 1.115 | 1.056 | 1.009 | 0.970 | 0.944 | 0.923 | 0.908 |
| **90** | **0** | 0.979 | 0.885 | 0.874 | 0.878 | 0.883 | 0.889 | 0.896 | 0.904 | 0.907 | 0.912 |
| | **0.1** | 1.230 | 1.079 | 1.007 | 0.965 | 0.936 | 0.927 | 0.919 | 0.914 | 0.913 | 0.913 |
| | **0.2** | 1.340 | 1.171 | 1.074 | 1.020 | 0.982 | 0.952 | 0.938 | 0.925 | 0.917 | 0.913 |
| | **0.3** | 1.406 | 1.229 | 1.128 | 1.055 | 1.015 | 0.976 | 0.952 | 0.936 | 0.920 | 0.913 |
| | **0.4** | 1.451 | 1.280 | 1.173 | 1.093 | 1.034 | 0.999 | 0.968 | 0.944 | 0.925 | 0.912 |
| | **0.5** | 1.484 | 1.318 | 1.204 | 1.123 | 1.066 | 1.015 | 0.976 | 0.949 | 0.928 | 0.912 |
| **100** | **0** | 0.972 | 0.879 | 0.875 | 0.883 | 0.888 | 0.893 | 0.900 | 0.907 | 0.914 | 0.916 |
| | **0.1** | 1.232 | 1.080 | 1.007 | 0.976 | 0.941 | 0.930 | 0.924 | 0.921 | 0.918 | 0.915 |
| | **0.2** | 1.338 | 1.171 | 1.078 | 1.023 | 0.982 | 0.959 | 0.944 | 0.931 | 0.921 | 0.918 |
| | **0.3** | 1.409 | 1.233 | 1.131 | 1.063 | 1.016 | 0.981 | 0.959 | 0.939 | 0.927 | 0.916 |
| | **0.4** | 1.454 | 1.285 | 1.176 | 1.100 | 1.045 | 1.003 | 0.972 | 0.947 | 0.932 | 0.916 |
| | **0.5** | 1.487 | 1.320 | 1.210 | 1.128 | 1.072 | 1.020 | 0.981 | 0.954 | 0.931 | 0.917 |

TABLE 5.7

ESTIMATION OF $Q_{0.05}$ IN TERMS OF $I_1$, $I_2$, $I_4$ (CASE A)

| | | I2 | | | | | | | | | |
|---|---|---|---|---|---|---|---|---|---|---|---|
| **I1** | **I4** | **0.1** | **0.2** | **0.3** | **0.4** | **0.5** | **0.6** | **0.7** | **0.8** | **0.9** | **1** |
| **10** | **0** | 0.905 | 0.835 | 0.792 | 0.787 | 0.763 | 0.761 | 0.769 | 0.777 | 0.783 | 0.794 |
| | **0.1** | 0.930 | 0.850 | 0.817 | 0.791 | 0.764 | 0.765 | 0.770 | 0.779 | 0.783 | 0.794 |
| | **0.2** | 0.946 | 0.877 | 0.830 | 0.804 | 0.783 | 0.789 | 0.783 | 0.775 | 0.787 | 0.793 |
| | **0.3** | 0.961 | 0.904 | 0.849 | 0.824 | 0.806 | 0.806 | 0.783 | 0.777 | 0.783 | 0.793 |
| | **0.4** | 0.978 | 0.907 | 0.866 | 0.845 | 0.808 | 0.804 | 0.803 | 0.792 | 0.782 | 0.793 |
| | **0.5** | 0.983 | 0.922 | 0.873 | 0.856 | 0.819 | 0.817 | 0.805 | 0.796 | 0.786 | 0.793 |
| **20** | **0** | 0.940 | 0.864 | 0.828 | 0.818 | 0.815 | 0.812 | 0.819 | 0.830 | 0.837 | 0.843 |
| | **0.1** | 1.016 | 0.918 | 0.869 | 0.849 | 0.825 | 0.828 | 0.836 | 0.830 | 0.837 | 0.842 |
| | **0.2** | 1.056 | 0.955 | 0.902 | 0.877 | 0.851 | 0.853 | 0.843 | 0.838 | 0.836 | 0.843 |
| | **0.3** | 1.082 | 0.987 | 0.932 | 0.890 | 0.871 | 0.860 | 0.854 | 0.851 | 0.837 | 0.845 |
| | **0.4** | 1.100 | 1.012 | 0.955 | 0.916 | 0.881 | 0.875 | 0.860 | 0.849 | 0.844 | 0.845 |
| | **0.5** | 1.121 | 1.028 | 0.972 | 0.929 | 0.897 | 0.883 | 0.866 | 0.856 | 0.846 | 0.845 |
| **30** | **0** | 0.971 | 0.872 | 0.842 | 0.838 | 0.835 | 0.838 | 0.848 | 0.856 | 0.864 | 0.867 |
| | **0.1** | 1.065 | 0.951 | 0.905 | 0.883 | 0.858 | 0.858 | 0.857 | 0.864 | 0.866 | 0.867 |
| | **0.2** | 1.122 | 1.004 | 0.950 | 0.907 | 0.886 | 0.880 | 0.871 | 0.871 | 0.869 | 0.868 |
| | **0.3** | 1.161 | 1.043 | 0.977 | 0.932 | 0.907 | 0.889 | 0.883 | 0.877 | 0.869 | 0.870 |
| | **0.4** | 1.193 | 1.067 | 1.001 | 0.953 | 0.921 | 0.906 | 0.891 | 0.880 | 0.875 | 0.868 |
| | **0.5** | 1.212 | 1.086 | 1.023 | 0.972 | 0.934 | 0.914 | 0.894 | 0.884 | 0.875 | 0.870 |
| **40** | **0** | 0.968 | 0.879 | 0.851 | 0.852 | 0.852 | 0.854 | 0.864 | 0.874 | 0.878 | 0.884 |
| | **0.1** | 1.093 | 0.974 | 0.925 | 0.901 | 0.878 | 0.881 | 0.880 | 0.877 | 0.879 | 0.881 |
| | **0.2** | 1.160 | 1.028 | 0.966 | 0.937 | 0.907 | 0.898 | 0.891 | 0.888 | 0.883 | 0.885 |
| | **0.3** | 1.207 | 1.073 | 1.000 | 0.958 | 0.929 | 0.913 | 0.902 | 0.894 | 0.887 | 0.885 |
| | **0.4** | 1.246 | 1.102 | 1.025 | 0.984 | 0.938 | 0.925 | 0.907 | 0.899 | 0.887 | 0.882 |
| | **0.5** | 1.260 | 1.127 | 1.052 | 0.996 | 0.960 | 0.940 | 0.919 | 0.901 | 0.891 | 0.884 |
| **50** | **0** | 0.968 | 0.886 | 0.859 | 0.863 | 0.863 | 0.867 | 0.878 | 0.883 | 0.889 | 0.893 |
| | **0.1** | 1.112 | 0.991 | 0.935 | 0.917 | 0.896 | 0.890 | 0.890 | 0.890 | 0.891 | 0.893 |
| | **0.2** | 1.191 | 1.055 | 0.989 | 0.955 | 0.924 | 0.913 | 0.904 | 0.898 | 0.894 | 0.895 |
| | **0.3** | 1.244 | 1.100 | 1.029 | 0.978 | 0.948 | 0.926 | 0.913 | 0.906 | 0.894 | 0.894 |
| | **0.4** | 1.274 | 1.140 | 1.053 | 0.998 | 0.963 | 0.939 | 0.922 | 0.911 | 0.901 | 0.895 |
| | **0.5** | 1.307 | 1.164 | 1.075 | 1.019 | 0.978 | 0.953 | 0.932 | 0.917 | 0.906 | 0.893 |
| **60** | **0** | 0.962 | 0.881 | 0.867 | 0.870 | 0.874 | 0.878 | 0.886 | 0.893 | 0.898 | 0.903 |
| | **0.1** | 1.131 | 1.009 | 0.956 | 0.922 | 0.908 | 0.902 | 0.900 | 0.902 | 0.901 | 0.903 |
| | **0.2** | 1.216 | 1.071 | 1.002 | 0.966 | 0.932 | 0.923 | 0.917 | 0.908 | 0.904 | 0.903 |
| | **0.3** | 1.264 | 1.119 | 1.040 | 0.991 | 0.955 | 0.943 | 0.928 | 0.913 | 0.907 | 0.902 |
| | **0.4** | 1.308 | 1.154 | 1.068 | 1.011 | 0.975 | 0.953 | 0.936 | 0.922 | 0.910 | 0.902 |
| | **0.5** | 1.338 | 1.183 | 1.095 | 1.032 | 0.989 | 0.965 | 0.943 | 0.925 | 0.910 | 0.903 |
| **70** | **0** | 0.955 | 0.884 | 0.872 | 0.878 | 0.881 | 0.885 | 0.893 | 0.902 | 0.904 | 0.906 |
| | **0.1** | 1.144 | 1.018 | 0.959 | 0.935 | 0.918 | 0.912 | 0.910 | 0.910 | 0.908 | 0.908 |
| | **0.2** | 1.238 | 1.086 | 1.019 | 0.973 | 0.945 | 0.932 | 0.923 | 0.917 | 0.911 | 0.908 |
| | **0.3** | 1.287 | 1.135 | 1.052 | 0.997 | 0.966 | 0.944 | 0.932 | 0.924 | 0.913 | 0.908 |
| | **0.4** | 1.328 | 1.169 | 1.081 | 1.026 | 0.984 | 0.963 | 0.941 | 0.927 | 0.918 | 0.909 |
| | **0.5** | 1.367 | 1.201 | 1.108 | 1.045 | 1.000 | 0.972 | 0.948 | 0.931 | 0.918 | 0.909 |
| **80** | **0** | 0.958 | 0.889 | 0.878 | 0.881 | 0.888 | 0.892 | 0.900 | 0.906 | 0.910 | 0.914 |
| | **0.1** | 1.155 | 1.029 | 0.969 | 0.942 | 0.925 | 0.918 | 0.915 | 0.914 | 0.913 | 0.916 |
| | **0.2** | 1.245 | 1.099 | 1.022 | 0.980 | 0.952 | 0.938 | 0.929 | 0.921 | 0.917 | 0.912 |
| | **0.3** | 1.312 | 1.150 | 1.061 | 1.010 | 0.977 | 0.953 | 0.941 | 0.930 | 0.919 | 0.916 |
| | **0.4** | 1.353 | 1.189 | 1.091 | 1.033 | 0.994 | 0.970 | 0.945 | 0.934 | 0.923 | 0.914 |
| | **0.5** | 1.384 | 1.215 | 1.117 | 1.052 | 1.008 | 0.981 | 0.958 | 0.938 | 0.923 | 0.913 |
| **90** | **0** | 0.963 | 0.892 | 0.883 | 0.888 | 0.891 | 0.896 | 0.903 | 0.909 | 0.914 | 0.917 |
| | **0.1** | 1.162 | 1.034 | 0.982 | 0.950 | 0.930 | 0.926 | 0.923 | 0.919 | 0.917 | 0.919 |
| | **0.2** | 1.267 | 1.106 | 1.032 | 0.990 | 0.963 | 0.949 | 0.932 | 0.929 | 0.922 | 0.917 |
| | **0.3** | 1.332 | 1.160 | 1.073 | 1.019 | 0.981 | 0.959 | 0.945 | 0.933 | 0.925 | 0.918 |
| | **0.4** | 1.378 | 1.195 | 1.102 | 1.043 | 1.001 | 0.973 | 0.955 | 0.941 | 0.926 | 0.919 |
| | **0.5** | 1.415 | 1.228 | 1.127 | 1.060 | 1.017 | 0.986 | 0.963 | 0.943 | 0.930 | 0.920 |
| **100** | **0** | 0.954 | 0.891 | 0.886 | 0.893 | 0.897 | 0.903 | 0.908 | 0.914 | 0.919 | 0.920 |
| | **0.1** | 1.181 | 1.040 | 0.983 | 0.957 | 0.934 | 0.930 | 0.924 | 0.925 | 0.922 | 0.922 |
| | **0.2** | 1.275 | 1.116 | 1.037 | 0.995 | 0.968 | 0.949 | 0.939 | 0.931 | 0.925 | 0.922 |
| | **0.3** | 1.339 | 1.164 | 1.077 | 1.025 | 0.988 | 0.967 | 0.950 | 0.937 | 0.928 | 0.924 |
| | **0.4** | 1.385 | 1.203 | 1.109 | 1.048 | 1.007 | 0.979 | 0.959 | 0.944 | 0.932 | 0.923 |
| | **0.5** | 1.425 | 1.236 | 1.138 | 1.071 | 1.023 | 0.994 | 0.966 | 0.950 | 0.933 | 0.922 |

TABLE 5.8

ESTIMATION OF $Q_{0.05}$ IN TERMS OF $I_1$, $I_2$, $I_4$ (CASE B)

| | | I2 | | | | | | | | | |
|---|---|---|---|---|---|---|---|---|---|---|---|
| **I1** | **I4** | **0.1** | **0.2** | **0.3** | **0.4** | **0.5** | **0.6** | **0.7** | **0.8** | **0.9** | **1** |
| **10** | **0** | 0.890 | 0.715 | 0.619 | 0.583 | 0.542 | 0.543 | 0.523 | 0.522 | 0.527 | 0.546 |
| | **0.1** | 0.925 | 0.757 | 0.660 | 0.580 | 0.546 | 0.543 | 0.524 | 0.525 | 0.534 | 0.543 |
| | **0.2** | 0.959 | 0.789 | 0.680 | 0.618 | 0.581 | 0.581 | 0.577 | 0.528 | 0.528 | 0.546 |
| | **0.3** | 0.990 | 0.829 | 0.710 | 0.663 | 0.596 | 0.615 | 0.556 | 0.524 | 0.523 | 0.545 |
| | **0.4** | 1.044 | 0.832 | 0.745 | 0.670 | 0.599 | 0.605 | 0.586 | 0.555 | 0.527 | 0.543 |
| | **0.5** | 1.042 | 0.883 | 0.774 | 0.718 | 0.628 | 0.616 | 0.592 | 0.558 | 0.524 | 0.537 |
| **20** | **0** | 1.011 | 0.802 | 0.703 | 0.659 | 0.630 | 0.621 | 0.619 | 0.631 | 0.649 | 0.672 |
| | **0.1** | 1.121 | 0.899 | 0.770 | 0.721 | 0.664 | 0.662 | 0.652 | 0.631 | 0.644 | 0.662 |
| | **0.2** | 1.205 | 0.997 | 0.850 | 0.766 | 0.708 | 0.702 | 0.680 | 0.668 | 0.647 | 0.660 |
| | **0.3** | 1.272 | 1.034 | 0.894 | 0.818 | 0.753 | 0.712 | 0.688 | 0.680 | 0.648 | 0.658 |
| | **0.4** | 1.333 | 1.089 | 0.956 | 0.858 | 0.770 | 0.756 | 0.711 | 0.679 | 0.668 | 0.661 |
| | **0.5** | 1.361 | 1.136 | 0.986 | 0.887 | 0.807 | 0.765 | 0.731 | 0.690 | 0.670 | 0.656 |
| **30** | **0** | 1.046 | 0.822 | 0.729 | 0.694 | 0.677 | 0.670 | 0.673 | 0.688 | 0.711 | 0.715 |
| | **0.1** | 1.231 | 0.961 | 0.856 | 0.765 | 0.716 | 0.715 | 0.701 | 0.710 | 0.703 | 0.720 |
| | **0.2** | 1.344 | 1.070 | 0.929 | 0.836 | 0.788 | 0.761 | 0.730 | 0.722 | 0.719 | 0.718 |
| | **0.3** | 1.435 | 1.152 | 1.002 | 0.901 | 0.840 | 0.785 | 0.760 | 0.740 | 0.717 | 0.718 |
| | **0.4** | 1.511 | 1.212 | 1.059 | 0.936 | 0.856 | 0.822 | 0.772 | 0.750 | 0.732 | 0.721 |
| | **0.5** | 1.573 | 1.278 | 1.113 | 0.978 | 0.889 | 0.836 | 0.798 | 0.765 | 0.731 | 0.720 |
| **40** | **0** | 1.039 | 0.822 | 0.733 | 0.707 | 0.705 | 0.701 | 0.713 | 0.726 | 0.748 | 0.757 |
| | **0.1** | 1.279 | 1.003 | 0.884 | 0.825 | 0.770 | 0.759 | 0.753 | 0.741 | 0.745 | 0.754 |
| | **0.2** | 1.437 | 1.134 | 0.977 | 0.894 | 0.825 | 0.802 | 0.772 | 0.767 | 0.759 | 0.749 |
| | **0.3** | 1.537 | 1.222 | 1.063 | 0.952 | 0.878 | 0.837 | 0.806 | 0.782 | 0.767 | 0.749 |
| | **0.4** | 1.632 | 1.303 | 1.118 | 0.996 | 0.908 | 0.864 | 0.822 | 0.796 | 0.765 | 0.756 |
| | **0.5** | 1.682 | 1.370 | 1.163 | 1.038 | 0.947 | 0.891 | 0.836 | 0.806 | 0.776 | 0.750 |
| **50** | **0** | 1.030 | 0.823 | 0.744 | 0.729 | 0.732 | 0.724 | 0.738 | 0.756 | 0.766 | 0.780 |
| | **0.1** | 1.317 | 1.049 | 0.913 | 0.852 | 0.801 | 0.777 | 0.766 | 0.769 | 0.764 | 0.773 |
| | **0.2** | 1.504 | 1.176 | 1.021 | 0.931 | 0.868 | 0.827 | 0.808 | 0.791 | 0.784 | 0.780 |
| | **0.3** | 1.616 | 1.284 | 1.096 | 0.989 | 0.916 | 0.869 | 0.825 | 0.801 | 0.787 | 0.781 |
| | **0.4** | 1.699 | 1.369 | 1.165 | 1.033 | 0.943 | 0.898 | 0.851 | 0.818 | 0.792 | 0.779 |
| | **0.5** | 1.780 | 1.426 | 1.216 | 1.074 | 0.986 | 0.927 | 0.867 | 0.830 | 0.802 | 0.776 |
| **60** | **0** | 1.018 | 0.807 | 0.754 | 0.743 | 0.742 | 0.749 | 0.759 | 0.777 | 0.788 | 0.801 |
| | **0.1** | 1.351 | 1.071 | 0.946 | 0.868 | 0.822 | 0.806 | 0.794 | 0.802 | 0.794 | 0.798 |
| | **0.2** | 1.536 | 1.203 | 1.043 | 0.953 | 0.892 | 0.849 | 0.833 | 0.813 | 0.806 | 0.798 |
| | **0.3** | 1.669 | 1.316 | 1.132 | 1.014 | 0.944 | 0.882 | 0.852 | 0.823 | 0.812 | 0.800 |
| | **0.4** | 1.757 | 1.399 | 1.200 | 1.063 | 0.979 | 0.918 | 0.878 | 0.842 | 0.816 | 0.796 |
| | **0.5** | 1.833 | 1.467 | 1.249 | 1.111 | 1.016 | 0.945 | 0.895 | 0.854 | 0.822 | 0.799 |
| **70** | **0** | 1.011 | 0.814 | 0.764 | 0.754 | 0.767 | 0.759 | 0.775 | 0.790 | 0.801 | 0.814 |
| | **0.1** | 1.379 | 1.089 | 0.960 | 0.889 | 0.841 | 0.826 | 0.824 | 0.807 | 0.807 | 0.814 |
| | **0.2** | 1.584 | 1.241 | 1.081 | 0.971 | 0.909 | 0.873 | 0.850 | 0.831 | 0.821 | 0.815 |
| | **0.3** | 1.720 | 1.357 | 1.157 | 1.034 | 0.961 | 0.907 | 0.872 | 0.846 | 0.825 | 0.810 |
| | **0.4** | 1.806 | 1.443 | 1.220 | 1.096 | 0.992 | 0.940 | 0.892 | 0.856 | 0.833 | 0.811 |
| | **0.5** | 1.886 | 1.508 | 1.288 | 1.146 | 1.039 | 0.962 | 0.915 | 0.866 | 0.832 | 0.810 |
| **80** | **0** | 0.996 | 0.820 | 0.770 | 0.768 | 0.779 | 0.775 | 0.791 | 0.804 | 0.814 | 0.821 |
| | **0.1** | 1.380 | 1.101 | 0.983 | 0.903 | 0.860 | 0.836 | 0.833 | 0.829 | 0.819 | 0.828 |
| | **0.2** | 1.607 | 1.267 | 1.100 | 0.986 | 0.922 | 0.884 | 0.864 | 0.840 | 0.833 | 0.821 |
| | **0.3** | 1.739 | 1.373 | 1.183 | 1.059 | 0.971 | 0.916 | 0.887 | 0.858 | 0.834 | 0.823 |
| | **0.4** | 1.834 | 1.474 | 1.260 | 1.113 | 1.014 | 0.958 | 0.905 | 0.869 | 0.846 | 0.822 |
| | **0.5** | 1.914 | 1.538 | 1.310 | 1.157 | 1.061 | 0.982 | 0.924 | 0.880 | 0.846 | 0.823 |
| **90** | **0** | 0.990 | 0.812 | 0.775 | 0.776 | 0.793 | 0.788 | 0.799 | 0.812 | 0.824 | 0.836 |
| | **0.1** | 1.414 | 1.125 | 0.986 | 0.919 | 0.873 | 0.855 | 0.842 | 0.834 | 0.832 | 0.833 |
| | **0.2** | 1.636 | 1.291 | 1.107 | 1.012 | 0.937 | 0.896 | 0.871 | 0.856 | 0.843 | 0.835 |
| | **0.3** | 1.773 | 1.400 | 1.205 | 1.070 | 0.991 | 0.931 | 0.897 | 0.865 | 0.851 | 0.834 |
| | **0.4** | 1.887 | 1.486 | 1.259 | 1.130 | 1.027 | 0.970 | 0.916 | 0.885 | 0.851 | 0.837 |
| | **0.5** | 1.976 | 1.561 | 1.335 | 1.175 | 1.070 | 0.994 | 0.937 | 0.891 | 0.858 | 0.834 |
| **100** | **0** | 0.981 | 0.807 | 0.775 | 0.789 | 0.795 | 0.796 | 0.814 | 0.821 | 0.834 | 0.843 |
| | **0.1** | 1.427 | 1.127 | 1.004 | 0.938 | 0.880 | 0.858 | 0.849 | 0.845 | 0.836 | 0.842 |
| | **0.2** | 1.654 | 1.305 | 1.122 | 1.021 | 0.947 | 0.908 | 0.879 | 0.860 | 0.850 | 0.841 |
| | **0.3** | 1.809 | 1.416 | 1.214 | 1.082 | 1.002 | 0.944 | 0.904 | 0.875 | 0.856 | 0.840 |
| | **0.4** | 1.923 | 1.508 | 1.287 | 1.144 | 1.045 | 0.980 | 0.928 | 0.891 | 0.860 | 0.842 |
| | **0.5** | 1.996 | 1.573 | 1.351 | 1.185 | 1.082 | 1.006 | 0.943 | 0.899 | 0.867 | 0.841 |

TABLE 5.9

ESTIMATION OF $Q_{0.05}$ IN TERMS OF $I_1$, $I_2$, $I_4$ (CASE C)

| | | I2 | | | | | | | | | |
|---|---|---|---|---|---|---|---|---|---|---|---|
| **I1** | **I4** | **0.1** | **0.2** | **0.3** | **0.4** | **0.5** | **0.6** | **0.7** | **0.8** | **0.9** | **1** |
| **10** | **0** | 1.800 | 1.598 | 1.474 | 1.418 | 1.344 | 1.338 | 1.308 | 1.291 | 1.269 | 1.258 |
| | **0.1** | 1.841 | 1.636 | 1.507 | 1.424 | 1.339 | 1.341 | 1.313 | 1.285 | 1.276 | 1.262 |
| | **0.2** | 1.872 | 1.662 | 1.541 | 1.456 | 1.372 | 1.373 | 1.338 | 1.290 | 1.272 | 1.260 |
| | **0.3** | 1.904 | 1.693 | 1.557 | 1.475 | 1.402 | 1.398 | 1.339 | 1.290 | 1.269 | 1.254 |
| | **0.4** | 1.897 | 1.709 | 1.585 | 1.501 | 1.394 | 1.398 | 1.361 | 1.320 | 1.281 | 1.258 |
| | **0.5** | 1.924 | 1.720 | 1.590 | 1.518 | 1.417 | 1.427 | 1.367 | 1.320 | 1.274 | 1.261 |
| **20** | **0** | 1.654 | 1.458 | 1.354 | 1.309 | 1.252 | 1.237 | 1.222 | 1.208 | 1.197 | 1.183 |
| | **0.1** | 1.739 | 1.532 | 1.416 | 1.349 | 1.278 | 1.258 | 1.238 | 1.206 | 1.195 | 1.187 |
| | **0.2** | 1.786 | 1.588 | 1.458 | 1.374 | 1.302 | 1.287 | 1.249 | 1.218 | 1.197 | 1.179 |
| | **0.3** | 1.825 | 1.615 | 1.480 | 1.397 | 1.333 | 1.302 | 1.261 | 1.229 | 1.195 | 1.188 |
| | **0.4** | 1.858 | 1.651 | 1.515 | 1.428 | 1.346 | 1.322 | 1.271 | 1.235 | 1.209 | 1.184 |
| | **0.5** | 1.875 | 1.673 | 1.541 | 1.447 | 1.366 | 1.334 | 1.280 | 1.239 | 1.208 | 1.189 |
| **30** | **0** | 1.577 | 1.384 | 1.300 | 1.252 | 1.207 | 1.193 | 1.178 | 1.168 | 1.157 | 1.151 |
| | **0.1** | 1.691 | 1.480 | 1.376 | 1.301 | 1.239 | 1.226 | 1.193 | 1.180 | 1.160 | 1.150 |
| | **0.2** | 1.757 | 1.532 | 1.420 | 1.334 | 1.274 | 1.242 | 1.210 | 1.187 | 1.169 | 1.151 |
| | **0.3** | 1.800 | 1.578 | 1.453 | 1.366 | 1.300 | 1.258 | 1.229 | 1.192 | 1.168 | 1.150 |
| | **0.4** | 1.828 | 1.619 | 1.483 | 1.391 | 1.314 | 1.277 | 1.231 | 1.197 | 1.172 | 1.150 |
| | **0.5** | 1.856 | 1.647 | 1.510 | 1.418 | 1.340 | 1.291 | 1.242 | 1.207 | 1.171 | 1.150 |
| **40** | **0** | 1.521 | 1.339 | 1.255 | 1.224 | 1.184 | 1.168 | 1.155 | 1.145 | 1.138 | 1.133 |
| | **0.1** | 1.652 | 1.444 | 1.344 | 1.278 | 1.216 | 1.201 | 1.174 | 1.153 | 1.138 | 1.131 |
| | **0.2** | 1.724 | 1.508 | 1.387 | 1.318 | 1.251 | 1.222 | 1.190 | 1.168 | 1.143 | 1.133 |
| | **0.3** | 1.773 | 1.557 | 1.428 | 1.343 | 1.278 | 1.240 | 1.198 | 1.171 | 1.149 | 1.130 |
| | **0.4** | 1.804 | 1.593 | 1.460 | 1.368 | 1.300 | 1.258 | 1.213 | 1.184 | 1.152 | 1.132 |
| | **0.5** | 1.840 | 1.623 | 1.489 | 1.395 | 1.318 | 1.269 | 1.220 | 1.183 | 1.152 | 1.130 |
| **50** | **0** | 1.472 | 1.317 | 1.231 | 1.194 | 1.163 | 1.149 | 1.138 | 1.131 | 1.125 | 1.118 |
| | **0.1** | 1.618 | 1.421 | 1.322 | 1.262 | 1.202 | 1.175 | 1.153 | 1.141 | 1.121 | 1.118 |
| | **0.2** | 1.690 | 1.492 | 1.377 | 1.297 | 1.238 | 1.201 | 1.168 | 1.147 | 1.128 | 1.118 |
| | **0.3** | 1.751 | 1.538 | 1.410 | 1.329 | 1.263 | 1.223 | 1.184 | 1.156 | 1.133 | 1.118 |
| | **0.4** | 1.790 | 1.580 | 1.447 | 1.352 | 1.281 | 1.240 | 1.198 | 1.161 | 1.135 | 1.116 |
| | **0.5** | 1.824 | 1.610 | 1.478 | 1.378 | 1.302 | 1.253 | 1.208 | 1.170 | 1.141 | 1.118 |
| **60** | **0** | 1.433 | 1.277 | 1.207 | 1.178 | 1.151 | 1.137 | 1.126 | 1.118 | 1.114 | 1.108 |
| | **0.1** | 1.595 | 1.403 | 1.309 | 1.238 | 1.193 | 1.166 | 1.142 | 1.131 | 1.118 | 1.107 |
| | **0.2** | 1.678 | 1.470 | 1.355 | 1.281 | 1.228 | 1.193 | 1.163 | 1.140 | 1.120 | 1.106 |
| | **0.3** | 1.736 | 1.521 | 1.398 | 1.318 | 1.257 | 1.212 | 1.175 | 1.143 | 1.125 | 1.106 |
| | **0.4** | 1.774 | 1.563 | 1.434 | 1.349 | 1.273 | 1.227 | 1.188 | 1.152 | 1.127 | 1.103 |
| | **0.5** | 1.808 | 1.598 | 1.463 | 1.367 | 1.294 | 1.245 | 1.195 | 1.160 | 1.128 | 1.106 |
| **70** | **0** | 1.406 | 1.250 | 1.195 | 1.164 | 1.140 | 1.126 | 1.117 | 1.114 | 1.106 | 1.096 |
| | **0.1** | 1.577 | 1.397 | 1.291 | 1.232 | 1.182 | 1.158 | 1.138 | 1.118 | 1.110 | 1.096 |
| | **0.2** | 1.668 | 1.463 | 1.345 | 1.270 | 1.216 | 1.183 | 1.154 | 1.131 | 1.109 | 1.100 |
| | **0.3** | 1.727 | 1.513 | 1.389 | 1.306 | 1.246 | 1.202 | 1.165 | 1.138 | 1.114 | 1.098 |
| | **0.4** | 1.769 | 1.549 | 1.422 | 1.333 | 1.264 | 1.221 | 1.175 | 1.145 | 1.118 | 1.100 |
| | **0.5** | 1.802 | 1.583 | 1.456 | 1.359 | 1.289 | 1.234 | 1.188 | 1.150 | 1.122 | 1.097 |
| **80** | **0** | 1.384 | 1.247 | 1.178 | 1.152 | 1.130 | 1.117 | 1.108 | 1.103 | 1.099 | 1.092 |
| | **0.1** | 1.566 | 1.386 | 1.283 | 1.223 | 1.176 | 1.152 | 1.130 | 1.115 | 1.103 | 1.089 |
| | **0.2** | 1.655 | 1.452 | 1.341 | 1.267 | 1.210 | 1.174 | 1.147 | 1.125 | 1.107 | 1.094 |
| | **0.3** | 1.715 | 1.505 | 1.379 | 1.299 | 1.239 | 1.192 | 1.157 | 1.129 | 1.109 | 1.092 |
| | **0.4** | 1.759 | 1.545 | 1.417 | 1.326 | 1.259 | 1.210 | 1.170 | 1.139 | 1.111 | 1.092 |
| | **0.5** | 1.791 | 1.580 | 1.449 | 1.352 | 1.281 | 1.227 | 1.182 | 1.144 | 1.114 | 1.094 |
| **90** | **0** | 1.363 | 1.227 | 1.172 | 1.147 | 1.122 | 1.114 | 1.106 | 1.098 | 1.090 | 1.086 |
| | **0.1** | 1.556 | 1.373 | 1.277 | 1.217 | 1.166 | 1.144 | 1.125 | 1.110 | 1.096 | 1.086 |
| | **0.2** | 1.653 | 1.447 | 1.330 | 1.260 | 1.203 | 1.172 | 1.140 | 1.120 | 1.100 | 1.086 |
| | **0.3** | 1.707 | 1.495 | 1.373 | 1.290 | 1.234 | 1.187 | 1.151 | 1.123 | 1.103 | 1.087 |
| | **0.4** | 1.753 | 1.539 | 1.411 | 1.322 | 1.249 | 1.207 | 1.165 | 1.132 | 1.105 | 1.086 |
| | **0.5** | 1.785 | 1.577 | 1.442 | 1.344 | 1.276 | 1.222 | 1.174 | 1.137 | 1.110 | 1.087 |
| **100** | **0** | 1.340 | 1.212 | 1.160 | 1.136 | 1.118 | 1.105 | 1.097 | 1.090 | 1.086 | 1.082 |
| | **0.1** | 1.543 | 1.358 | 1.266 | 1.212 | 1.161 | 1.137 | 1.120 | 1.104 | 1.088 | 1.083 |
| | **0.2** | 1.640 | 1.436 | 1.322 | 1.252 | 1.197 | 1.161 | 1.136 | 1.114 | 1.096 | 1.083 |
| | **0.3** | 1.701 | 1.490 | 1.367 | 1.287 | 1.225 | 1.180 | 1.149 | 1.122 | 1.099 | 1.081 |
| | **0.4** | 1.744 | 1.531 | 1.406 | 1.317 | 1.249 | 1.201 | 1.159 | 1.126 | 1.101 | 1.081 |
| | **0.5** | 1.781 | 1.564 | 1.438 | 1.340 | 1.273 | 1.216 | 1.167 | 1.132 | 1.103 | 1.083 |

TABLE 5.10

ESTIMATION OF $Q_{0.95}$ IN TERMS OF $I_1$, $I_2$, $I_4$ (CASE A)

| | | I2 | | | | | | | | | |
|---|---|---|---|---|---|---|---|---|---|---|---|
| I1 | I4 | 0.1 | 0.2 | 0.3 | 0.4 | 0.5 | 0.6 | 0.7 | 0.8 | 0.9 | 1 |
| 10 | 0 | 2.246 | 1.818 | 1.604 | 1.519 | 1.410 | 1.394 | 1.355 | 1.337 | 1.298 | 1.284 |
| | 0.1 | 2.342 | 1.870 | 1.664 | 1.515 | 1.400 | 1.398 | 1.361 | 1.323 | 1.307 | 1.287 |
| | 0.2 | 2.321 | 1.932 | 1.704 | 1.551 | 1.455 | 1.448 | 1.381 | 1.326 | 1.306 | 1.286 |
| | 0.3 | 2.415 | 1.969 | 1.728 | 1.590 | 1.482 | 1.473 | 1.391 | 1.334 | 1.301 | 1.291 |
| | 0.4 | 2.426 | 2.006 | 1.751 | 1.619 | 1.459 | 1.475 | 1.427 | 1.366 | 1.298 | 1.283 |
| | 0.5 | 2.462 | 2.023 | 1.761 | 1.632 | 1.495 | 1.522 | 1.426 | 1.356 | 1.308 | 1.293 |
| 20 | 0 | 2.027 | 1.612 | 1.437 | 1.360 | 1.290 | 1.264 | 1.248 | 1.226 | 1.212 | 1.203 |
| | 0.1 | 2.147 | 1.716 | 1.520 | 1.406 | 1.316 | 1.295 | 1.267 | 1.231 | 1.219 | 1.204 |
| | 0.2 | 2.229 | 1.790 | 1.576 | 1.448 | 1.346 | 1.329 | 1.272 | 1.240 | 1.211 | 1.199 |
| | 0.3 | 2.324 | 1.850 | 1.608 | 1.469 | 1.394 | 1.353 | 1.292 | 1.252 | 1.217 | 1.206 |
| | 0.4 | 2.360 | 1.888 | 1.644 | 1.514 | 1.403 | 1.366 | 1.312 | 1.264 | 1.223 | 1.200 |
| | 0.5 | 2.393 | 1.932 | 1.689 | 1.533 | 1.417 | 1.377 | 1.311 | 1.263 | 1.227 | 1.207 |
| 30 | 0 | 1.890 | 1.496 | 1.359 | 1.296 | 1.235 | 1.215 | 1.196 | 1.182 | 1.170 | 1.165 |
| | 0.1 | 2.071 | 1.634 | 1.458 | 1.353 | 1.273 | 1.247 | 1.217 | 1.194 | 1.175 | 1.164 |
| | 0.2 | 2.160 | 1.718 | 1.518 | 1.388 | 1.309 | 1.271 | 1.226 | 1.201 | 1.184 | 1.163 |
| | 0.3 | 2.269 | 1.778 | 1.558 | 1.428 | 1.344 | 1.287 | 1.252 | 1.218 | 1.186 | 1.166 |
| | 0.4 | 2.315 | 1.834 | 1.595 | 1.462 | 1.358 | 1.303 | 1.255 | 1.214 | 1.188 | 1.162 |
| | 0.5 | 2.351 | 1.872 | 1.631 | 1.481 | 1.378 | 1.330 | 1.265 | 1.225 | 1.188 | 1.162 |
| 40 | 0 | 1.765 | 1.420 | 1.296 | 1.245 | 1.206 | 1.187 | 1.172 | 1.158 | 1.153 | 1.142 |
| | 0.1 | 1.993 | 1.576 | 1.403 | 1.312 | 1.242 | 1.222 | 1.190 | 1.164 | 1.148 | 1.139 |
| | 0.2 | 2.120 | 1.669 | 1.469 | 1.364 | 1.280 | 1.249 | 1.203 | 1.181 | 1.153 | 1.139 |
| | 0.3 | 2.200 | 1.727 | 1.520 | 1.396 | 1.309 | 1.267 | 1.215 | 1.185 | 1.159 | 1.140 |
| | 0.4 | 2.281 | 1.775 | 1.553 | 1.427 | 1.329 | 1.278 | 1.232 | 1.194 | 1.164 | 1.139 |
| | 0.5 | 2.307 | 1.829 | 1.590 | 1.453 | 1.349 | 1.297 | 1.238 | 1.202 | 1.165 | 1.140 |
| 50 | 0 | 1.672 | 1.375 | 1.270 | 1.215 | 1.181 | 1.165 | 1.151 | 1.140 | 1.130 | 1.124 |
| | 0.1 | 1.909 | 1.536 | 1.373 | 1.294 | 1.224 | 1.198 | 1.164 | 1.150 | 1.128 | 1.126 |
| | 0.2 | 2.065 | 1.629 | 1.433 | 1.333 | 1.259 | 1.221 | 1.186 | 1.157 | 1.134 | 1.123 |
| | 0.3 | 2.162 | 1.703 | 1.489 | 1.369 | 1.293 | 1.237 | 1.199 | 1.165 | 1.140 | 1.127 |
| | 0.4 | 2.211 | 1.749 | 1.532 | 1.406 | 1.311 | 1.259 | 1.211 | 1.173 | 1.145 | 1.124 |
| | 0.5 | 2.257 | 1.799 | 1.572 | 1.429 | 1.333 | 1.274 | 1.221 | 1.181 | 1.152 | 1.121 |
| 60 | 0 | 1.623 | 1.342 | 1.244 | 1.197 | 1.174 | 1.149 | 1.134 | 1.127 | 1.119 | 1.115 |
| | 0.1 | 1.878 | 1.500 | 1.354 | 1.269 | 1.210 | 1.186 | 1.153 | 1.137 | 1.123 | 1.112 |
| | 0.2 | 2.030 | 1.604 | 1.418 | 1.320 | 1.245 | 1.204 | 1.178 | 1.149 | 1.129 | 1.115 |
| | 0.3 | 2.108 | 1.673 | 1.476 | 1.353 | 1.273 | 1.227 | 1.186 | 1.151 | 1.134 | 1.112 |
| | 0.4 | 2.153 | 1.730 | 1.514 | 1.380 | 1.290 | 1.238 | 1.199 | 1.159 | 1.135 | 1.113 |
| | 0.5 | 2.225 | 1.769 | 1.544 | 1.408 | 1.319 | 1.260 | 1.206 | 1.163 | 1.135 | 1.112 |
| 70 | 0 | 1.544 | 1.310 | 1.221 | 1.181 | 1.152 | 1.137 | 1.126 | 1.118 | 1.107 | 1.100 |
| | 0.1 | 1.838 | 1.488 | 1.328 | 1.255 | 1.197 | 1.173 | 1.146 | 1.128 | 1.113 | 1.105 |
| | 0.2 | 1.983 | 1.582 | 1.404 | 1.305 | 1.232 | 1.198 | 1.165 | 1.136 | 1.120 | 1.101 |
| | 0.3 | 2.090 | 1.647 | 1.452 | 1.335 | 1.265 | 1.209 | 1.176 | 1.145 | 1.117 | 1.105 |
| | 0.4 | 2.166 | 1.699 | 1.493 | 1.372 | 1.284 | 1.232 | 1.182 | 1.151 | 1.129 | 1.103 |
| | 0.5 | 2.200 | 1.756 | 1.532 | 1.390 | 1.296 | 1.243 | 1.196 | 1.156 | 1.127 | 1.103 |
| 80 | 0 | 1.504 | 1.281 | 1.203 | 1.166 | 1.142 | 1.123 | 1.116 | 1.106 | 1.103 | 1.098 |
| | 0.1 | 1.804 | 1.471 | 1.323 | 1.246 | 1.194 | 1.162 | 1.135 | 1.120 | 1.107 | 1.099 |
| | 0.2 | 1.961 | 1.556 | 1.391 | 1.290 | 1.224 | 1.185 | 1.154 | 1.130 | 1.113 | 1.095 |
| | 0.3 | 2.065 | 1.631 | 1.442 | 1.325 | 1.249 | 1.203 | 1.165 | 1.136 | 1.117 | 1.097 |
| | 0.4 | 2.120 | 1.688 | 1.481 | 1.358 | 1.274 | 1.221 | 1.176 | 1.141 | 1.117 | 1.094 |
| | 0.5 | 2.178 | 1.735 | 1.516 | 1.383 | 1.289 | 1.236 | 1.188 | 1.146 | 1.121 | 1.096 |
| 90 | 0 | 1.481 | 1.272 | 1.192 | 1.162 | 1.132 | 1.121 | 1.110 | 1.101 | 1.095 | 1.090 |
| | 0.1 | 1.784 | 1.461 | 1.316 | 1.234 | 1.179 | 1.151 | 1.132 | 1.115 | 1.099 | 1.093 |
| | 0.2 | 1.954 | 1.551 | 1.378 | 1.284 | 1.218 | 1.179 | 1.146 | 1.124 | 1.105 | 1.092 |
| | 0.3 | 2.036 | 1.621 | 1.428 | 1.322 | 1.242 | 1.196 | 1.157 | 1.131 | 1.109 | 1.090 |
| | 0.4 | 2.129 | 1.676 | 1.465 | 1.351 | 1.262 | 1.209 | 1.169 | 1.138 | 1.109 | 1.091 |
| | 0.5 | 2.162 | 1.713 | 1.504 | 1.373 | 1.287 | 1.227 | 1.178 | 1.142 | 1.116 | 1.092 |
| 100 | 0 | 1.446 | 1.248 | 1.182 | 1.148 | 1.124 | 1.115 | 1.104 | 1.098 | 1.092 | 1.087 |
| | 0.1 | 1.771 | 1.430 | 1.302 | 1.230 | 1.168 | 1.147 | 1.124 | 1.106 | 1.096 | 1.089 |
| | 0.2 | 1.929 | 1.537 | 1.366 | 1.273 | 1.210 | 1.168 | 1.137 | 1.118 | 1.100 | 1.087 |
| | 0.3 | 2.028 | 1.615 | 1.417 | 1.307 | 1.232 | 1.190 | 1.154 | 1.124 | 1.102 | 1.086 |
| | 0.4 | 2.094 | 1.657 | 1.457 | 1.337 | 1.256 | 1.202 | 1.163 | 1.130 | 1.107 | 1.091 |
| | 0.5 | 2.145 | 1.702 | 1.497 | 1.367 | 1.276 | 1.216 | 1.172 | 1.136 | 1.108 | 1.084 |

**TABLE 5.11**

**ESTIMATION OF $Q_{0.95}$ IN TERMS OF $I_1$, $I_2$, $I_4$ (CASE B)**

| I1 | I4 | I2: 0.1 | 0.2 | 0.3 | 0.4 | 0.5 | 0.6 | 0.7 | 0.8 | 0.9 | 1 |
|---|---|---|---|---|---|---|---|---|---|---|---|
| 10 | 0 | 3.262 | 2.555 | 2.190 | 2.015 | 1.808 | 1.809 | 1.718 | 1.651 | 1.595 | 1.564 |
| | 0.1 | 3.365 | 2.650 | 2.295 | 1.971 | 1.827 | 1.809 | 1.714 | 1.643 | 1.609 | 1.564 |
| | 0.2 | 3.483 | 2.756 | 2.289 | 2.075 | 1.897 | 1.873 | 1.790 | 1.646 | 1.604 | 1.558 |
| | 0.3 | 3.517 | 2.799 | 2.367 | 2.145 | 1.940 | 1.946 | 1.784 | 1.650 | 1.609 | 1.578 |
| | 0.4 | 3.588 | 2.876 | 2.478 | 2.193 | 1.935 | 1.933 | 1.823 | 1.706 | 1.608 | 1.557 |
| | 0.5 | 3.592 | 2.966 | 2.485 | 2.244 | 1.963 | 1.992 | 1.821 | 1.718 | 1.599 | 1.569 |
| 20 | 0 | 2.780 | 2.143 | 1.877 | 1.705 | 1.592 | 1.537 | 1.478 | 1.449 | 1.411 | 1.400 |
| | 0.1 | 3.053 | 2.342 | 2.004 | 1.820 | 1.619 | 1.586 | 1.520 | 1.433 | 1.409 | 1.395 |
| | 0.2 | 3.219 | 2.480 | 2.083 | 1.871 | 1.700 | 1.646 | 1.550 | 1.478 | 1.410 | 1.402 |
| | 0.3 | 3.336 | 2.579 | 2.160 | 1.921 | 1.765 | 1.666 | 1.558 | 1.516 | 1.413 | 1.385 |
| | 0.4 | 3.414 | 2.672 | 2.250 | 2.007 | 1.778 | 1.732 | 1.605 | 1.505 | 1.439 | 1.397 |
| | 0.5 | 3.471 | 2.720 | 2.305 | 2.041 | 1.822 | 1.761 | 1.612 | 1.516 | 1.442 | 1.391 |
| 30 | 0 | 2.557 | 1.949 | 1.694 | 1.584 | 1.466 | 1.421 | 1.380 | 1.353 | 1.339 | 1.308 |
| | 0.1 | 2.834 | 2.198 | 1.886 | 1.684 | 1.532 | 1.490 | 1.413 | 1.375 | 1.333 | 1.327 |
| | 0.2 | 3.031 | 2.320 | 1.993 | 1.763 | 1.610 | 1.531 | 1.456 | 1.404 | 1.346 | 1.308 |
| | 0.3 | 3.203 | 2.433 | 2.075 | 1.832 | 1.682 | 1.571 | 1.479 | 1.409 | 1.350 | 1.320 |
| | 0.4 | 3.358 | 2.545 | 2.152 | 1.885 | 1.708 | 1.610 | 1.506 | 1.421 | 1.367 | 1.324 |
| | 0.5 | 3.395 | 2.650 | 2.216 | 1.957 | 1.740 | 1.641 | 1.525 | 1.438 | 1.363 | 1.321 |
| 40 | 0 | 2.364 | 1.816 | 1.589 | 1.499 | 1.415 | 1.361 | 1.331 | 1.302 | 1.286 | 1.272 |
| | 0.1 | 2.726 | 2.070 | 1.795 | 1.619 | 1.499 | 1.429 | 1.367 | 1.333 | 1.282 | 1.265 |
| | 0.2 | 2.985 | 2.246 | 1.883 | 1.713 | 1.554 | 1.474 | 1.396 | 1.356 | 1.306 | 1.276 |
| | 0.3 | 3.119 | 2.374 | 2.014 | 1.768 | 1.601 | 1.518 | 1.424 | 1.363 | 1.312 | 1.269 |
| | 0.4 | 3.265 | 2.473 | 2.092 | 1.833 | 1.638 | 1.552 | 1.456 | 1.389 | 1.310 | 1.273 |
| | 0.5 | 3.319 | 2.562 | 2.139 | 1.885 | 1.693 | 1.582 | 1.461 | 1.394 | 1.325 | 1.264 |
| 50 | 0 | 2.212 | 1.735 | 1.523 | 1.425 | 1.368 | 1.323 | 1.283 | 1.273 | 1.256 | 1.243 |
| | 0.1 | 2.632 | 2.035 | 1.736 | 1.588 | 1.446 | 1.377 | 1.332 | 1.283 | 1.254 | 1.243 |
| | 0.2 | 2.889 | 2.198 | 1.844 | 1.659 | 1.524 | 1.431 | 1.370 | 1.308 | 1.263 | 1.245 |
| | 0.3 | 3.045 | 2.322 | 1.951 | 1.734 | 1.569 | 1.478 | 1.387 | 1.333 | 1.274 | 1.240 |
| | 0.4 | 3.172 | 2.432 | 2.028 | 1.792 | 1.605 | 1.508 | 1.417 | 1.348 | 1.292 | 1.249 |
| | 0.5 | 3.268 | 2.513 | 2.098 | 1.855 | 1.656 | 1.551 | 1.440 | 1.348 | 1.297 | 1.240 |
| 60 | 0 | 2.102 | 1.653 | 1.461 | 1.393 | 1.327 | 1.292 | 1.266 | 1.253 | 1.230 | 1.224 |
| | 0.1 | 2.546 | 1.958 | 1.699 | 1.531 | 1.427 | 1.352 | 1.304 | 1.274 | 1.245 | 1.226 |
| | 0.2 | 2.809 | 2.136 | 1.815 | 1.627 | 1.490 | 1.404 | 1.346 | 1.284 | 1.248 | 1.225 |
| | 0.3 | 2.998 | 2.264 | 1.908 | 1.695 | 1.534 | 1.440 | 1.373 | 1.298 | 1.261 | 1.227 |
| | 0.4 | 3.101 | 2.374 | 2.006 | 1.748 | 1.583 | 1.473 | 1.395 | 1.319 | 1.269 | 1.216 |
| | 0.5 | 3.201 | 2.472 | 2.059 | 1.807 | 1.635 | 1.515 | 1.412 | 1.329 | 1.268 | 1.224 |
| 70 | 0 | 2.007 | 1.584 | 1.432 | 1.361 | 1.312 | 1.272 | 1.246 | 1.229 | 1.218 | 1.202 |
| | 0.1 | 2.497 | 1.919 | 1.665 | 1.500 | 1.392 | 1.335 | 1.295 | 1.251 | 1.226 | 1.207 |
| | 0.2 | 2.771 | 2.099 | 1.794 | 1.591 | 1.460 | 1.390 | 1.330 | 1.278 | 1.231 | 1.207 |
| | 0.3 | 2.950 | 2.241 | 1.881 | 1.665 | 1.529 | 1.425 | 1.341 | 1.290 | 1.235 | 1.206 |
| | 0.4 | 3.069 | 2.347 | 1.966 | 1.736 | 1.567 | 1.461 | 1.373 | 1.301 | 1.252 | 1.203 |
| | 0.5 | 3.159 | 2.428 | 2.040 | 1.787 | 1.615 | 1.495 | 1.394 | 1.322 | 1.253 | 1.205 |
| 80 | 0 | 1.942 | 1.560 | 1.403 | 1.331 | 1.283 | 1.250 | 1.233 | 1.214 | 1.195 | 1.189 |
| | 0.1 | 2.451 | 1.900 | 1.640 | 1.488 | 1.372 | 1.314 | 1.270 | 1.241 | 1.213 | 1.191 |
| | 0.2 | 2.721 | 2.063 | 1.755 | 1.572 | 1.450 | 1.365 | 1.304 | 1.255 | 1.221 | 1.192 |
| | 0.3 | 2.899 | 2.206 | 1.871 | 1.655 | 1.500 | 1.409 | 1.332 | 1.272 | 1.227 | 1.194 |
| | 0.4 | 3.036 | 2.315 | 1.938 | 1.712 | 1.545 | 1.440 | 1.348 | 1.295 | 1.235 | 1.188 |
| | 0.5 | 3.142 | 2.417 | 2.011 | 1.775 | 1.593 | 1.473 | 1.380 | 1.295 | 1.238 | 1.191 |
| 90 | 0 | 1.888 | 1.523 | 1.377 | 1.306 | 1.274 | 1.235 | 1.220 | 1.198 | 1.191 | 1.181 |
| | 0.1 | 2.418 | 1.870 | 1.611 | 1.465 | 1.361 | 1.298 | 1.261 | 1.226 | 1.196 | 1.177 |
| | 0.2 | 2.693 | 2.053 | 1.745 | 1.565 | 1.433 | 1.356 | 1.291 | 1.246 | 1.205 | 1.179 |
| | 0.3 | 2.893 | 2.170 | 1.844 | 1.631 | 1.481 | 1.386 | 1.317 | 1.260 | 1.214 | 1.175 |
| | 0.4 | 3.012 | 2.295 | 1.920 | 1.693 | 1.531 | 1.430 | 1.340 | 1.272 | 1.220 | 1.179 |
| | 0.5 | 3.148 | 2.395 | 1.992 | 1.755 | 1.576 | 1.455 | 1.352 | 1.279 | 1.221 | 1.177 |
| 100 | 0 | 1.833 | 1.485 | 1.357 | 1.290 | 1.252 | 1.219 | 1.209 | 1.192 | 1.178 | 1.172 |
| | 0.1 | 2.393 | 1.834 | 1.597 | 1.454 | 1.347 | 1.287 | 1.246 | 1.208 | 1.183 | 1.168 |
| | 0.2 | 2.677 | 2.024 | 1.720 | 1.543 | 1.419 | 1.341 | 1.280 | 1.234 | 1.198 | 1.171 |
| | 0.3 | 2.862 | 2.162 | 1.827 | 1.614 | 1.478 | 1.380 | 1.297 | 1.248 | 1.202 | 1.168 |
| | 0.4 | 2.994 | 2.270 | 1.912 | 1.685 | 1.528 | 1.409 | 1.328 | 1.264 | 1.211 | 1.170 |
| | 0.5 | 3.109 | 2.364 | 1.989 | 1.733 | 1.575 | 1.442 | 1.354 | 1.274 | 1.216 | 1.166 |

**TABLE 5.12**

**ESTIMATION OF $Q_{0.95}$ IN TERMS OF $I_1$, $I_2$, $I_4$ (CASE C)**

| | LogNormal | | | | Normal | | | | Exponential | | | |
|---|---|---|---|---|---|---|---|---|---|---|---|---|
| | Mean | Q0.95 | S.E. | Q0.05 | Mean | Q0.95 | S.E. | Q0.05 | Mean | Q0.95 | S.E. | Q0.05 |
| Initial Variance | 0.0300 | 0.0921 | 0.0065 | 0.0138 | 0.0255 | 0.0381 | 0.0017 | 0.0237 | 0.1290 | 0.2957 | 0.0161 | 0.0743 |
| r2 | 71.0% | 73.4% | 74.6% | 68.5% | 77.4% | 81.3% | 82.0% | 72.9% | 73.6% | 76.2% | 78.5% | 70.4% |
| k0 | 1.31693 | 1.90226 | 0.31313 | 0.93316 | 1.30662 | 1.64949 | 0.20283 | 0.98028 | 1.68756 | 2.67120 | 0.53648 | 0.93459 |
| k1 | 0.00011 | -0.00276 | -0.00150 | 0.00176 | 0.00005 | -0.00191 | -0.00116 | 0.00193 | 0.00009 | -0.00513 | -0.00276 | 0.00374 |
| k2 | -0.47811 | -0.83634 | -0.18888 | -0.25242 | -0.45483 | -0.55923 | -0.05810 | -0.36493 | -1.01064 | -1.51959 | -0.27518 | -0.63025 |
| k3 | 0.28623 | 0.34725 | 0.03280 | 0.23544 | 0.29961 | 0.26720 | -0.01972 | 0.33050 | 0.59678 | 0.66230 | 0.04298 | 0.51870 |

**tk=k0+k1.I1+k2.I2+k3.I4**

| | LogNormal | | | | Normal | | | | Exponential | | | |
|---|---|---|---|---|---|---|---|---|---|---|---|---|
| | Mean | Q0.95 | S.E. | Q0.05 | Mean | Q0.95 | S.E. | Q0.05 | Mean | Q0.95 | S.E. | Q0.05 |
| Initial Variance | 0.0300 | 0.0921 | 0.0065 | 0.0138 | 0.0255 | 0.0381 | 0.0017 | 0.0237 | 0.1290 | 0.2957 | 0.0161 | 0.0743 |
| r2 | 93.5% | 94.3% | 94.2% | 93.7% | 96.1% | 97.1% | 96.4% | 95.6% | 82.6% | 95.5% | 95.0% | 94 3% |
| k0 | 1.45578 | 2.31298 | 0.48176 | 0.87116 | 1.38007 | 1.84810 | 0.27931 | 0.93047 | 1.80429 | 3.36660 | 0.79055 | 0.83683 |
| k1 | -0.00091 | -0.00959 | -0.00516 | 0.00513 | 0.00007 | -0.00623 | -0.00359 | 0.00556 | -0.00186 | -0.01811 | -0.00891 | 0.01109 |
| k2 | -1.33529 | -2.56606 | -0 65784 | -0.61351 | -1.10714 | -1.39066 | -0.15215 | -0.89647 | -1.14819 | -4.35894 | -0.87546 | -1.55116 |
| k3 | 0.62788 | 1 10637 | 0.12477 | 0.67911 | 0.65437 | 0.75719 | -0.06647 | 0.97727 | 0.03123 | 2.00279 | 0.03432 | 1.42786 |
| k4 | 0.00000 | 0.00004 | 0.00003 | -0.00003 | 0.00000 | 0.00003 | 0.00002 | -0.00003 | 0.00000 | 0.00009 | 0 00004 | -0.00006 |
| k5 | 0.97086 | 1.66591 | 0.36701 | 0.55952 | 0.77400 | 0.87656 | 0.05262 | 0.70844 | 0.59960 | 2.79227 | 0.45831 | 1.31699 |
| k6 | 0.00506 | -0.52876 | -0.04793 | -0.36335 | 0.00482 | -0.37038 | 0.04004 | -0.50112 | 0.28253 | -0.90839 | 0 22813 | -0.73132 |
| k7 | 0.00002 | 0.00297 | 0.00171 | -0.00158 | 0.00002 | 0.00083 | 0 00042 | -0.00048 | -0.00333 | 0.00474 | 0.00246 | -0.00293 |
| k8 | 0.00222 | 0.00165 | -0 00008 | 0 00194 | 0.00152 | 0.00160 | -0.00003 | 0.00165 | 0.01614 | 0.00359 | -0.00350 | 0.00479 |
| k9 | -0.84799 | -1.06465 | -0 11607 | -0.67055 | -0.80127 | -0.71432 | 0.05185 | -0.88561 | -1.16852 | -1.97055 | -0.15600 | -1.46677 |

**tk=k0+k1.I1+k2.I2+k3.I4+k4.I1^2+k5.I2^2+k6.I4^2+k7.I1.I2+k8.I1.I4+k9.I2.I4**

**TABLE 5.13**

**LINEAR AND QUADRATIC MODELS ESTIMATED**

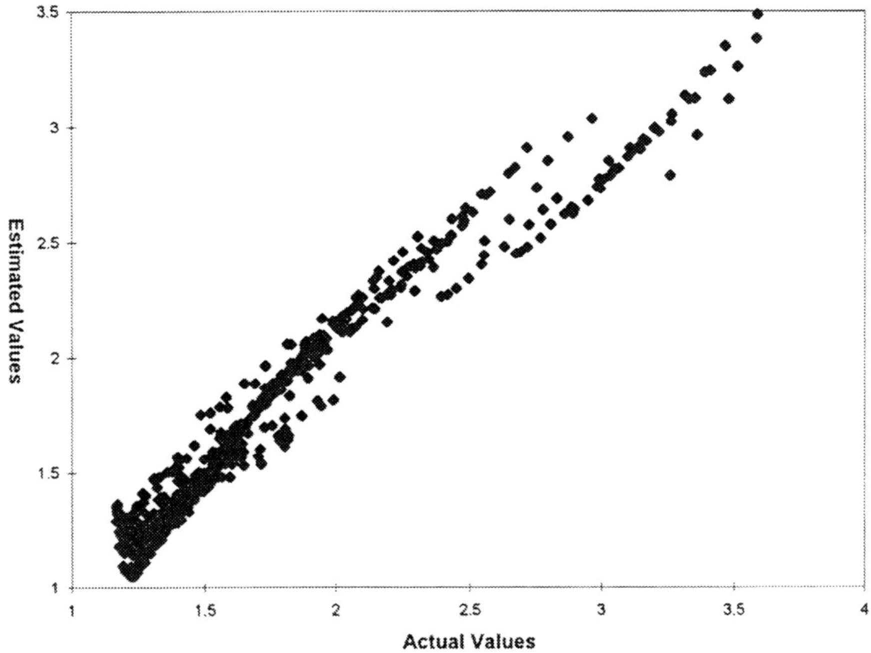


FIGURE 5.15
ACTUAL AND ESTIMATED VALUES FOR MEAN (CASE A)
USING THE QUADRATIC MODEL PRESENTED IN TABLE 5.13

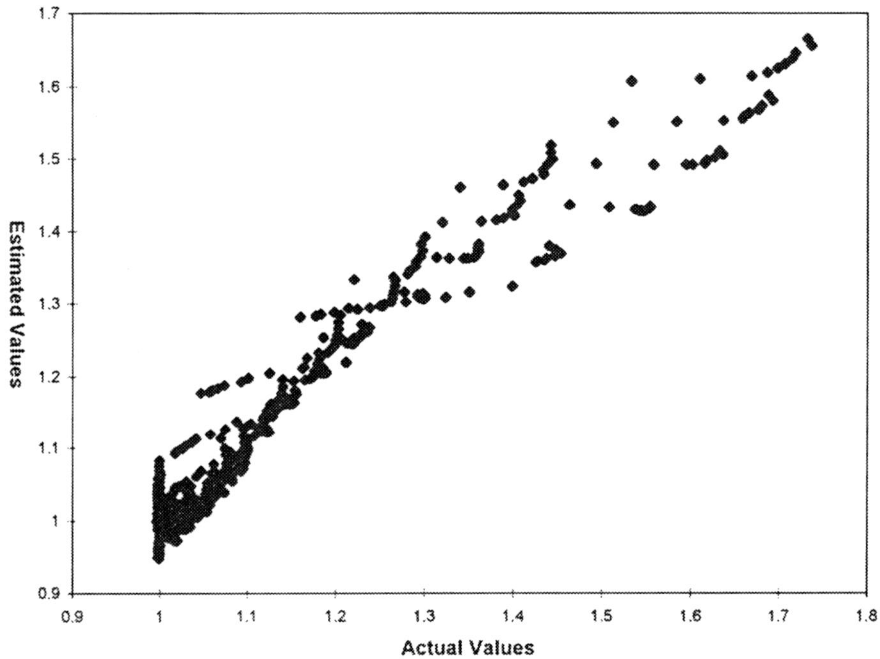


FIGURE 5.16
ACTUAL AND ESTIMATED VALUES FOR MEAN (CASE B)
USING THE QUADRATIC MODEL PRESENTED IN TABLE 5.13

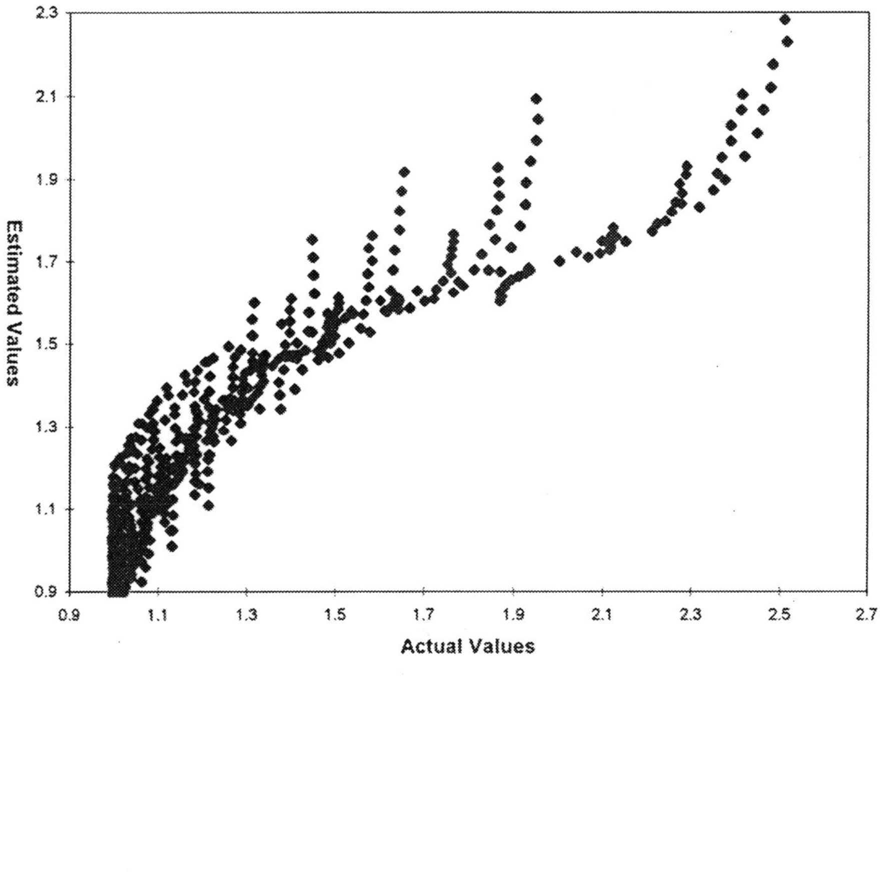


FIGURE 5.17
ACTUAL AND ESTIMATED VALUES FOR MEAN (CASE C)
USING THE QUADRATIC MODEL PRESENTED IN TABLE 5.13

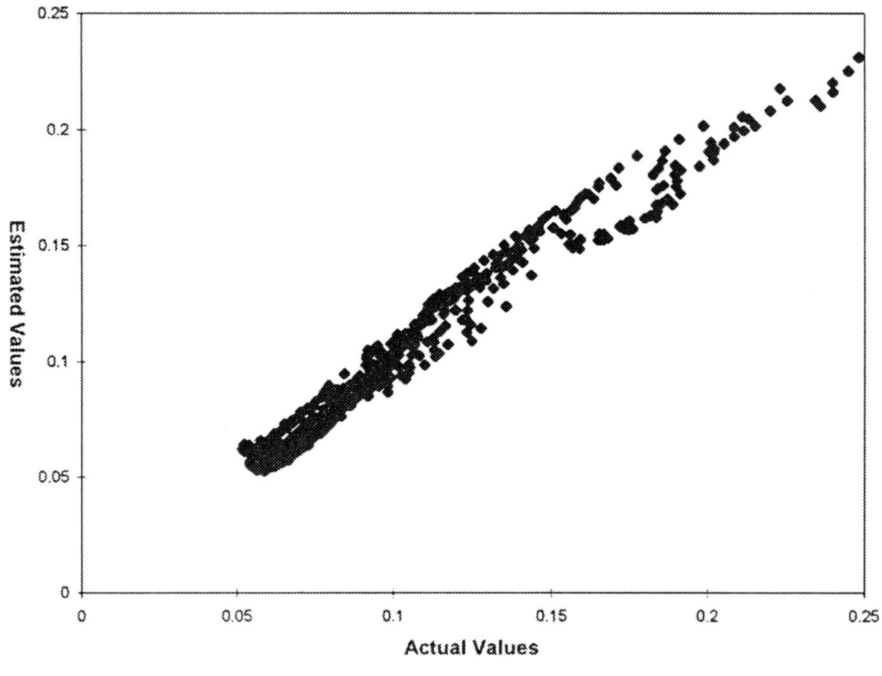


**FIGURE 5.18**
**ACTUAL AND ESTIMATED VALUES FOR S. E. (CASE A) USING THE QUADRATIC MODEL PRESENTED IN TABLE 5.13**

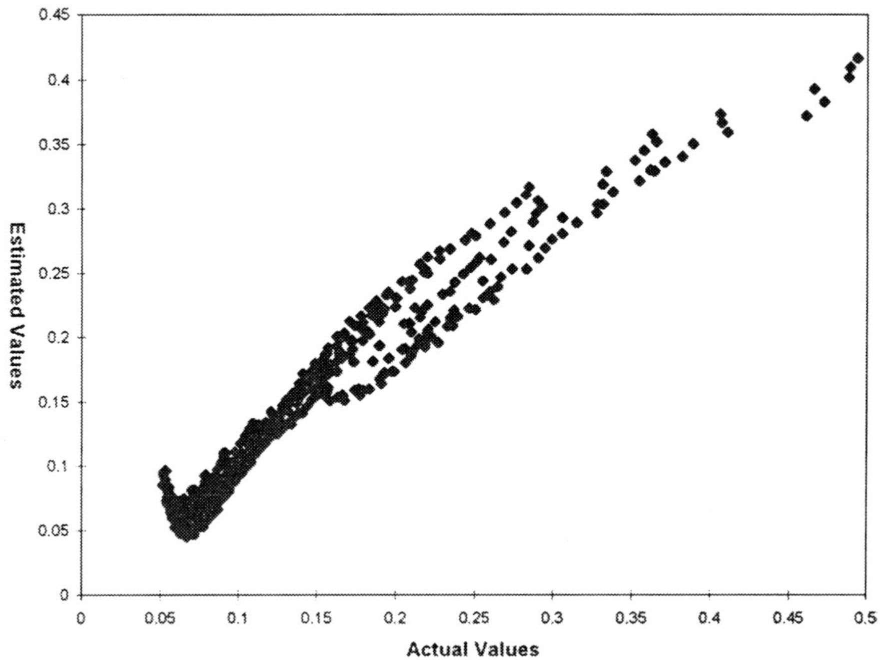


FIGURE 5.19
ACTUAL AND ESTIMATED VALUES FOR S. E. (CASE B)
USING THE QUADRATIC MODEL PRESENTED IN TABLE 5.13

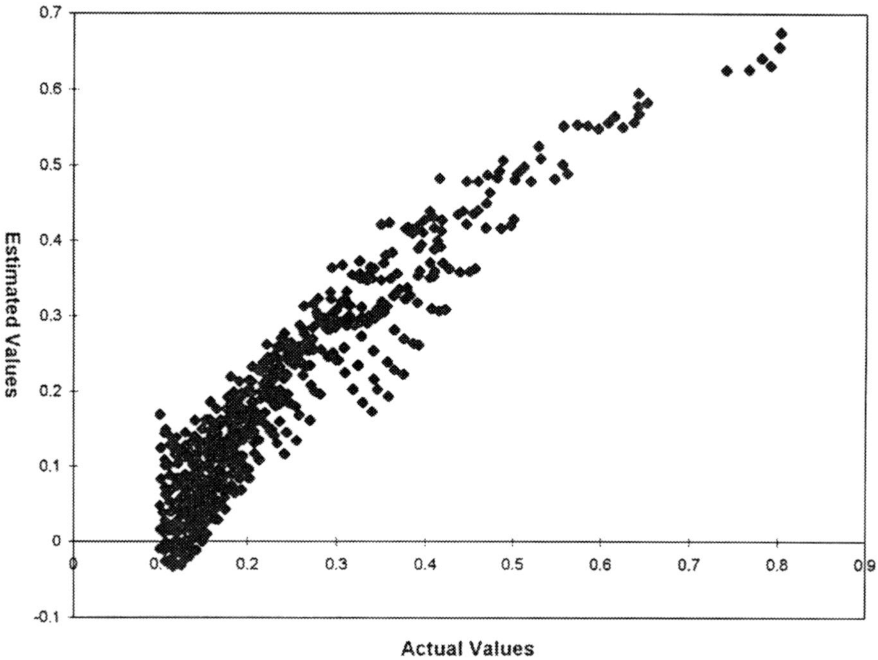


**FIGURE 5.20**
**ACTUAL AND ESTIMATED VALUES FOR S. E. (CASE C)**
**USING THE QUADRATIC MODEL**

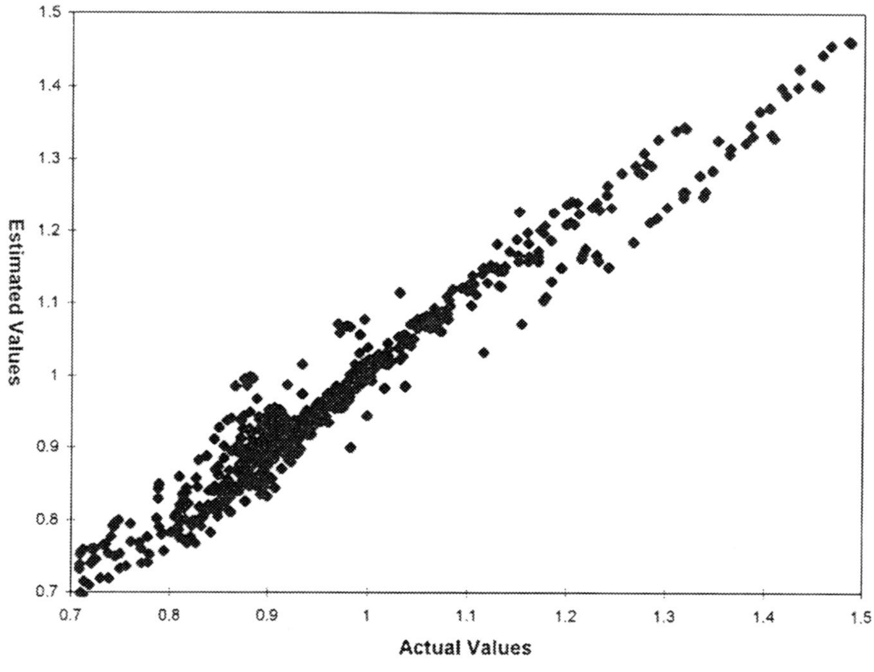


**FIGURE 5.21**
**ACTUAL AND ESTIMATED VALUES FOR $Q_{0.05}$ (CASE A) USING THE QUADRATIC MODEL PRESENTED IN TABLE 5.13**

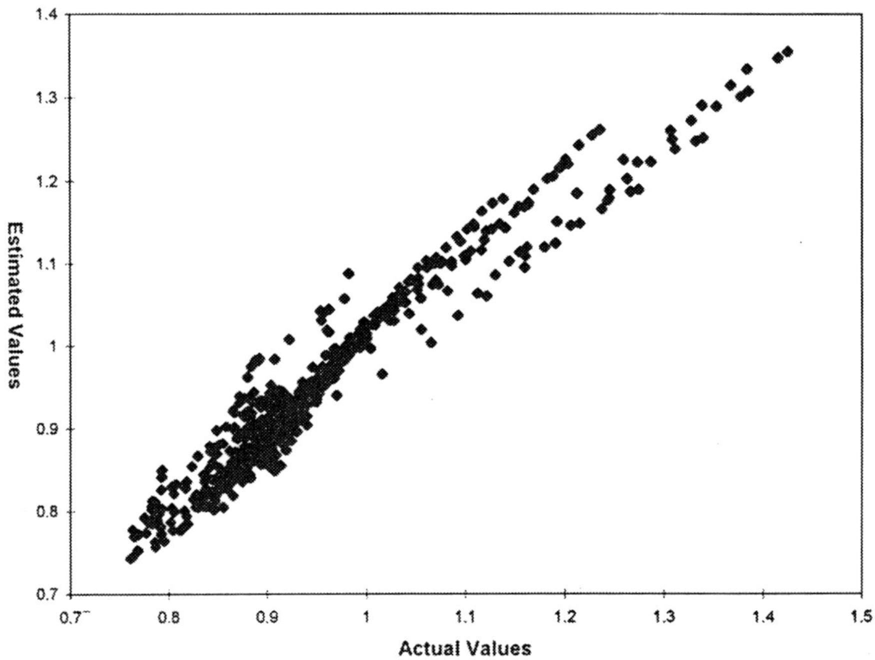


FIGURE 5.22
ACTUAL AND ESTIMATED VALUES FOR $Q_{0.05}$ (CASE B)
USING THE QUADRATIC MODEL PRESENTED IN TABLE 5.13

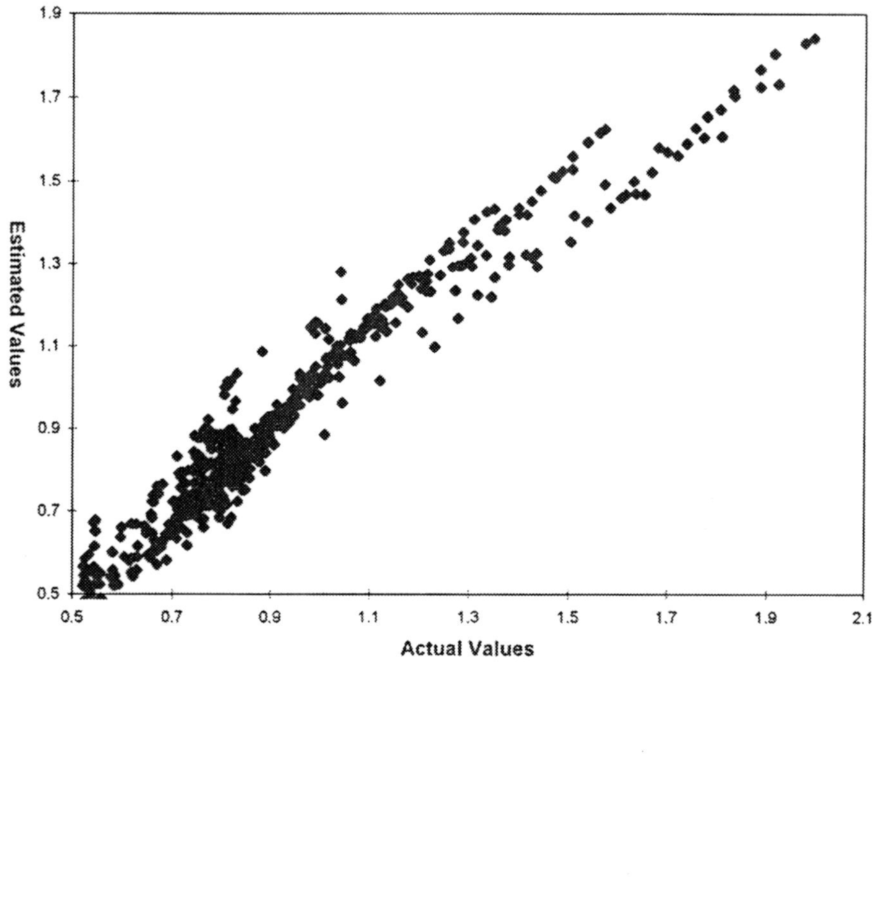


**FIGURE 5.23**
**ACTUAL AND ESTIMATED VALUES FOR $Q_{0.05}$ (CASE C)**
**USING THE QUADRATIC MODEL**

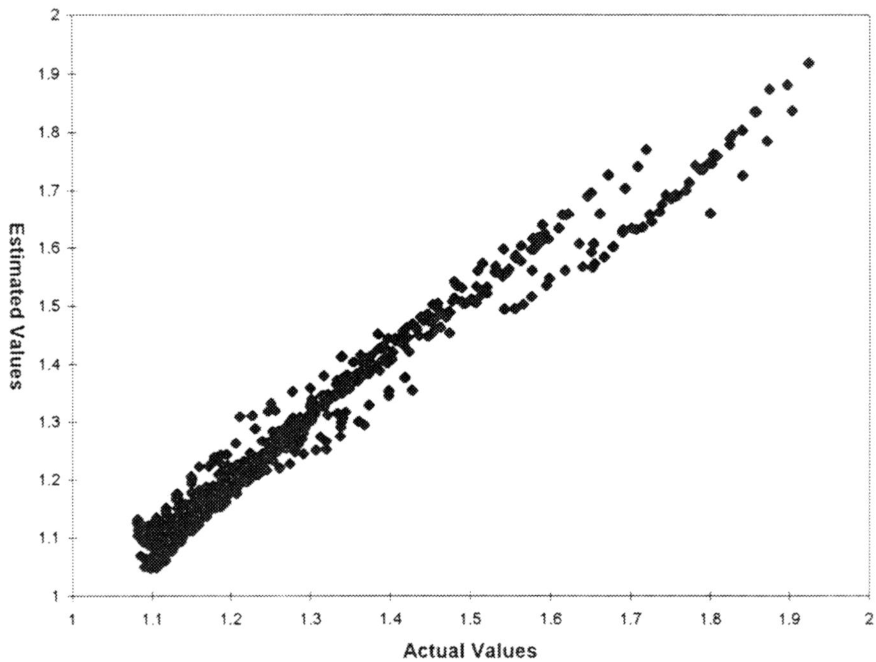


FIGURE 5.24
ACTUAL AND ESTIMATED VALUES FOR $Q_{0.95}$ (CASE A)
USING THE QUADRATIC MODEL PRESENTED IN TABLE 5.13

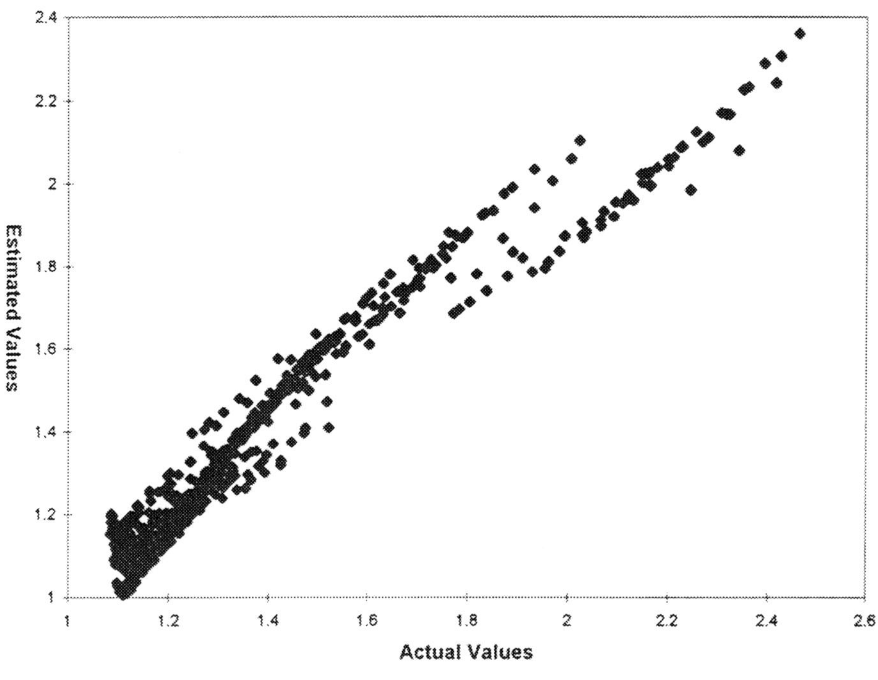


FIGURE 5.25
ACTUAL AND ESTIMATED VALUES FOR $Q_{0.95}$ (CASE B)
USING THE QUADRATIC MODEL PRESENTED IN TABLE 5.13

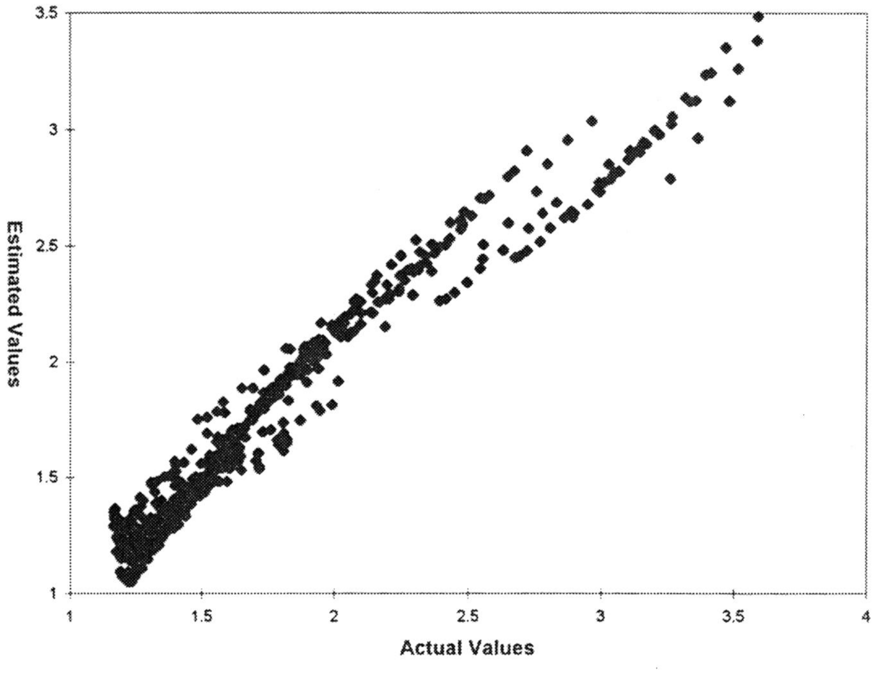


FIGURE 5.26
ACTUAL AND ESTIMATED VALUES FOR $Q_{0.95}$ (CASE C)
USING THE QUADRATIC MODEL

# 6

# SCHEDULING OF PROJECT NETWORKS

## *1. General formulation and typology*

The problem of scheduling is a basic and central topic of Project Management. This problem concerns, basically, the selection of the **starting times** and the **allocation of resources** (types, levels, nodes, etc.) needed to implement each activity of the project.

The typology of the scheduling is determined by three classes of entities: the **activities**, the **restrictions** and the **decisional criteria**.

A - The **activities** can be classified in terms of:

a) **Pre-emption**

If the activities can be interrupted when being implemented, they are called pre-emptive. Otherwise, there is no pre-emption.

b) **Mode**

Activities can have just one or multiple modes of implementation and this diversity can be continuous or discrete.

c) **Structure of Precedences**

Different types of precedences have been discussed already.

d) **Uncertainty**

The features of the activities can have a **deterministic** , **random** or **stochastic** nature.

B - The diversity of **restrictions** is quite wide as they can include:

a) The precedence conditions

The most common case corresponds to

$$t^f(j) \le t^s(i) \text{ for any } j \in J(i) \text{ with } i = 1,\dots,N.$$

b) The resources conditions

The **cumulative restriction** can be expressed by:

$$\overbrace{\sum_{i \in I_1} R_k(i) + \sum_{i \in I_2} R_k(i) \frac{t - t^s(i)}{t^f(i) - t^s(i)}}^{A_k(t)} \le LA_k(t)$$

for resource k = 1, ..., K and with $O \le t \le H$,

where $I_1 = \{i : t^f(i) \le t\}, I_2 = \{i : t^s(i) \le t \cap t^f(i) > t\}$,

H is the studied horizon, $R_k(i)$ the total consumption of resource k to carry out i and $LA_k(t)$ the maximal available amount of k until time t. This restriction assumes that the consumption is proportional to the fraction of the duration of i taking place before t.

The **non-cumulative restriction** can be given by:

$$\overbrace{\sum_{i \in I_3} r_k(i)}^{a_k(t)} \le La_k(t) \text{ for } O \le t \le H \text{ with } I_3 = \{i : t^s(i) \le t \cap t^f(i) \ge t\}$$

where $r_k$ (i) is the consumption of k by per time unit and where $La_k$ (t) is the maximal available amount of resource k at time unit t. It is assumed that the starting or finishing times of i are integer variables measured in the same time units than t (Fig. 6.1).

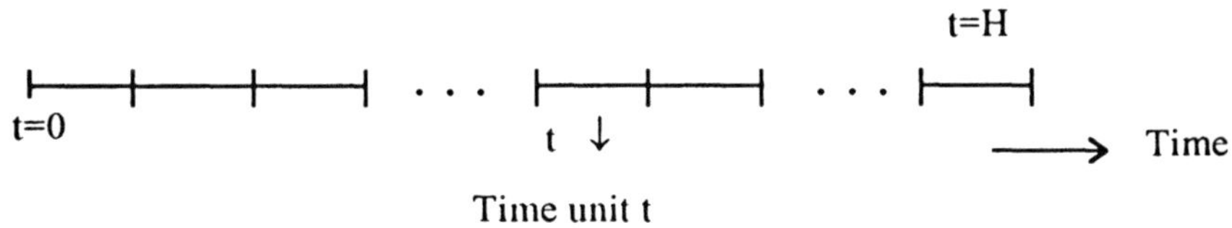


**FIGURE 6.1**
**THE ADOPTED DISCRETE TIME SCALE**

The function $a_k(t)$ can also represent a measure of the use of the capacity of any facility at time unit t.

c) The **cost-benefit (expenditure-income)** conditions

Quite often, the concept of cost (or expenditure) is used to synthetize all the requirements of resources as was pointed out before. Also, the cost-benefit or expenditure-income analysis has a crucial importance in carrying out any economic or financial study.
The formulation of the cost condition can be done in cumulative terms by:

$$\overbrace{\sum_{i \in I_1} C(i) + \sum_{i \in I_2} C(i) \frac{t - t^s(i)}{t^f(i) - t^s(i)}}^{M(t)} \leq LM(t)$$

where C(i) is the cost of i and LM(t) is the maximal cost acceptable until time t. In non-cumulative terms one has:

$$\overbrace{\sum_{i \in I_3} C(i)}^{m(t)} \leq Lm(t)$$

where C(i) is the cost of i per time unit and Lm (t) is the total cost per time unit at t. Again, it is assumed that the starting and finishing times are expressed by integer variables.

A similar formulation can be presented for the benefits $B_p$ generated by the completion of each sub-set of activities, $\beta_p$ with p=1, ..., P and with $\beta_p = \left\{i: t^f(i) \le t\right\}$.

$$\overbrace{\sum_p B_p}^{Q(t)} > LQ(t)$$

where LQ(t) is the minimal total benefit until t.

Also, the benefit generated during time unit t assuming again integer values for the finishing times will be given by:

$$\overbrace{\sum_p B_p}^{q(t)} > Lq(t)$$

where $B_p$ is the benefit generated by $\beta'_p = \left\{i: \underset{i}{Max}\, t^f(i) = t\right\}$

and where Lq(t) is the minimal benefit to be generated during time unit t.

Usually, the restrictions include both the perspectives of cost and benefit (or expenditure and income):

$$\begin{cases} M(t) - Q(t) \le S(t) \\ m(t) - q(t) \le s(t) \end{cases}$$

where S(t) (or s (t)) is the maximal cumulative (or non-cumulative) net cost for time t. It should be noted that [M(t)-Q(t)] or [m(t)-q(t)] represent the cumulative or non-cumulative cash-flow if using expenditures and income and therefore these magnitudes are very useful to study the financial features of the project.

However, if the scale is sufficiently large, a discounted sum of costs and benefits has to be defined in terms of an appropriate discounting factor, $f = \frac{1}{1+\alpha}$ where $\alpha$ is the discount rate:

$$V(t) = \sum_{\tau=0}^{t} f^{\tau}\left[q(\tau) - m(\tau)\right]$$

where V(t) is the Net Present Value of the project until time t.

Several restrictions then can be used

$$\underset{t}{\mathrm{Min}}\, V(t) \geq L_1$$

or

$$V(T) \geq L_2$$

where T is the project duration

d) the relationship cost-duration

The duration of an activity, D(i), can be considered a function of the allocated resources or of their cost:

$$R_k(i) = f_k[D(i)] \text{ with } k = 1,\ldots,K$$

or

$$C(i) = g[D(i)]$$

where $f_k$ and g are appropriate functions (usually, decreasing functions, as was discussed before).

Also, the benefit generated by any sub-set $\beta_p$, also may be considered a function of its accomplishment time.

C - After presenting the major types of constraints, the decisional criteria should be studied:

a) Minimization of the total duration , T:

$$\mathrm{Min}\,T = \mathrm{Min}\left\{\underset{i}{\mathrm{Max}}\left[t^f(i)\right]\right\}$$

b) Minimization of the tardiness of the total duration or of any important milestone. If the target duration is given, $T^o$, then one should have

$$\mathrm{Min}\{\mathrm{Max}[T - T^o;0]\}$$

Special importance may be given to the completion of some sub-sets of activities, $\beta_p$ with p=1, ..., P and then the maximal tardiness of each of these events can be defined by

$$\mathrm{Max}\left\{\left[\underset{\substack{i\\ i\in\beta_p}}{\mathrm{Max}}\, t^f(i)\right]-T^o{}_p\,;0\right\}$$

and therefore a weighted function can be minimized:

$$\mathrm{Min}\sum_{p=1}^{P}\lambda_p\left\{\mathrm{Max}\left\{\left[\underset{\substack{i\\ i\in\beta_p}}{\mathrm{Max}}\, t^f(i)\right]-T^o{}_p\right\};0\right\}$$

where $\lambda_p$ are the appropriate weights with $\sum_{p=1}^{P}\lambda_p = 1$

c) Minimization of the value of the consumed resources until time t

$$\mathrm{Min}\sum_k h_k \cdot A_k(t)$$

where $h_k$ is the unit value of resource k.

d) Minimization of the maximal value per time unit of consumed resources:

$$\mathrm{Min}\left\{\underset{t}{\mathrm{Max}}\left\{\sum_k h_k\, a_k(t)\right\}\right\}$$

e) Minimization of a measure of the fluctuations of the consumption of each resource (resource levelling)

$$W_k\sum_{t=0}^{T-1}\left[a_k(t+1)-a_k(t)\right]^2$$

f) Maximizing the NPV

Max V(T)

g) Minimizing the maximal project cash-flow deficit

in cumulative terms

$$\mathrm{Min}\left\{\underset{0\le t\le T}{\mathrm{Max}}\left[M(t)-Q(t)\right]\right\}$$

and in non cumulative terms:

$$\mathrm{Min}\left\{\underset{0\le t\le T}{\mathrm{Max}}\left[m(t)-q(t)\right]\right\}.$$

Obviously, the objective-function can be an aggregation of some of these criteria and those not included in the objective-function can be formulated as additional restrictions. The multi-criteria study of different criteria is a very important line of development which will be studied in Chapter 7.

The presented general formulation can be used to propose a typology of the different scheduling problems (Fig. 6.2). This classification should consider the properties of three major entities: **activities, resources, criteria**. In Fig. 6.2, the major types of assumptions about each one are given and it should be noted that the data about activities durations or resources can have a deterministic or random nature. Similarly, the criteria can be measured by deterministic functions or by stochastic magnitudes in terms of expectations or extreme quantiles (risk measures).

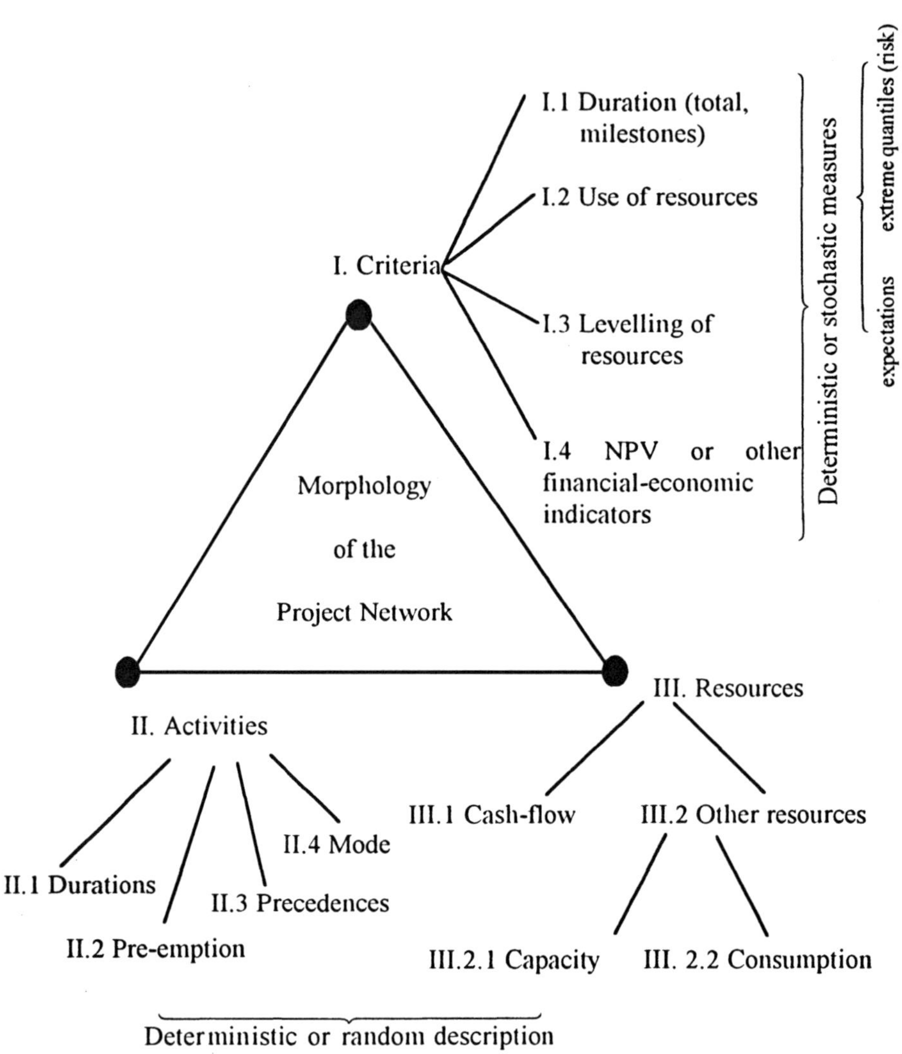


**FIG. 6.2**
**TYPOLOGY OF THE SCHEDULING PROBLEMS**

## 2. The binary formulation

The formulation of the presented problems of scheduling tends to require a two-dimension model based on

$\{X_{it}\}$ with $i=1, ..., N; t = 1, ..., T$

where $X_{it}$ represents the decision about the activity i at time unit t (Fig. 6.3). This is due to the need of building the precedence relations which are expressed in terms of time t and the resources constraints which require summing up the consumption of resources used by sets of activities which are active at time t (non-cumulative constraints) or until time t (cumulative constraints).

A large number of models (Pritsker, Watters and Wolfe 1969, Davies, 1972, Talbot, 1982 and Patterson, 1984) has been based on the definition of:

$X_{it} = 1$ or 0 if i is active or if it is not active at time t, respectively.

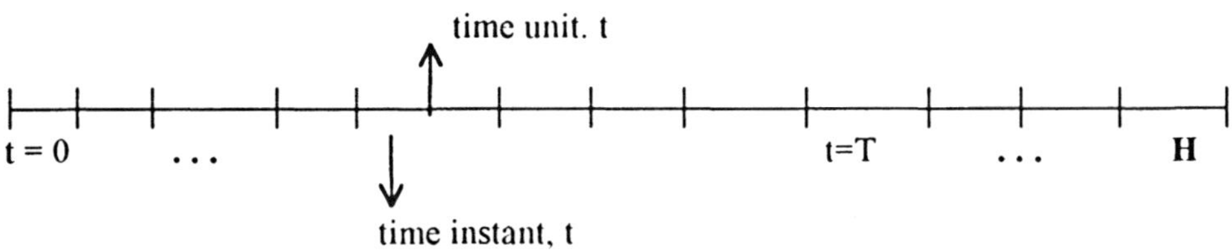


**Example of time scale with a project duration equal to (T+1) studied within an horizon of (H+1) units**

FIGURE 6.3

Then, one has

$$\sum_{t=0}^{H} X_{it} = D(i) \longrightarrow \text{completion of each activity}$$

$$X_{it} \leq \sum_{j \in J(i)} \sum_{\tau=0}^{t-1} X_{j\tau} / D(j) \longrightarrow \text{precedence constraints for any i with } J(i) \neq \emptyset$$

with $t = 1, ..., [H - D(i) + 1]$

This restriction implies that the right side is just one if, before t, j was active during D(j) units of time for all $j \in J(i)$ which means that all elements of J(i) were completed.

$$\sum_{i=1}^{N} X_{it}\, a_{ik} \leq La_k(t) \longrightarrow \text{resource constraint}$$

and/or $\sum_{\tau=0}^{t} \sum_{i=1}^{N} X_{i\tau}\, a_{ik} \leq LA_k(t)$ with k=1, ..., K and t=0, ..., H.

D(j) is the duration of j and $a_{ik}$ is the amount of resource k = 1, ..., K consumed by i during each time until.

In this formulation, the pre-emptive assumption is implicitly adopted as each activity can be interrupted any time. However, if this is not the case, one should add:

$$\sum_{\tau=t}^{t+D(i)-1} X_{i\tau} \geq X_{it} \cdot D(i) \text{ for } t = 0,\ldots,[H - D(i) + 1]$$

As was said before, several objectives can be considered such as levelling the resource usage, minimizing the cost of the resource usage, maximizing the net present value of the project but the most common objective is minimizing the project duration.

In this last case, one should find the minimal L for which:

with $\sum_{i=1}^{N} \sum_{t=0}^{T} X_{it} \geq \sum_{i=1}^{N} D(i)$ being $T+1 \leq L$.

However, this is not a convenient formulation for a binary model as T is implicitly defined by the number of terms in the summation. If no pre-emption is assumed, an alternative formulation can be used:

Min T

where the total duration is (T+1) subject to:

$$\frac{1}{D(i)} \sum_{t=0}^{H} X_{it} \cdot t + \frac{D(i)-1}{2} \leq T$$

where H is the maximal studied horizon and for any i with K(i)=Ø.

This condition stems from

$$\sum_{t=0}^{H} X_{it} \cdot t = D(i) \cdot T - \frac{D(i)[D(i)-1]}{2}$$

as is shown in Fig. 6.4.

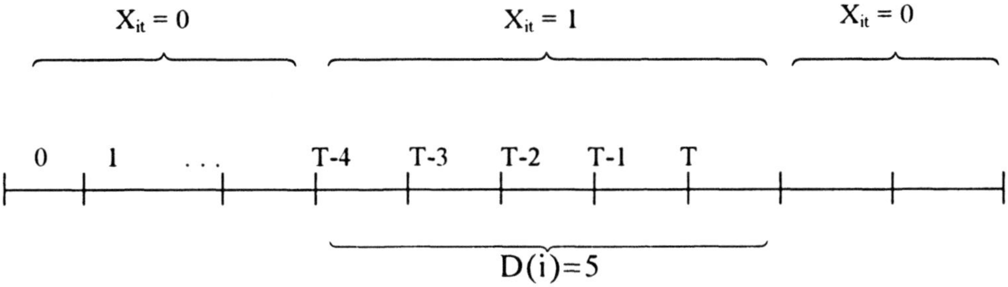


$$\sum_{t=0}^{H} X_{it} \cdot t = T + (T-1) + \ldots + [T - [D(i)-1]] = T \cdot D(i) - \left[ \underbrace{0 + 1 + \ldots + D(i) - 1}_{\frac{D(i) \cdot D(i)-1}{2}} \right]$$

**FIG. 6.4**
**A FORMULATION OF THE DURATION OF ANY ACTIVITY I**

An alternative formulation can define $X_{it}$ = 1 or 0 if i starts at time unit t or not. In this case one has:

$$\sum_{t=0}^{H} X_{it} = 1 \text{(starting of each activity)}$$

$$\sum_{\tau=0}^{T-D(j)-D(i)+1} \tau X_{j\tau} + D(j) \leq \sum_{\tau=0}^{T-D(i)+1} \tau X_{i\tau} \text{ (precedence condition } j \rightarrow i)$$

and

$$\sum_{i=1}^{N} \left[ \sum_{\tau=\max[0,t-D(i)+1]}^{t} X_{i\tau} \right] \cdot a_{ik} \leq La_k(t) \text{ with k and } 0 \leq t \leq H$$

or the equivalent cumulative restriction

$$\sum_{\tau=0}^{t-D(i)+1} X_{i\tau} \cdot [a_{ik} \cdot D(i)] + \sum_{\tau=t-D(i)+2}^{t} X_{i\tau} \cdot a_{ik} [b - \tau + 1] \leq LA_k(t)$$

with $k = 1, \ldots, K$ and $0 \leq t \leq H$

The objective of minimizing the total duration can be easily expressed by MIN T with $\sum X_{it} \cdot [t + D(i) - 1] \leq (T + 1)$ for any i as the total duration is given by (T+1).

Obviously, an equivalent formulation can be developed using $X_{it}$ = 1 or 0 if i is completed (or not) at time unit t.

The presented models can be simplified by restricting the conditions and the variables to the feasible domain within the earliest and latest starting or finishing times of each activity. For instance, if $t_s(i)'$ and $t_s(i)''$ are the earliest and latest starting times of the activity i, $X_{it}$ will be defined as a binary variable describing the starting of i just for the interval $[t_s(i)', t_s(i)'']$. Several authors have also adopted decomposition techniques to simplify this problem (See De et al., 1994 and De et al., 1977).

Finally, it should be noted that instead of using k=1, ..., K resources, the adoption of the concept of cost can be used reducing the number of coefficients ($C_i$ instead of $a_{ik}$ with k=1, ..., K) and the number of restrictions (M (t) or m (t) instead $a_k$ (t) or $A_k$ (t) with k=1, ..., K).

A generalization of the presented binary models can be proposed considering that D(i) is a function of C(i). This function can be discrete and the problem is then called the **multi-mode scheduling** as each activity can be performed by P alternative modes $\left[C(i)_p;D(i)_p\right]$ with i=1, ..., N and p=1, ...P. The presented models can be adopted again but the number of variables is increased by (P-1). N.

## 3. The heuristic optimization of the binary model

Unfortunately, the studied problems (the resource constrained scheduling with or without pre-emption, single or multi-mode) are NP-hard (Elmaghraby, 1995).

The improvement of binary programming codes during the last years is significant but the size of the model is quite large even for networks of medium size.

Therefore, many algorithms have been proposed to schedule the activities and, every year, several papers are published claiming better results than the algorithms of the previous year (see, for instance, Wiest, 1967, Johnson, 1967, Pritsker, Watters and Wolfe, 1969, Davies and Heindorn, 1971; Patterson, J. H., 1973, Davies, 1973, Davies, 1974, Patterson and Roth, 1976, Doersch and Patterson, 1977; Talbot, 1982; Elmaghraby, 1977; Patterson et al, 1990, Elmaghraby, 1995, Herroelen, Van Dommelen and Demeulemeester, 1997) which also confirms the limitation of the binary models.

These algorithms are based on two different methods:

a) the **exact methods** exploring the full space of alternative scheduling solutions. Usually, a strategy of the type "**branch and bound**" is adopted to avoid full enumeration. Each node corresponds to a partial schedule (a schedule of a sub-set of activities) and if an activity cannot start, the node is branched into several scenarios of delays.

   Recent proposals have been presented by Demeulemeester and Herroelen, 1997; Brucker et al, 1998 and Sprecher and Drexl, 1998.

b) the **heuristic methods**

   based on inductive procedures to achieve optimal or "near optimal" solutions without full analysis of the solutions space. Heuristics (from the greek word "heuristike" meaning the "art of discovery or of invention") do not guarantee the obtention of the best schedule but are faster than the exact methods.

   The adopted heuristics can be based on two major approaches:

- **Serial approach**

In this case, a rule is adopted to rank sequentially the whole set of activities before starting the schedule. Then, the activities are started according to this rule and if a non-feasible situation occurs due to shortage of resources, the "difficult" activity is passed over for other activities respecting the constraints. A second rule is usually required to avoid the same ranking by several activities.

- **Parallel approach**

The parallel approach is based on a process of selection of activities to be started at each time t from the set of candidates being t a time clock ranging from 0 to the end of the project. Such selection is also based on arbitrary rules.

The value of t is then advanced to the next instant when a decision has to be made. This means that the priority index assigned to each activity is changing as along as the scheduling is progressing.

Boctor 1990, has compared the alternative scheduling rules using a set of test problems and obtained the results presented in Table 6.1.

**TABLE 6.1**
**BEST ALTERNATIVE HEURISTIC RULES**

**a) PARALLEL APPROACH**

| | |
|---|---|
| MINSLK | Ascending order of total float |
| RSM | At each time t, $X_{ij}$ = max (0, t + duration of i - latest start time of j) and schedule first the activity i with lowest $X_{ij}$ |
| MINLFT | Ascending order of latest finish time |
| RAN | Random choice |
| GRD | Ascending order of the amount of required resources by each activity |
| SA | Ascending order of duration |
| SRD | Inverse of GRD |
| LA | Descending order of duration |

| **b) SERIAL APPROACH** | **FIRST RULE (ASCENDING ORDER)** | **SECOND RULE (ASCENDING ORDER)** |
|---|---|---|
| J-I | Number of the final node | Number of the initial node |
| J-D | Number of the final node | Duration |
| J-R | Number of the final node | Total resource requirements* |
| J-S | Number of the final node | Total float |
| I-J | Number of the initial node | Number of the final node |

* descending order instead of ascending order

The obtained conclusions are:

- The most efficient rules are MINSLK, MINLFT and RSM

- If the results obtained by one of these rules are not satisfactory, another heuristic making an efficient combination with the first one can be used to re-solve the problem constructing another schedule. As RSM requires more computational time than the others, the following sequence is suggested:

  MINSLK, MINLFT, J-R, GRD, SA, J-I, RSM

- However, even with the best rule, the optimal solution was not found in more than 30% of the tested problems. The successive use of 4 or 5 heuristics can reduce this percentage to less than 5% but it increases the computational time as the same problem has to be re-computed 4 or 5 times.

The performance of heuristics is improving due to the evelopment of new search methods allowing more flexible rules and procedures.

These methods (See, e. g., Blazewicz et al, 1994 and Crawels, Poth, Van Wassenhove, 1996) are based on:

- arbitrary choice of an **initial solution**, $\overline{X}(s=0)$
- definition of an **objective function** in terms of $\overline{X}, F(\overline{X})$
- design of an **iteration "engine"** going from a solution $\overline{X}(s)$ to $\overline{X}(s+1)$ with s = 0, 1,...
- definition of a **stopping rule** to finish the search.

The definition of any solution for the scheduling problem includes the starting times of all activities or of a sub-set of activities and, if required, the allocation of resources to such activities. The objective function is one of the criteria already discussed, such as the total duration or the Net Present Value.

The traditional type of "engines" is based on comparing the value $F(\overline{X}(s))$ with F at alternative points, $\overline{X}'$, belonging to the neighborhood of $\overline{X}(s)$, $\Omega(\overline{X}(s))$. The point $\overline{X}'$ with the greatest improvement of F is then selected:

$$\overline{X}(s+1) = \overline{X}'$$

and if no significant improvement is made the search is stopped.

Several ways of implementing this procedure have been proposed, namely:

- locating $\overline{X}'$ on the direction of steepest descent (gradient method)
- selecting $\overline{X}'$ over the corners of a poliedric definition of $\Omega$.

The main shortcoming of this approach concerns its myopic nature: any local optimum can be responsible for stopping the search.

Three alternative approaches have been more recently proposed to avoid this difficulty: **simulated annealing**, **tabu search** and **genetic algorithms**.

A complete review of the simulated annealing is given by Eglese (1990), and this approach is based on the principle of:

a) accepting $\overline{X}'$ always if $F(\overline{X}')$ is better than $\overline{X}(s)$
b) accepting $\overline{X}'$ if $F(\overline{X}')$ is worse than $\overline{X}(s)$ with a probability of $\alpha(\sigma)\cdot\exp[-\sigma/t]$ where $\sigma$ is the increase (decrease) of F if F should be minimized (maximized) and t is a parameter called "temperature".

The tabu approach (see Glover, 1989) carries out a systematic search of $\Omega(\overline{X}(s))$ and $\overline{X}'$ also can be accepted even if $\overline{X}'$ is responsible for deteriorating F. However, in each iteration, a list of alternative moves is forbidden (Tabu list) in terms of specific features of $\overline{X}'$. For instance, this approach can be applied to the scheduling problem by making the "tabu list" in terms of "forbidden" starting times of several activities.

The genetic algorithms (Goldberg, 1989) include the definition of the "chromosomal" component of each solution. For instance, in scheduling problems, this component is usually defined in terms of early or late schedules for sub-sets of activities. At each iteration, a new population of solutions is generated in terms of the chromosomal components of the parental population of solutions by use of three different genetic operators reproduction (keeping the features), crossover (exchange between different chromosomal components) and mutation (changing the chromosomal features).

The application of genetic or tabu algorithms to scheduling problems is not preventing the development of other heuristics specifically oriented to cope with the scheduling problem (See, Ahn and Erengue, 1998 and Bianco, Dell'Olmo and Speranza, 1998).

Finally, it should be noted that:

a) the exact algorithms or the heuristic algorithms just produce a numerical solution for a numerical example and little understanding about the relationship between data and solution can be provided (No sensitivity results are available);

b) a general criticism can be made about any rule adopted by an heuristic algorithm: why should the same rule be used along the whole scheduling, from its beginning until its end?

c) the performance of heuristics may be rather unstable depending heavily on the numerical configuration of the problem;

d) the validity of the comparative study of performance of alternative algorithms depends on the morphology of the network but, unfortunately, no results have been published.

## 4. The proposed non-linear continuous model

The previous model is based on a binary formulation which means that at each time, t, the manager can just decide to make or not to make each activity. For networks with a reasonable size of more than 50 activities and assuming an average number of 30 different starting times for each of these activities, one needs 1500 binary variables falling into the class of large binary optimization problems.

This difficulty explains the frequent presentation of heuristics or variations of the Branch and Bound but, unfortunately, these techniques have difficulties discussed before.

Furthermore, this model has two other shortcomings:

- the introduction of objectives or restrictions in terms of NPV or other discounted measures is responsible for severe difficulties.

- the binary formulation is not easily compatible with the introduction of quite important degrees of freedom such as:

  - the variation of the duration of each activity in terms of the amount of resources allocated to its implementation;

  - the variation of the intensity of implementation of each activity along different time units.

Actually, any change concerning the pattern of implementation of each activity can be just considered by the binary formulation through the multi-mode formulation.

However, the first challenge requires to multiply by p the size of the problem if assuming p alternative durations and the second one is still more dramatic: a binary variable has to be defined for each level of intensity for each time unit and for each activity.

Therefore, a network with 50 activities and with an average number of 30 time units and 10 alternative levels of intensity, requires 15 000 binary variables!

This explains why an alternative **continuous model** should be studied.

Furthermore, the new model should be based on the **resources** (rather than on the **time**) which is allocated to each activity. This option stems from the previous comments about the priority concerning the decisions on resources and it has three major advantages:

- the relationship between **duration** and **resources** is easily formulated
- the **variability on time** of the **intensity of the implementation** of each activity is also easily formulated
- the introduction of an objective function or of restrictions expressed in terms of the **NPV** or other discount formula do not increase the difficulty of the problem.

The construction of a continuous model can benefit from the very significant advances achieved in the area of **Non-Linear Programming**.

Actually, software like **GAMS-MINOS** easily can solve large scale problems in very small computing times. However, it should be noted that its solution may correspond just to a local optimum instead of a global optimum.

The proposed model follows the lines proposed by Ferreira (1989), Tavares (1987) and Tavares (1989) to develop an improved continuous formulation. This model is based on $0 \leq X_{it} \leq 1$ where $X_{it}$ represents the **amount of implementation of the activity i at time unit t**. The value 1 corresponds to the maximal intensity of development of i at time t and 0 to no development at all. In this case, the completion of an activity requires that:

$$\sum_{t=0}^{H} X_{it}\, m_{it} = M_i$$

where $m_{it}$ is the maximal intensity at time t and $M_i$ is the required total quantity allocated to have i completed. $M_i$ can be also defined as the total cost of the activity i. Quite often $m_{it}$ is assumed constant on time, $m_i$, and so this assumption will be adopted. The minimal duration of i is given by $M_i/m_i$.

This model is much more flexible than the previous ones because:

- it allows maximal intensities of implementation per time unit and per activity;
- it implies a natural relationship between the allocated resources (or cost) and the duration of each activity.
- it allows an easy and direct formulation of financial objective functions or restrictions such as NPV, Present Cost, etc. as it is based on the explicit formulation of the cost (or of the resources) allocated to each time unit.

The restrictions concerning the use each resource, k, will be given by:

$$\sum_{i=1}^{N} X_{it} \cdot m_i \cdot W_{ik} \leq La_k(t)$$

and

$$\sum_{i=1}^{N} \sum_{\tau=0}^{t} X_{i\tau} \cdot m_i \cdot W_{ik} \leq LA_k(t)$$

with k = 1, ..., K and t = 0, ..., H, $W_{ik}$ being the coefficient measuring the use of the resource k due to one unit of intensity of activity i, and $La_k$ or $LA_k$ being the maximal bound for the use of k per time unit or accumulated until the time t.

Obviously, if just a monetary resource is used one has the previous constraints now expressed by:

$$\sum_{i=1}^{N} X_{it} \cdot m_i \leq Lm(t)$$

$$\sum_{i=1}^{N} X_{it} \cdot m_i \leq LM(t)$$

with t=0, ..., M and where Lm (t) is the maximal resource available at time unit t and LM(t) is the maximal cumulative resource available until t.

The major difficulty concerns the formulation of the precedence constraints but a model (Fig. 6.5) can be built in terms of $\Delta_{it} = M_i - AX_{it}$ where $AX_{it}$ is given by $AX_{it} = \sum_{\tau=0}^{t-1} m_i \, x_{i\tau}$ with t=1, ..., H and with $AX_{io} = 0$. Actually, $AX_{i\tau}$ represents the cumulative amount of implementation of i until the beginning of the time unit t and $\Delta_{it}$ measures the amount done from that instant of time until finishing i.

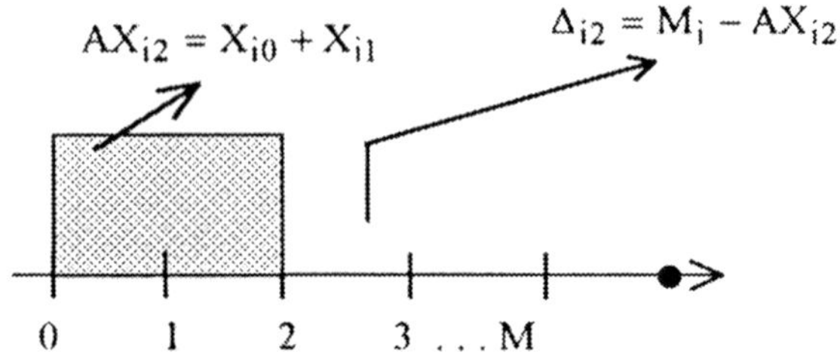


**FIG. 6.5**
**THE ADOPTED NOTATION**

Therefore, for any activity i with $J(i) \neq \emptyset$ one should satisfy the constraint:

$$\sum_{j \in J(i)} \sum_{\tau \in [S(i), F(i)]} \Delta_{j\tau} \cdot X_{i\tau} = 0$$

where S(i) is the earliest starting time of i and F(i) is the latest finishing time of i.

All these restrictions can be added up as all variables $\left(\Delta_{jt}, X_{it}\right)$ are non-negative:

$$\sum_{\substack{i=1 \\ \text{if } J(i) \neq \emptyset}}^{N} \sum_{j \in J(i)} \sum_{\tau \in [S(i), F(i)]} \Delta_{j\tau} \cdot X_{i\tau} = 0$$

This formulation can be easily solved by a **Linear Complementarity Method** if a linear objective function is adopted.

The previous restriction is a **complementarity restriction** and it can be relaxed by the inclusion of a penalty term in the objective function:

$$\text{Min}\left\{\pi \cdot \left[\sum_{\substack{i=1 \\ \text{if } J(i) \neq \emptyset}}^{N} \sum_{j \in J(i)} \sum_{\tau \in [S(i) F(i)]} \Delta_{j\tau} \cdot X_{i\tau}\right]\right\}$$

where $\pi$ is the **penalty coefficient**.

The objective function of this model can be easily constructed if it is a function of the allocated resources or of the cost related to each time unit. For instance, for the Present Cost problem one has:

$$\text{Min} \sum_{i=1}^{N} \sum_{t=0}^{H} X_{it}\, m_i\, f^t$$

where $m_i$ is defined as the maximal cost allocated to i per time unit. The factor f is the appropriate discounting factor.

The objective of levelling the resources can be also easily expressed by:

$$\text{Min} \sum_{t=1}^{H} \left\{ \left[ \sum_{i=1}^{N} X_{it}\, m_i \right] - \left[ \sum_{i=1}^{N} X_{it-1}\, m_i \right] \right\}^2 .$$

The objective of minimizing the total duration can be formulated by:

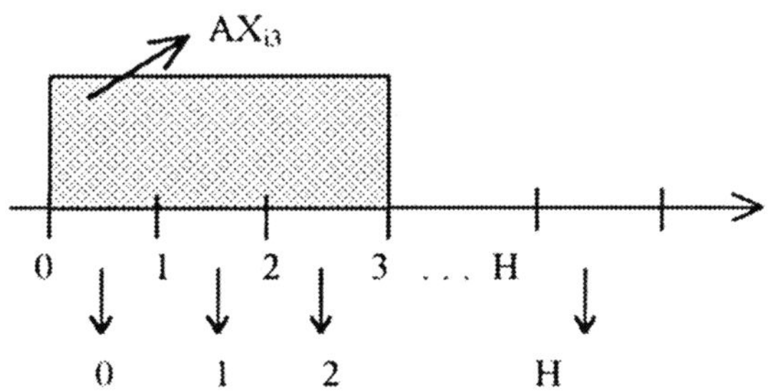


$AX_{i3} = M_i \Rightarrow \Delta_{i3} = 0 \Rightarrow Y_3, \ldots, Y_H = 1 \Rightarrow T = 3$

**FIG. 6.6**
**EXAMPLE OF THE FORMULATION ADOPTED TO STUDY THE TOTAL DURATION**

$$\sum_{\tau=1}^{H} \left( \sum_{i=1}^{N} \Delta_{i\tau} \right) \cdot Y_\tau = 0 \text{ with } 0 \le Y_\tau \le 1 \text{ and maximizing } \sum_{t=1}^{H} Y_t .$$

This means that $Y_\tau$ will be zero as long as the full set of activities was not completed until $\tau$ and that $Y_\tau$, $Y_{\tau+1}$, ..., $Y_H$ will be equal to one if such set was completed until the starting instant of the time unit, $\tau$ (Fig.6.6 ). Therefore, in this case, the total duration is equal to $\tau$ and $S = \sum_{t=1}^{H} Y_t = Y_\tau + \ldots + Y_H$. The maximization of this sum means to find $\tau$ as small as possible.

Thus,

$S = H - \tau + 1$ and $\tau = H - S + 1$

Alternatively, the objective of minimizing the total duration can be achieved by the inclusion of a penalty term in the objective function to be minimized:

$$\sum_{t=0}^{H} \sum_{i=1}^{N} a^t \cdot [X_{it}\ m_i]$$

where a is an appropriate penalty coefficient with a>1.

It should be noted that $a^H << \pi$ and therefore an automatic procedure to select a and $\pi$ can be based on:

a) arbitrary choice of the coefficient a near 1; for instance a=1.1
b) selection of $\pi$ much higher than $a^H$; for instance $\pi = 100 \times a^H$.

Another function can be used with the same purpose:

$\sum_{t=1}^{H} \sum_{i=1}^{N} \Delta_{i\tau}$ because any late implementation of an activity increases this summation.

The presented model can be easily implemented using a standard software for Non-linear Optimization like GAMS-MINOS (Brooke, Kendrick and Meeruns, 1996) et al, 1996). This package has the advantage of allowing the direct construction of the objective function and of restrictions using the standard algebraic notation.

Other formulations can enrich the presented model offered by this package:

a) starting time of each activity, i

$$t^s(i) = \mathrm{Max}\left\{0; \overbrace{1 + \mathrm{Max}_t \left\{\underbrace{t - \pi \cdot AX_{it}}_{①}\right\}}^{②}\right\}$$

where $\pi$ is a high penalty coefficient. Whenever $AX_{it} > 0$, ① will be very negative and so the selected t will be the maximal t, $t^\bullet$, with $aX_{it} = 0$. This means that $(1 + t^\bullet)$ will be equal to the starting time of i, in general, as it is shown in Fig. 6.7.

However, if the activity starts at t = 0, $t^\bullet$ will be equal to zero but ② will be very negative which means that if ② is negative one should adopt the lower bound $t^s(i) = 0$ (Fig. 6.7).

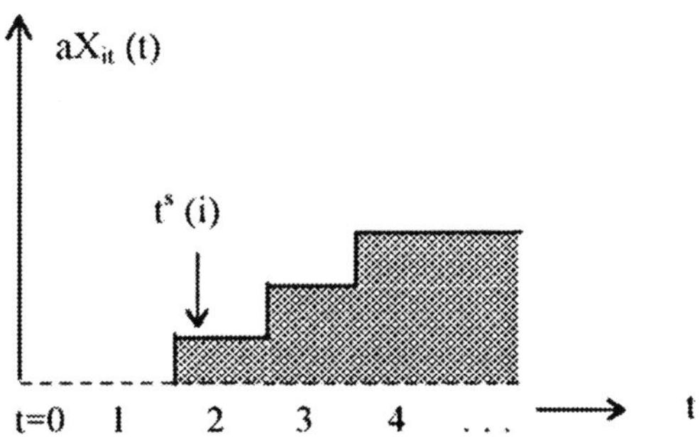


**FIG. 6.7**
**THE STARTING TIME, $T^S$ (I)**

b) finishing time of each activity, i.

$$t^f(i) = \mathrm{Min}\left\{H; \overbrace{-1 + \mathrm{Min}_t \left\{\underbrace{t + \pi \cdot \Delta_{it}}_{①}\right\}}^{②}\right\}$$

where $\pi$ is a high penalty coefficient. Whenever $\Delta_{it} > 0$, ① will be very positive and so ② will be equal to $t^f(i)$, in general, as is shown in Fig. 6.8.

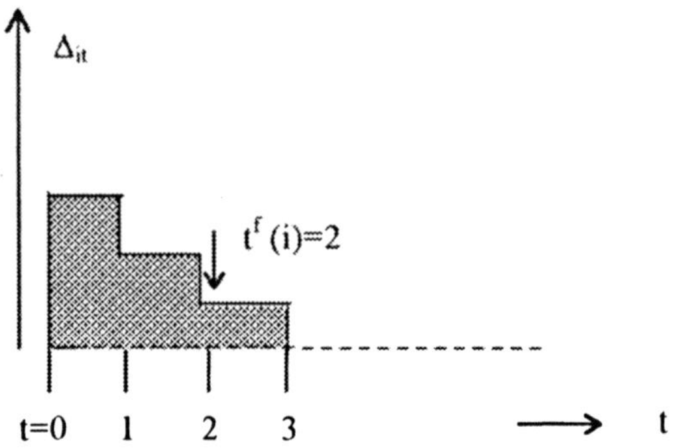


**FIG. 6.8**
**THE FINISHING TIME, $T^F$ (I)**

However, if $t^f$ (i) is equal to H, ② will be a high positive number which implies that $t^f$ (i) should have H as its upper bound.

c) duration of each activity, i

The duration of i, D(i), is given by:

$D(i) = t^f(i) - t^s(i) + 1$ if pre-emption is excluded

d) total duration of the project; T

Using a similar approach, one has:

$$T = \mathrm{Min}\left\{ H; \underset{t}{\mathrm{Min}}\left\{ t + \pi \cdot \sum_{i} \Delta_{it} \right\}\right\}$$

e) Occurrence time for a specific milestone, $t^*$

The occurrence time, $t^*$, of a milestone defined by the completion of the set I of activities can be defined by

$$t^* = \mathrm{Min}\left\{ H; \underset{t}{\mathrm{Min}}\left\{ t + \pi \cdot \sum_{i \in I} \Delta_{it} \right\}\right\}$$

f) Discounted value of the activities cost

As was mentioned before, the discounted value of costs or of benefits easily can be formulated.

The general formulation of the NPV can be expressed by:

$$NPV = \sum_p B_p f^{t_p^*} - \sum_t \left[ \sum_i X_{it} m_i \right] f^t$$

where $t_p^*$ is the generic occurrence time of the milestone p=1, ..., P associated to the benefit $B_p$ and where f is the discount factor.

Quite often, the cost associated to any activity i should be allocated to its starting time as its implementation requires providing all the corresponding resources at that time. Then, one has $NPV = \sum_p B_p f^{t_p^*} - \sum_t \left[ \sum_i X_{it} m_i \right] f^{t^s(i)}$ where $t^s$ (i) is the starting time already presented.

Another advantage of this model is allowing a direct formulation of precedence links of different types which is not the case of the previous models:

a) Precedence **Finish of j ⇒ Start of i with a time lag, d**. In this case, the precedence constraint should be now:

$$\sum_{j \in J(i)} \sum_{t=[S(i),F(i)]} \Delta_{j,t-d_{ij}} X_{it} = 0$$

where $\Delta_{j,t-d_{ij}} = M_j - \underset{\gamma}{AX_{j\,t-d_{ij}}}$ being $d_{ij}$ the delay between j and i.

b) Precedence [Finish of j ⇒ Start of i], with resource overlapping

This link means that i can start providing that $\gamma \cdot M_j$ has been spent, with $0<\gamma<1$.

The usual precedence conditions can be now adopted substituting

$$\Delta_{jt} \text{ by } \Delta'_{jt} = \gamma\, M_j - AX_{jt}$$

c) Precedence [**Start of j ⇒ Start of i**]

This condition can be expressed by $AX_{it} \le AX_{jt} \cdot \pi$ where $\pi$ is a very large coefficient. Actually, if j has not started, $AX_{jt}$ is nil and $AX_{it}$ has also to be zero. However, if $AX_{jt}$ is positive, even with a small magnitude, one can adopt then a positive $AX_{it}$.

d) Precedence **[Finish of j ⇒ Finish of i]**

In this case, one has $\pi \cdot \Delta_{it} \ge \Delta_{jt}$ $\pi$ being a very large coefficient. This means that $\Delta_{it}$ can just come to zero if $\Delta_{jt} = 0$.

e) Precedence **[Start j ⇒ Finish i]**

This condition implies selecting a small number $\delta$ to represent the minimal value given to $X_{jt}$ so that this magnitude will be significantly different of zero:

$$\Delta_{it} \ge \delta - AX_{jt} \cdot \pi$$

Therefore, $\Delta_{it}$ cannot be zero if $AX_{jt}$ is zero but it can be nil if j has started, even with a small magnitude.

## 5. *An application of the proposed model using RISKNET and GAMS-MINOS*

### 5.1 Model

The proposed model has required the development of a new software, RISKNET Module 7, (See Annex 1) which receives as input the generated or given network as well as:

- Scheduling horizon, H
- $M_i$ and $m_i$ for i=1, ..., N
- $Lm(t)$ and $LM(t)$ for $t = o,...,H$.

This software determines the minimal duration for each activity, i, $D_i = \frac{M_i}{m_i}$, and completes its earliest starting time, $S_i$, and its latest finishing time, $F_i$, adopting such duration. Then, this software generates automatically the formulation to be received by GAMS-MINOS.

Two case studies were developed to illustrate this approach: Case A with 75 activities and Case B with 150 activities.

Each network was generated and presented by RISKNET using AoN and the following notation.

a)

| X | Y |
|---|---|
| Z | W |

X → n° of the activity
Y → duration
Z → earliest starting time
W → latest finishing time

b) Square → activity without successor

c) Double line → critical activity

The Adjacency Matrix and the Level Adjacency Matrix have been also produced.

The earliest and the latest schedule of each activity is also computed by RISKNET assuming a minimal duration for each activity and the critical activities are indicated by dark lines.

The application of the optimization model is done using the minimal duration as objective function.

The maximal use of renewable resource per activity and time unit is equal to 10 units and therefore the amount of resource required by each activity is equal to Y multiplied by 10.

## 5.2 Case A

This application is based on one network generated by RISKNET with 75 activities presented in Figures 6.9 and 6.10. The level adjacency matrix is presented in Figure 6.11. The earliest and latest schedules are given in Figures 6.12 and 6.13.

This problem is studied for several levels of the maximal total use of resource per time unit:

a - constant, equal to 120
b - constant, equal to 100
c - constant, equal to 80
d - constant, equal to 75
e - constant, equal to 70
f - constant, equal to 65

The application of GAMS-MINOS produces the schedules presented in Figures 6.14 to 6.19 corresponding to examples from a to f. The empty diagrams corresponds to the maximal level of resource and the dashed diagram to the use of resource.

In Figure 6.20 the relationship between $L_m$ and the total duration is shown.

The scheduling with a variable maximal level is also studied for three different examples (Examples g, h and i) and the results are presented in Figs. 6.21, 6.22, 6.23.

Each case is solved in about 100 major iterations using less than 5 minutes CPU (Pentium 166).

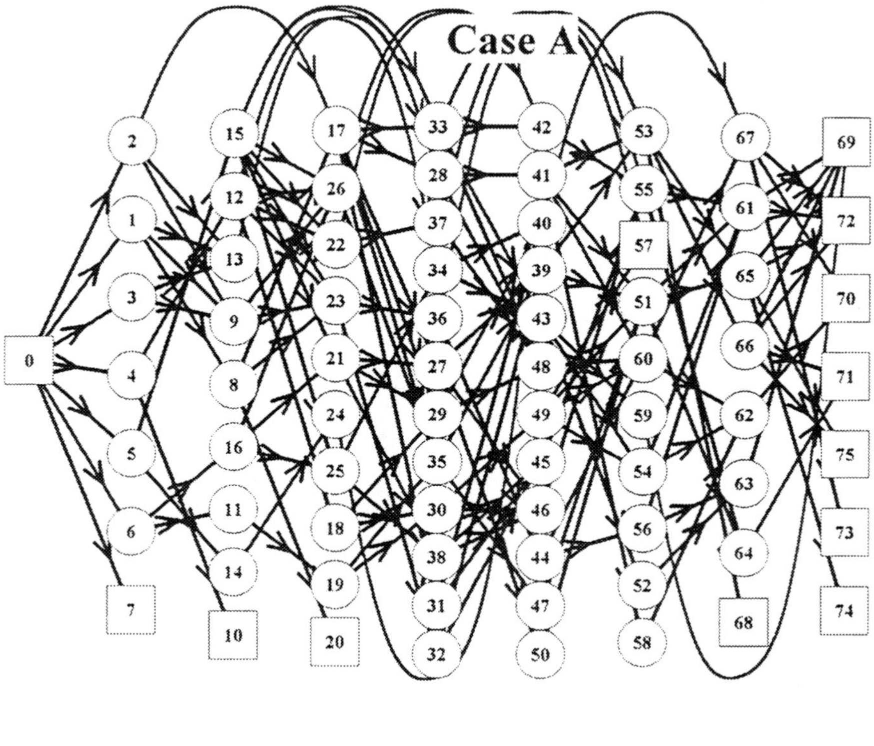


FIG. 6.9
NETWORK WITH 75 ACTIVITIES (CASE A)

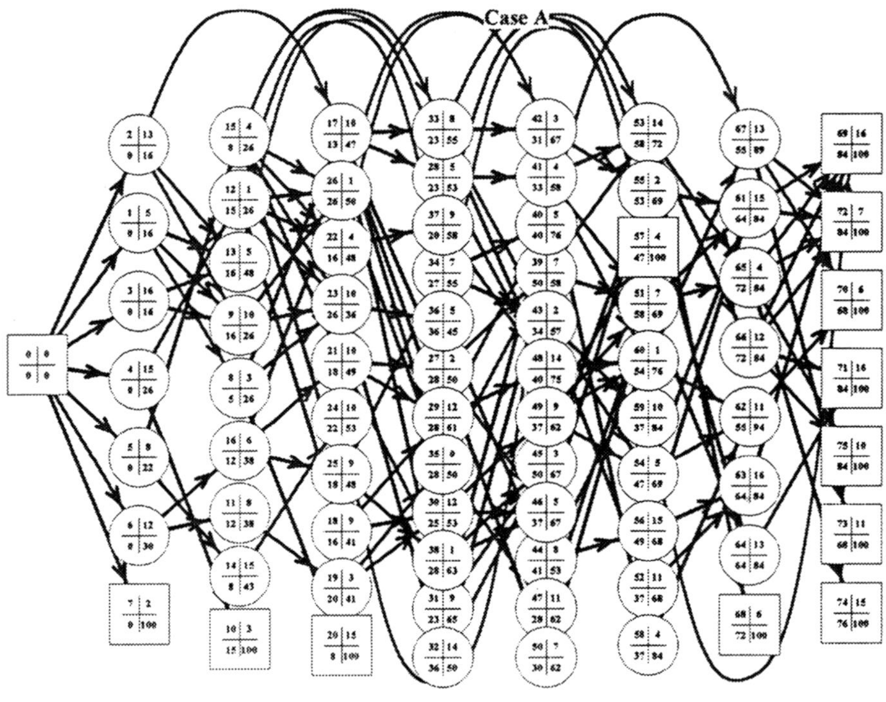


FIG. 6.10
ANALYSED NETWORK WITH 75 ACTIVITIES (CASE A)

| i \ k | 0 | 1 | 2 |
|---|---|---|---|
| 0 | 1 | 7 | |
| 1 | 7 | 13 | 1 |
| 2 | 9 | 19 | 4 |
| 3 | 10 | 19 | 3 |
| 4 | 12 | 24 | 4 |
| 5 | 12 | 21 | 1 |
| 6 | 10 | 16 | 1 |
| 7 | 8 | 16 | |
| 8 | 7 | | |

**Number of precedence links between each progressive level, i, and the level given by (i+k) with k not equal to 0. The element (i, k=0) represents the number of activities belonging to the level i.**

FIG. 6.11
LEVEL ADJACENCY MATRIX FOR THE CASE A

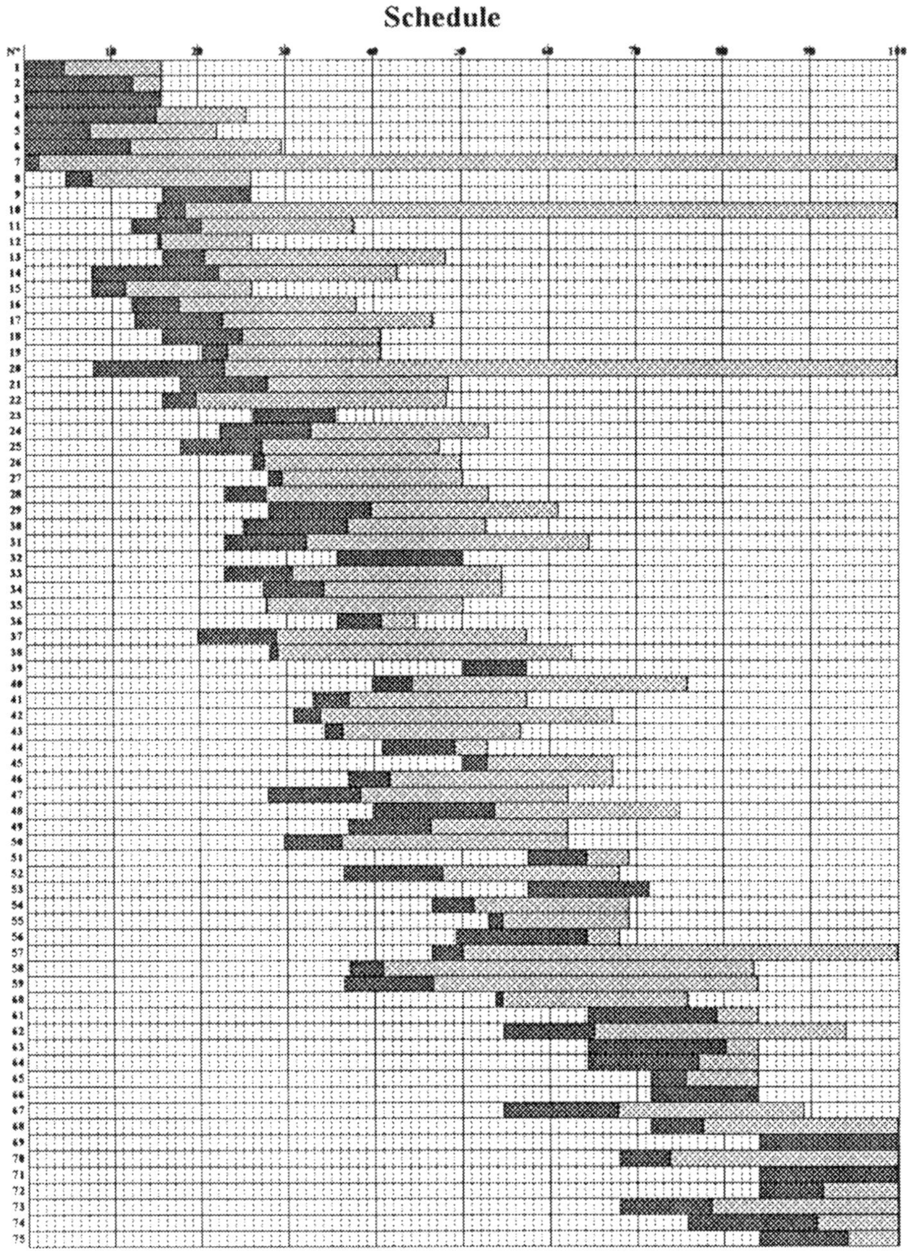

FIG. 6.12
EARLIEST SCHEDULE FOR CASE A

## Schedule

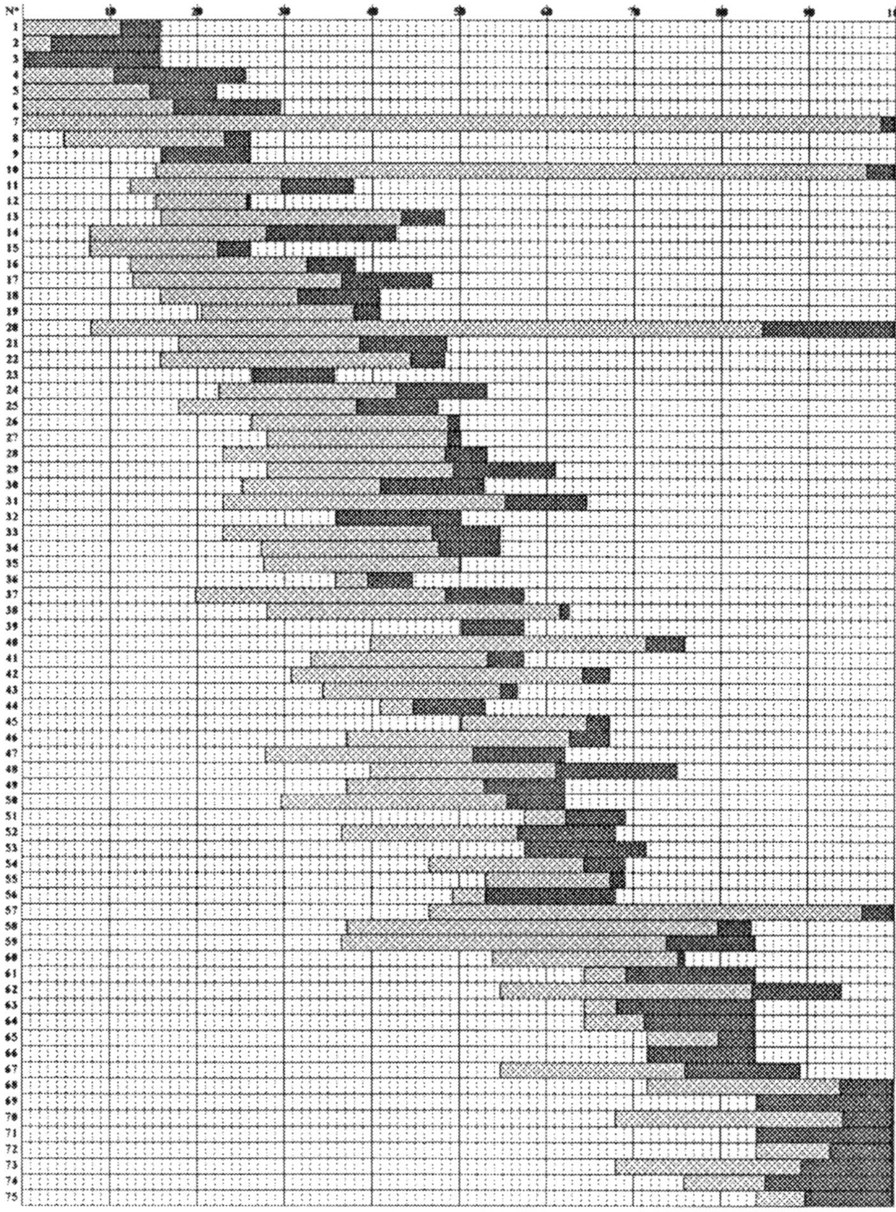

**FIG. 6.13**
**LATEST SCHEDULE FOR CASE A**

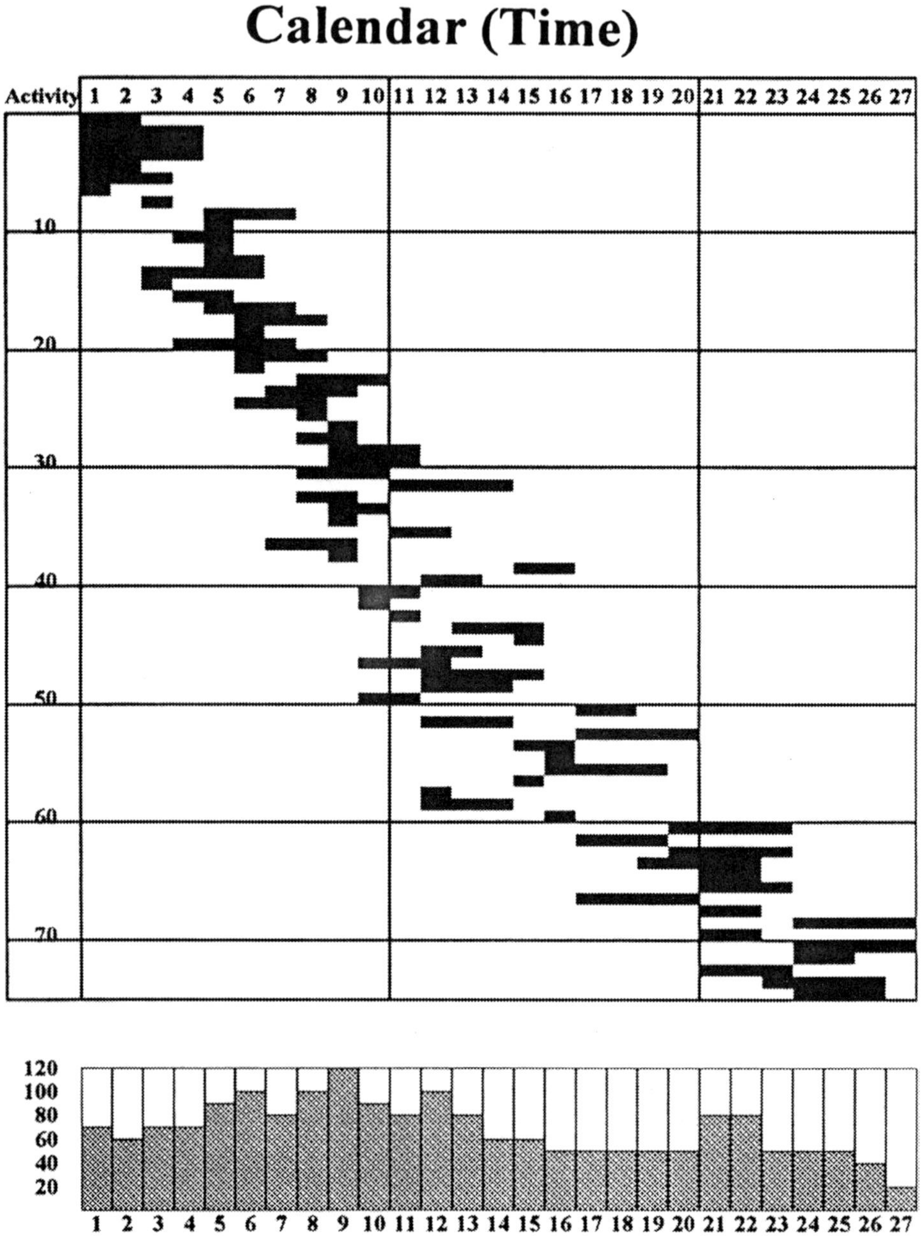


FIG. 6.14
COMPUTED SCHEDULE WITH Lm = 120 (CASE A)

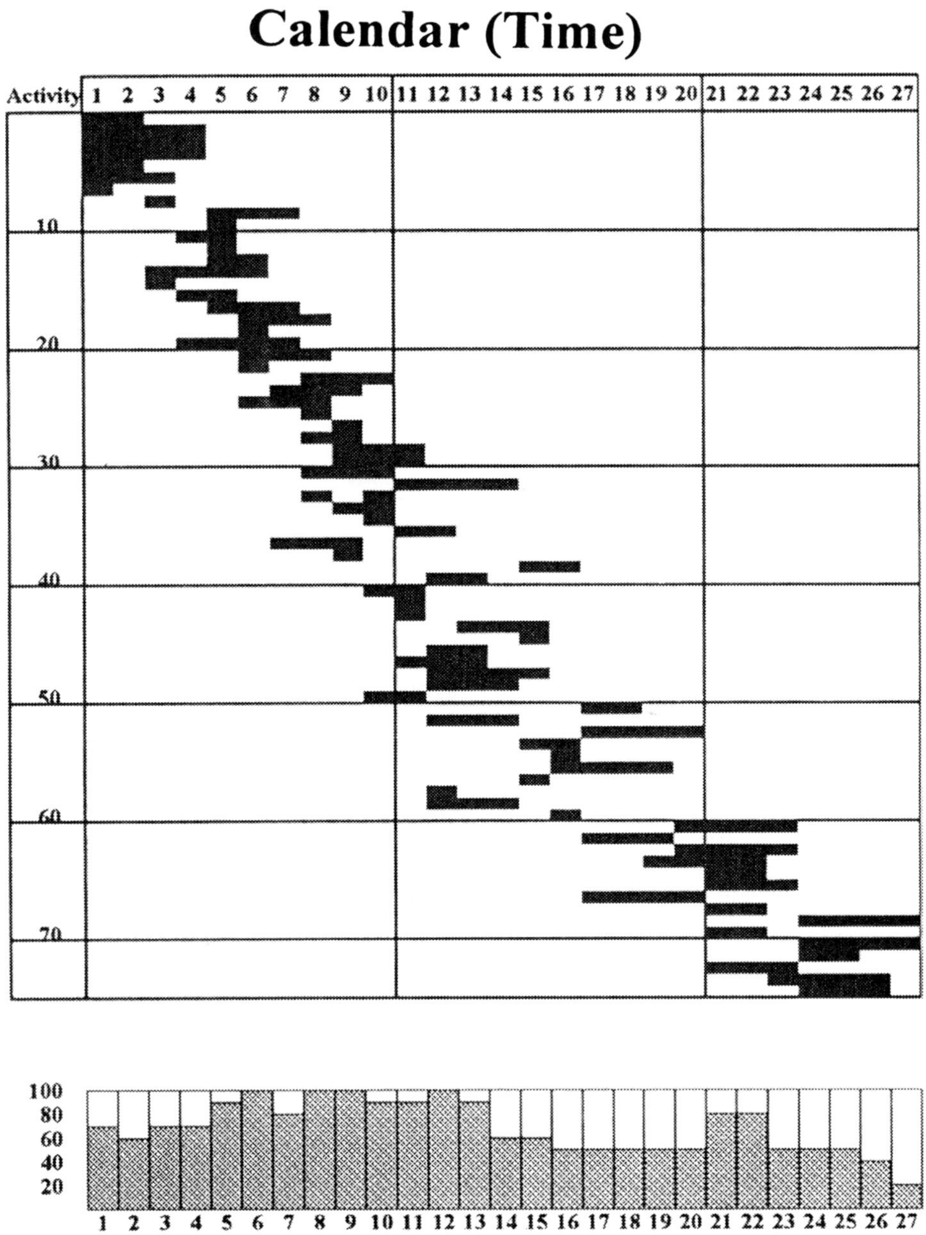


FIG. 6.15
COMPUTED SCHEDULE WITH Lm = 100 (CASE A)

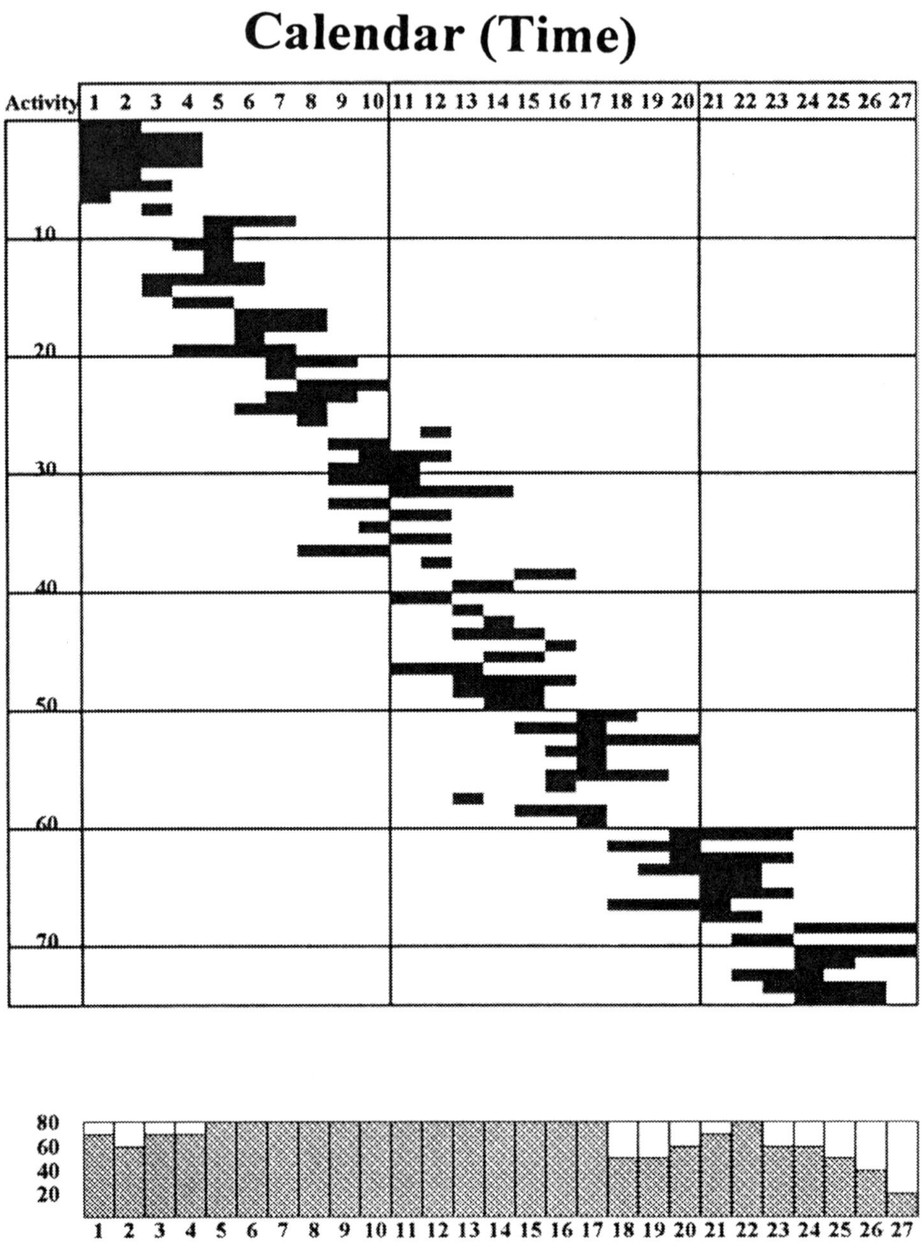


FIG. 6.16
COMPUTED SCHEDULE WITH Lm = 80 (CASE A)

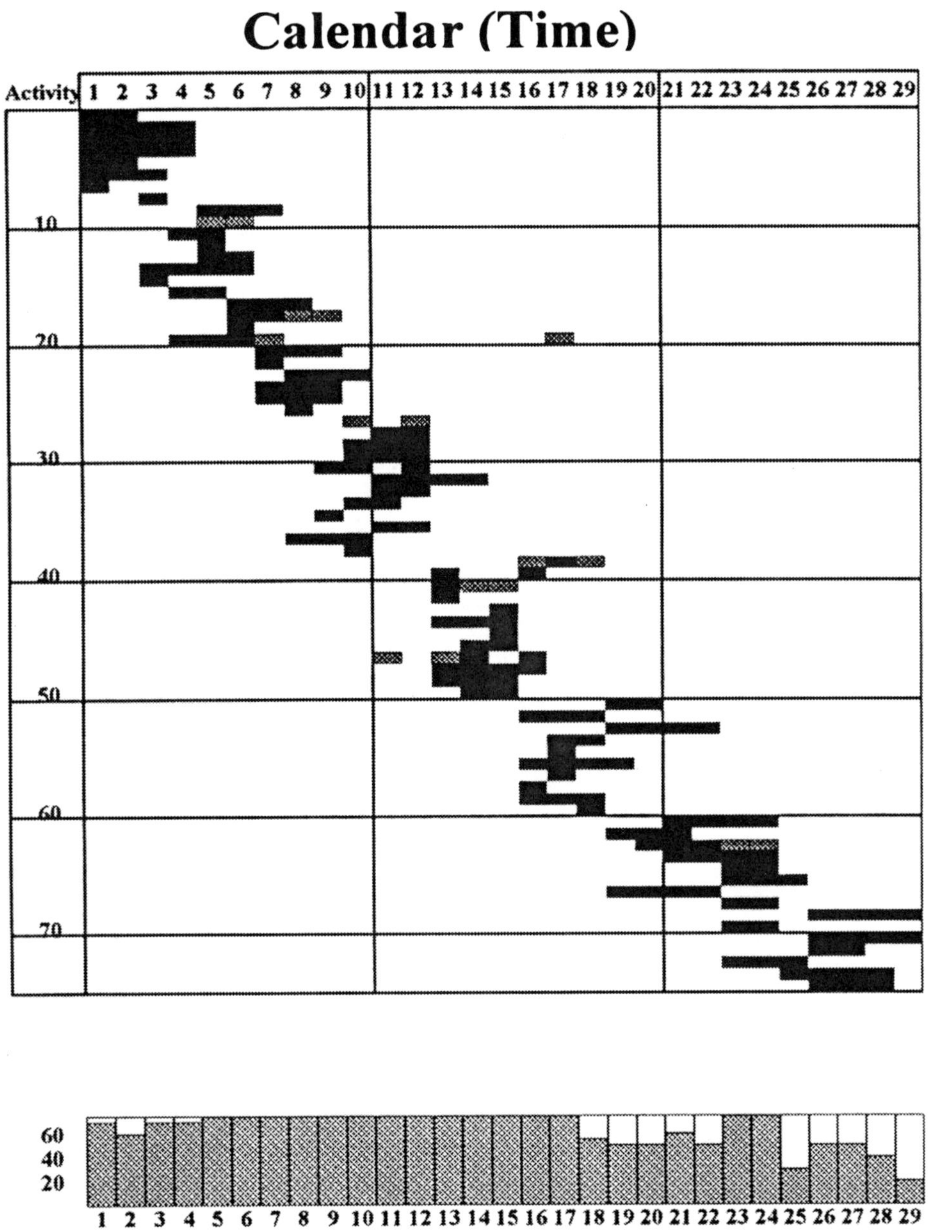


FIG. 6.17
COMPUTED SCHEDULE WITH Lm = 75 (CASE A)

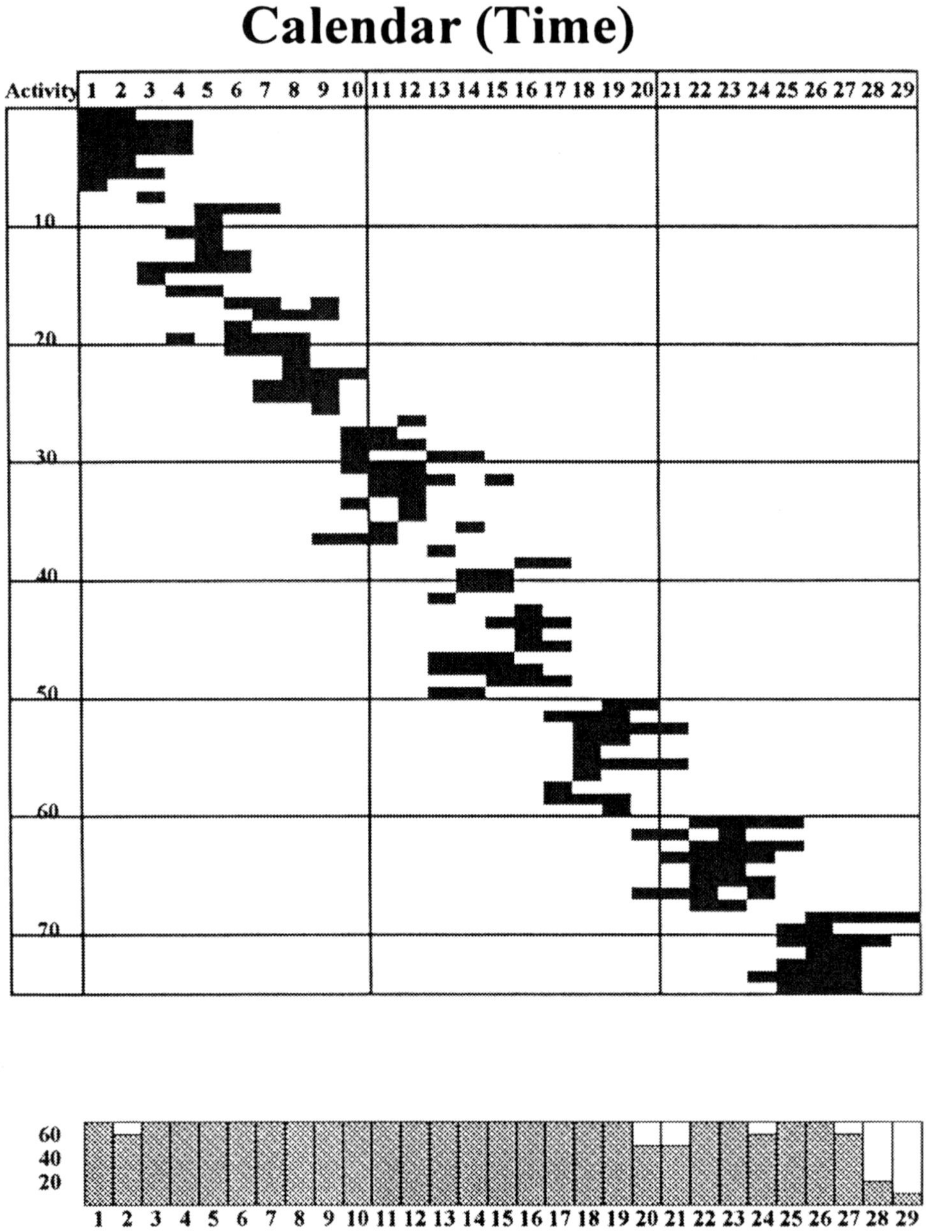


FIG. 6.18
COMPUTED SCHEDULE WITH Lm = 70 (CASE A)

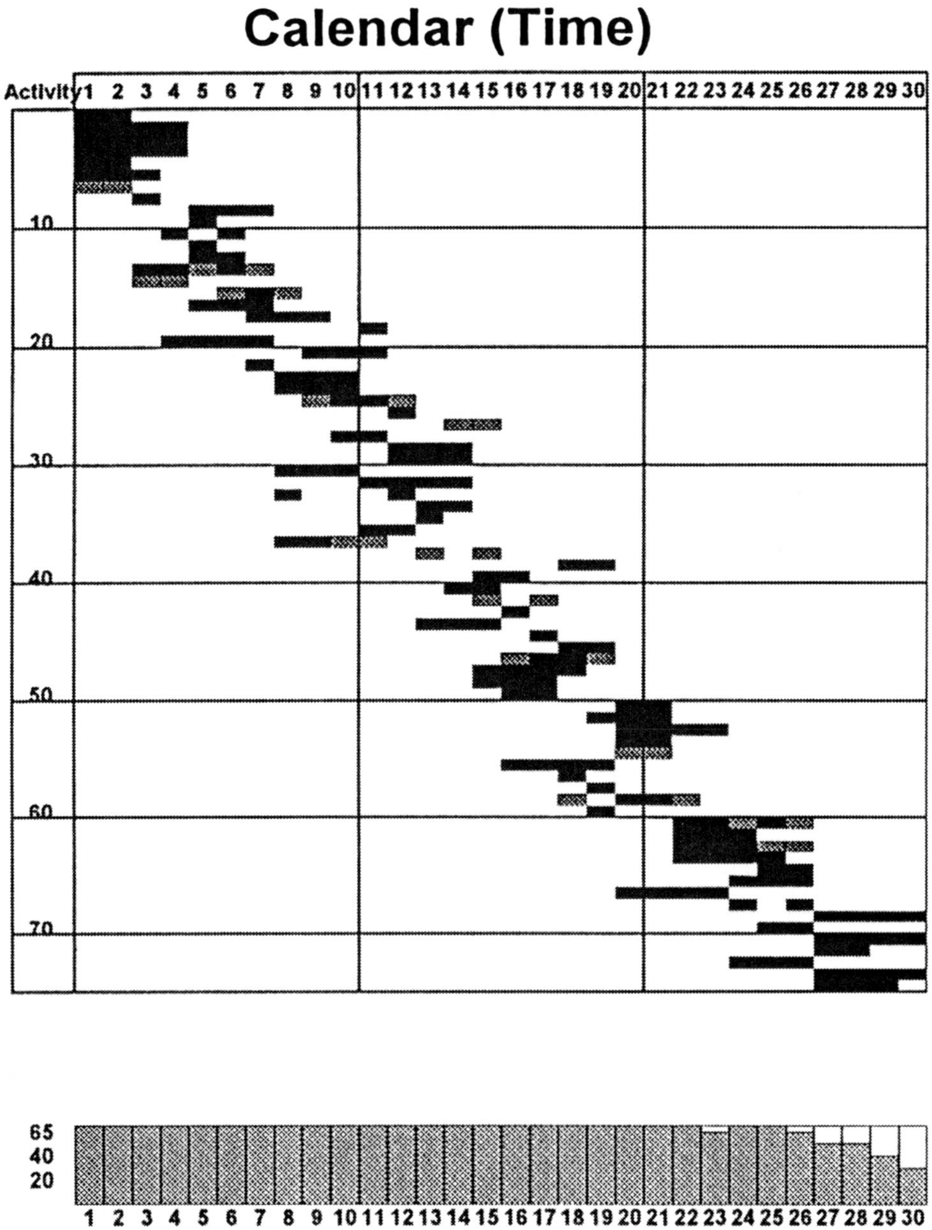


FIG. 6.19
COMPUTED SCHEDULE WITH Lm = 65 (CASE A)

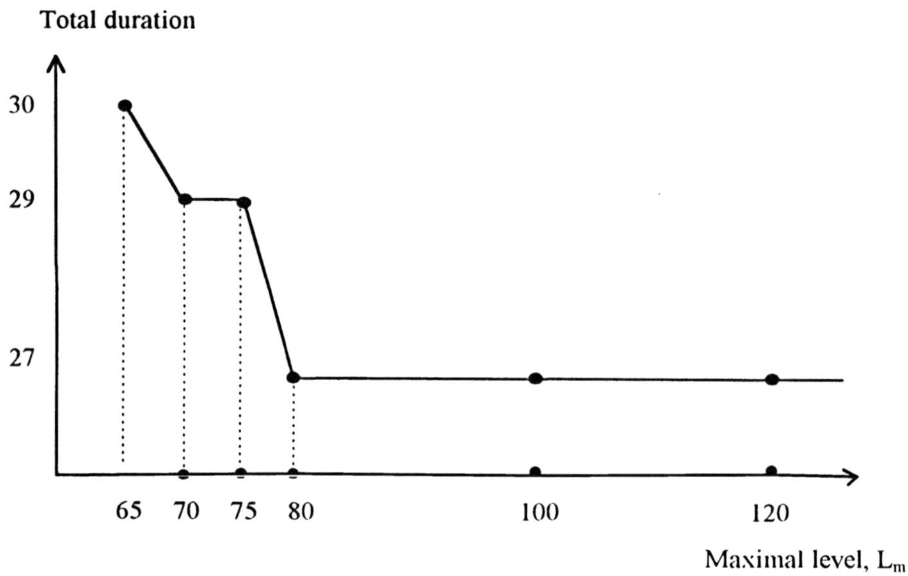


FIG. 6.20
TOTAL DURATION IN TERMS OF THE MAXIMAL LEVEL (CASE A)

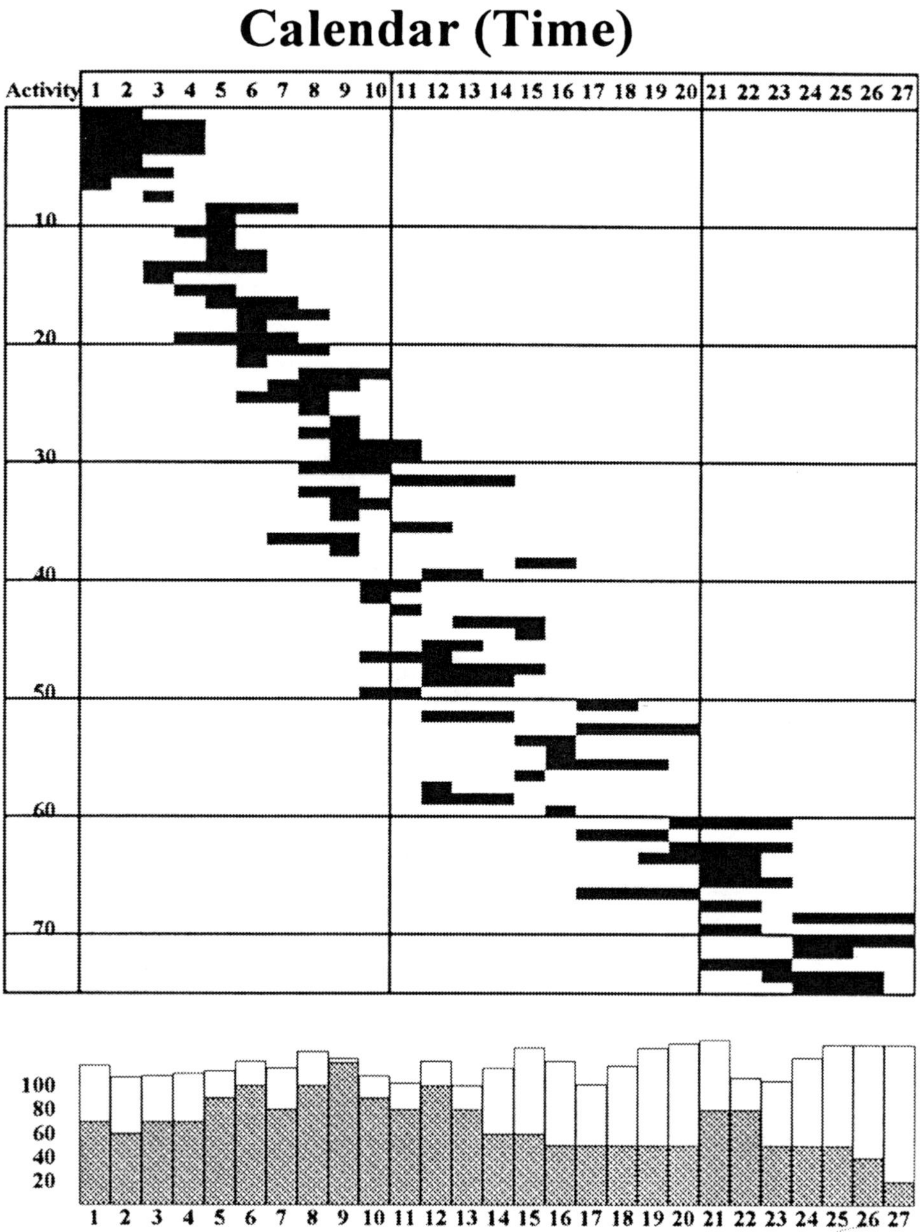


FIG. 6.21
COMPUTED SCHEDULE WITH VARIABLE Lm STARTING WITH Lm = 120 (CASE A)

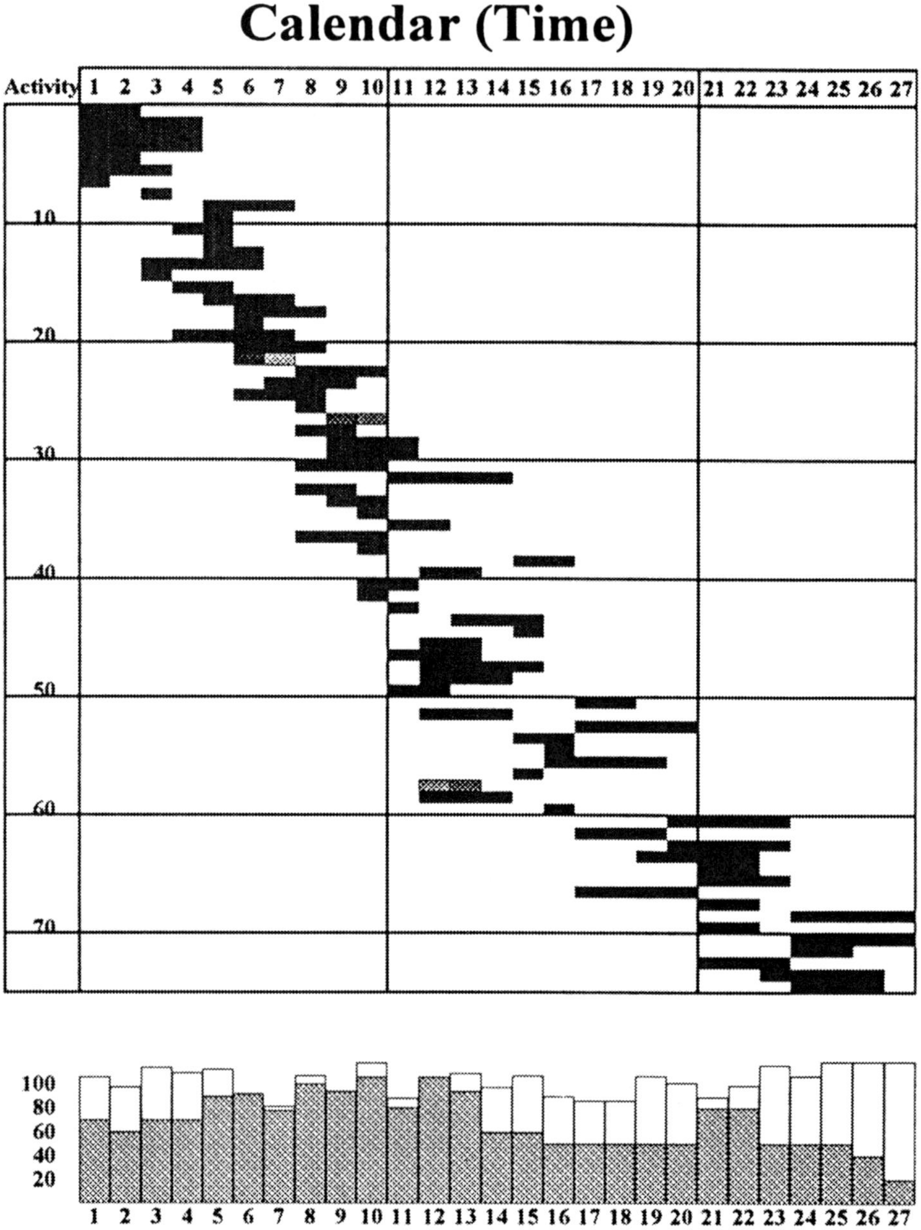


FIG. 6.22
COMPUTED SCHEDULE WITH VARIABLE Lm STARTING WITH Lm = 100 (CASE A)

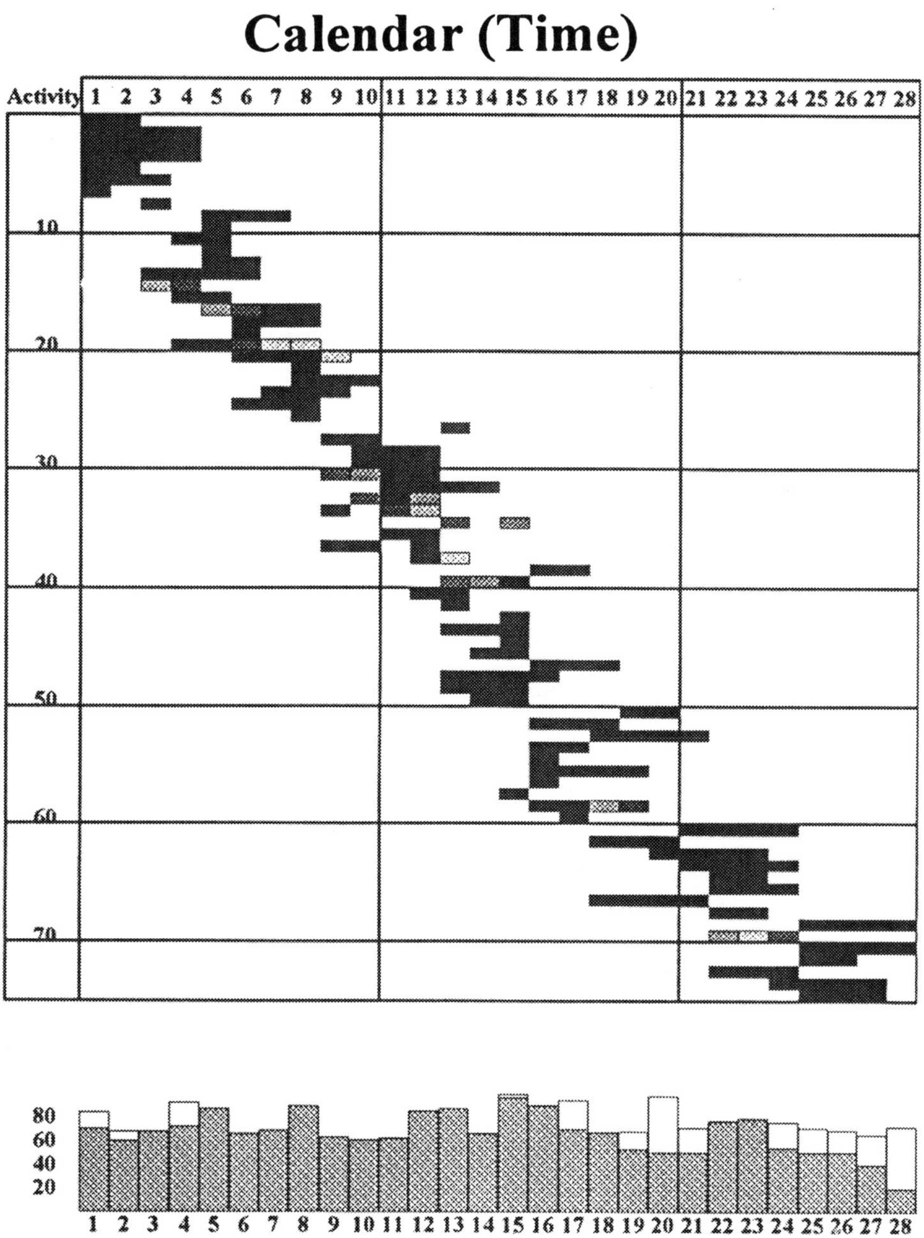


FIG. 6.23
COMPUTED SCHEDULE WITH VARIABLE Lm STARTING WITH Lm = 80 (CASE A)

### 5.3 Case B

This application has followed a path similar to Case A but is based on a network generated with 150 activities. The graphical representation, the level adjacency matrix and the earliest - latest schedules are now presented in Figs. 6.24 to 6.28.

Again, the required amount of resource per activity is equal to the maximal use per time unit (10 units) multiplied by Y.

This problem was studied for the following constant maximal levels of total use of resource per time unit (Figs. 6.29 to 6.32):

a - $L_m$ constant, equal to 240
b - $L_m$ constant, equal to 200
c - $L_m$ constant, equal to 180
d - $L_m$ constant, equal to 120

The relationship between the total duration and $L_m$ is shown in Fig. 6.33.

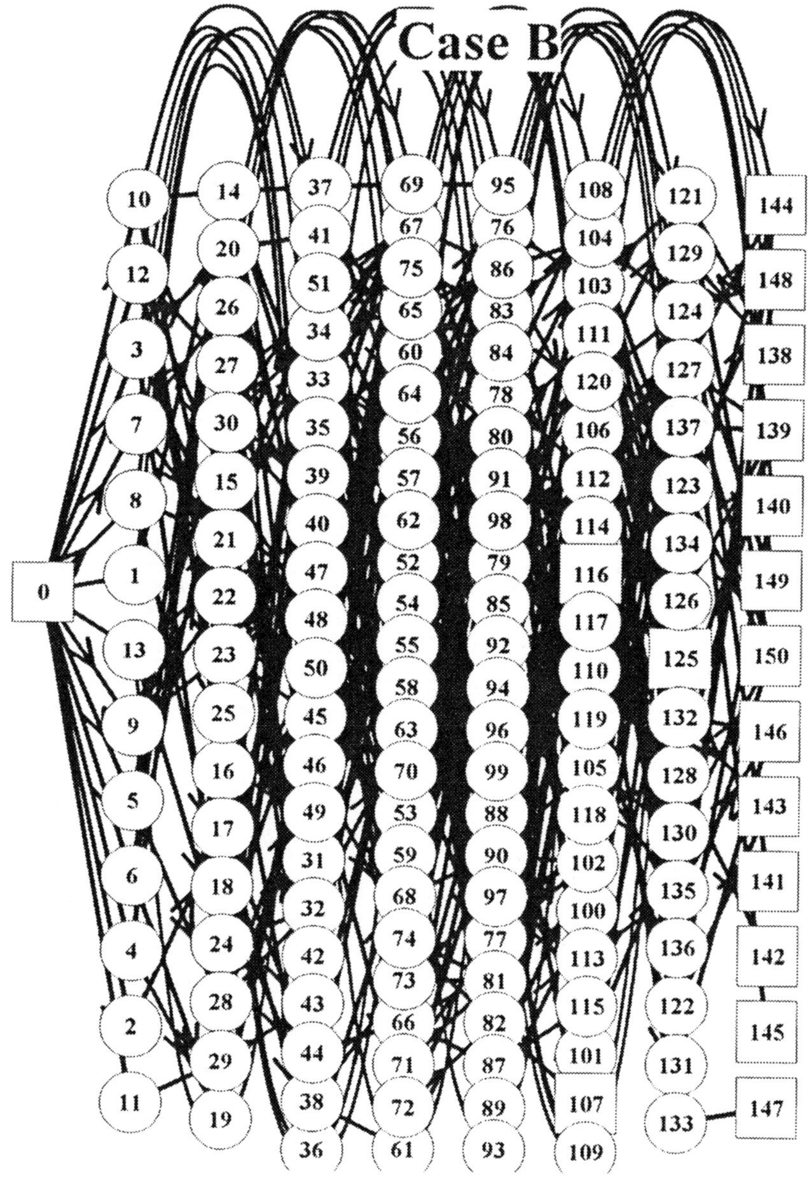


FIG. 6.24
NETWORK WITH 150 ACTIVITIES (CASE B)

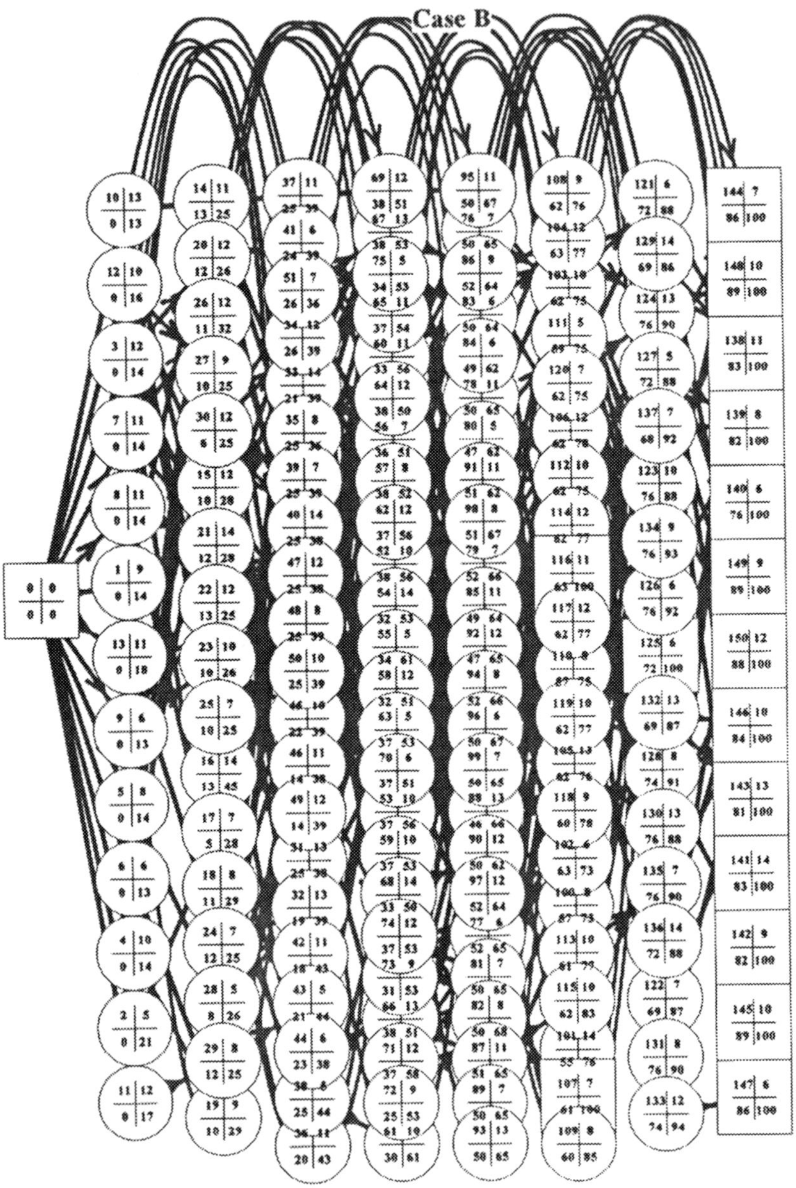

FIG. 6.25
ANALYZED NETWORK WITH 150 ACTIVITIES (CASE B)

# Case B, Level Adjacency Matrix

| i \ k | 0 | 1 | 2 |
|---|---|---|---|
| 0 | 1 | 13 | |
| 1 | 13 | 30 | 8 |
| 2 | 17 | 52 | 7 |
| 3 | 21 | 75 | 6 |
| 4 | 24 | 89 | 6 |
| 5 | 24 | 69 | 9 |
| 6 | 21 | 53 | 5 |
| 7 | 17 | 29 | |
| 8 | 13 | | |

**Number of precedence links between each progressive order, i, and the order given by (i+k) with k not equal to 0. The element (i,k=0) represents the number of activities belonging to the order i.**

FIG. 6.26
LEVEL ADJACENCY MATRIX (CASE B)

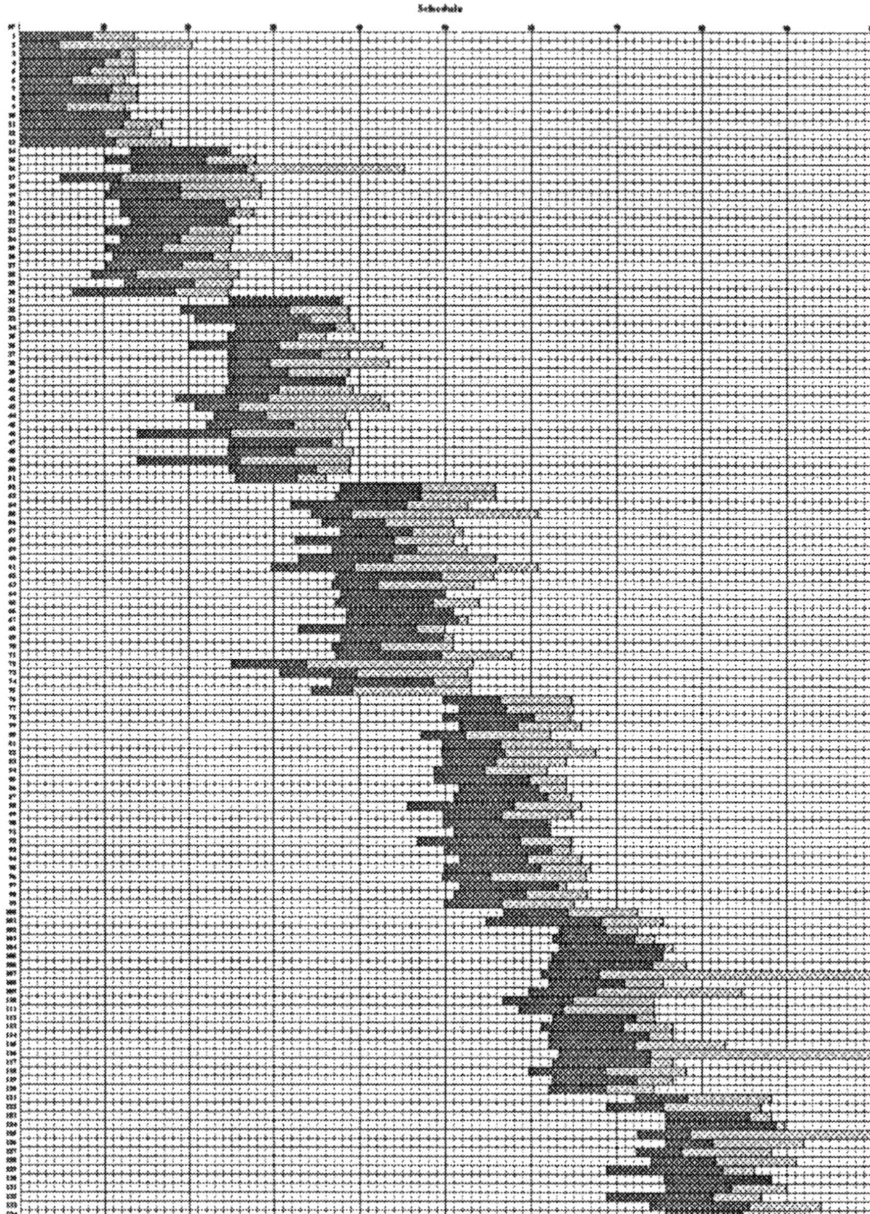

FIG. 6.27
EARLIEST SCHEDULE FOR CASE B

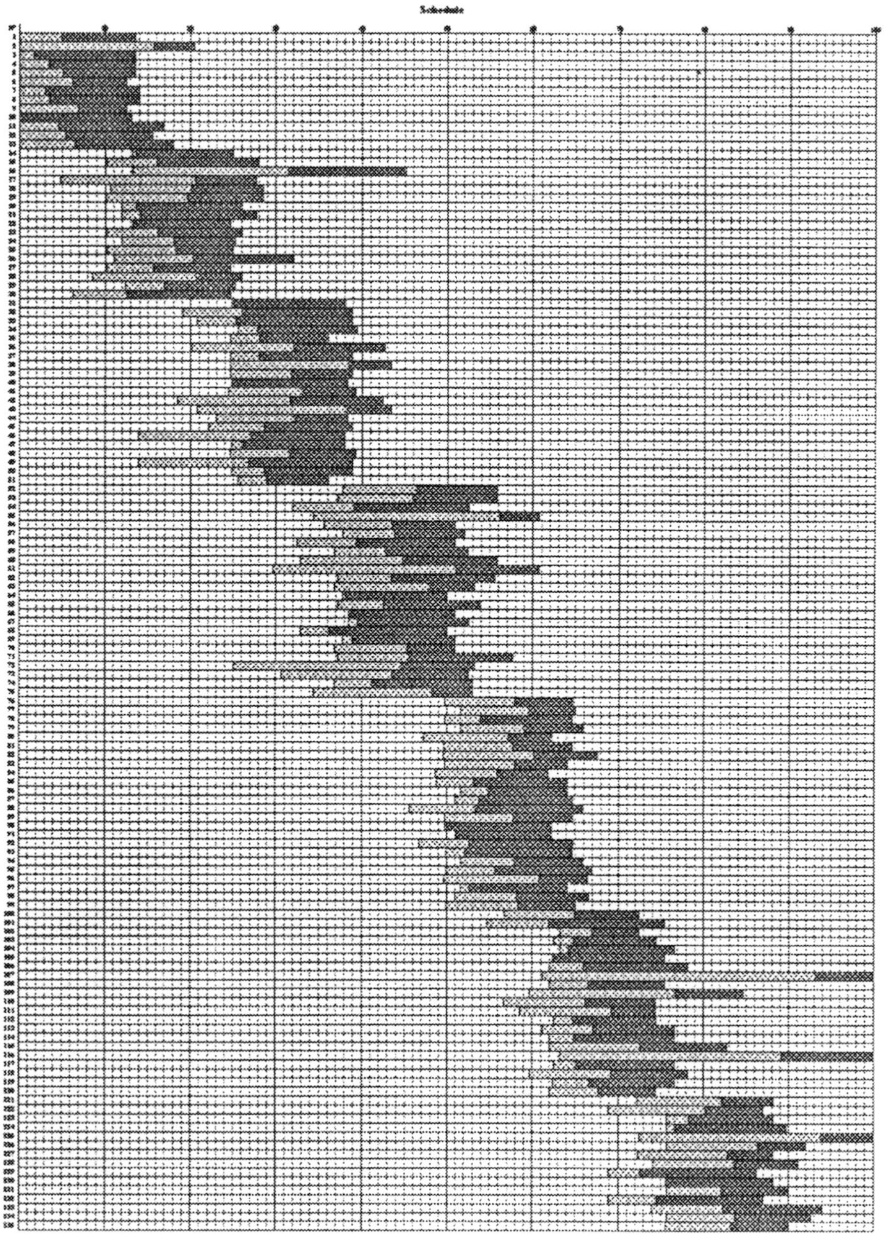

**FIG. 6.28**
**LATEST SCHEDULE FOR CASE B**

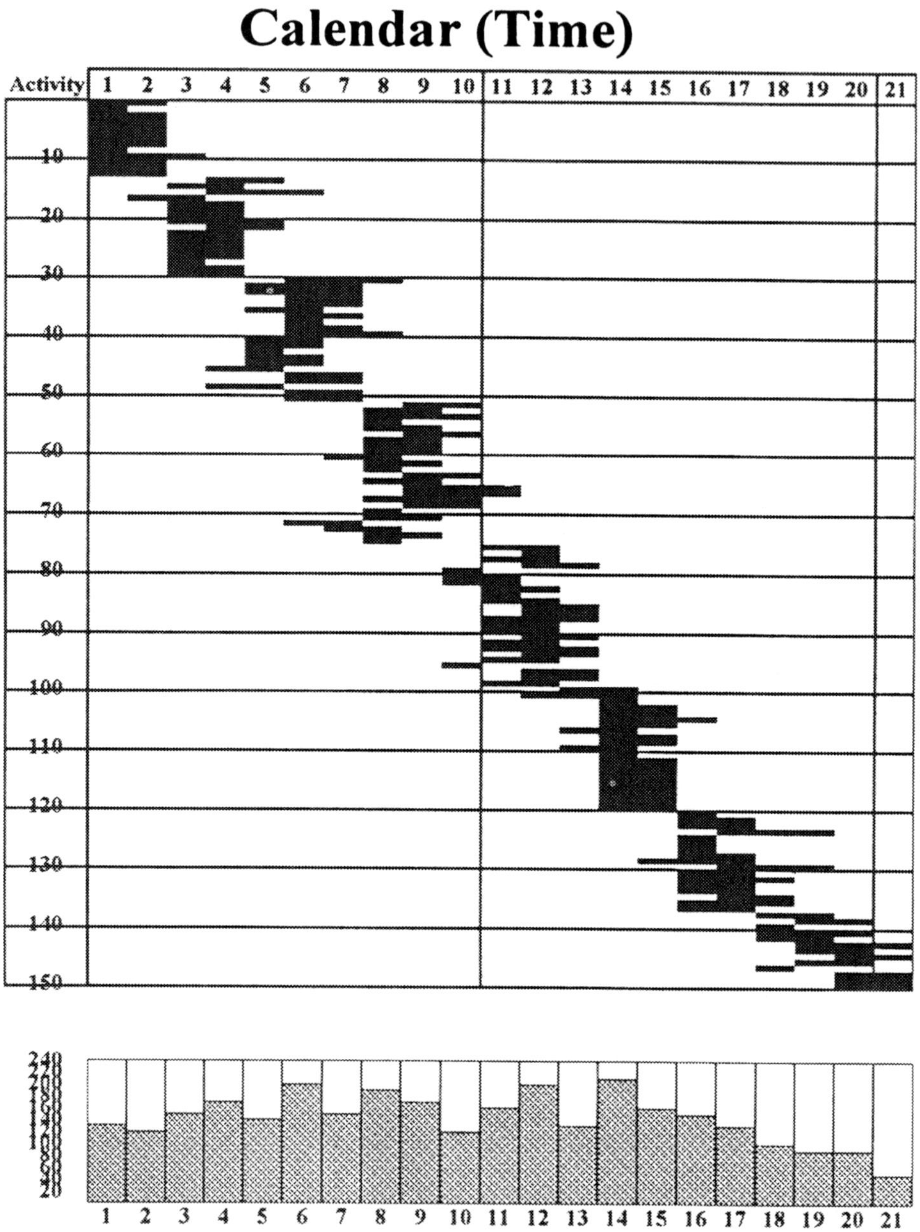


**FIG. 6.29**
**COMPUTED SCHEDULE WITH Lm = 240 (CASE B)**

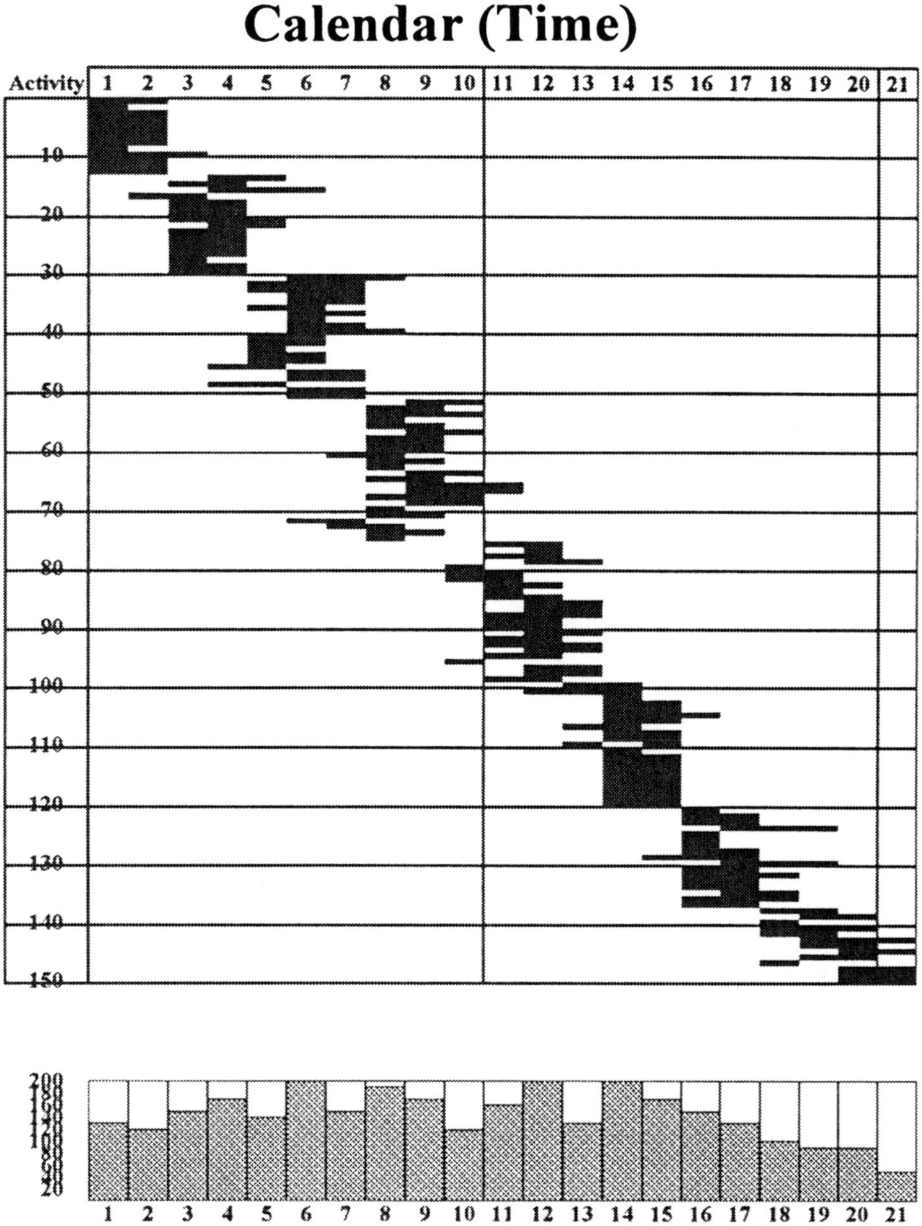


FIG. 6.30
COMPUTED SCHEDULE WITH Lm = 200 (CASE B)

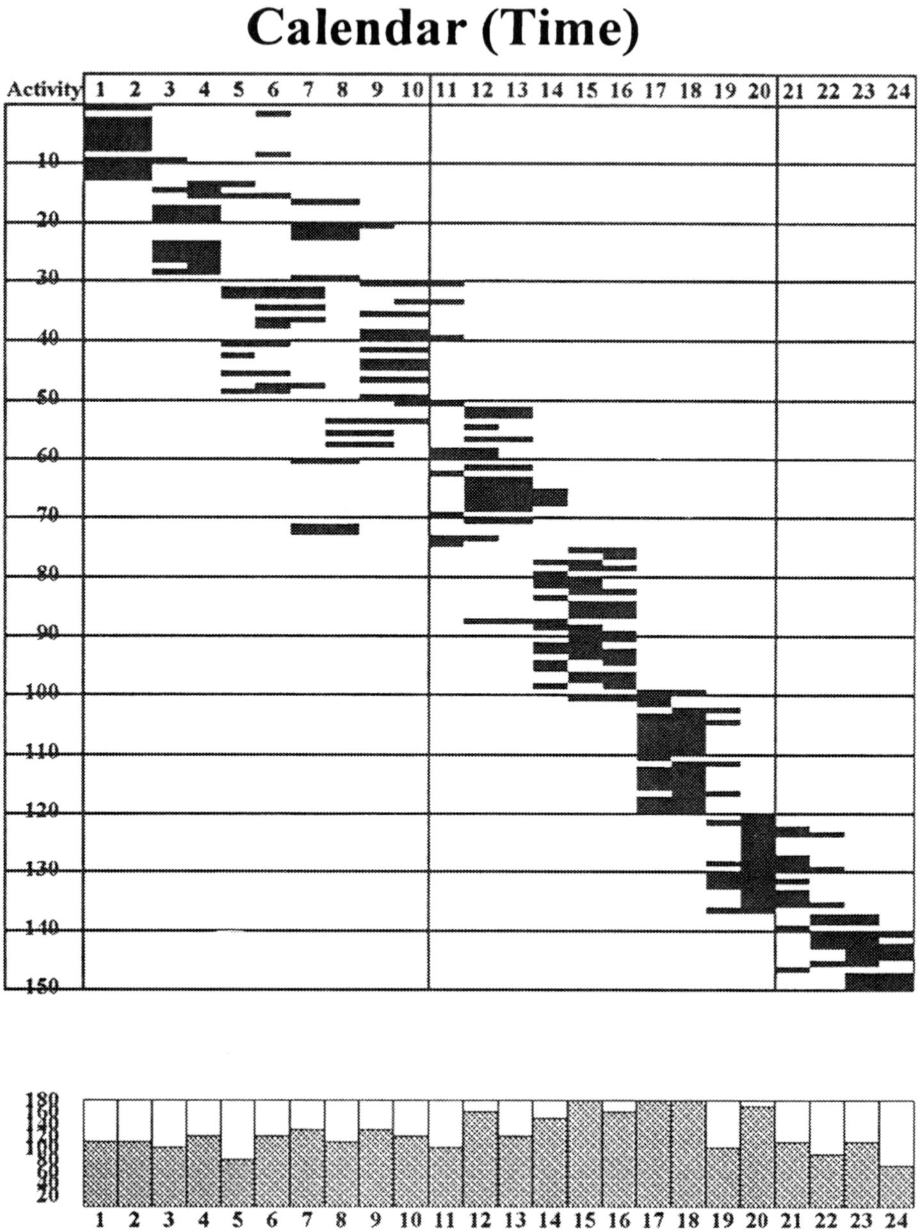


FIG. 6.31
COMPUTED SCHEDULE WITH Lm = 180 (CASE B)

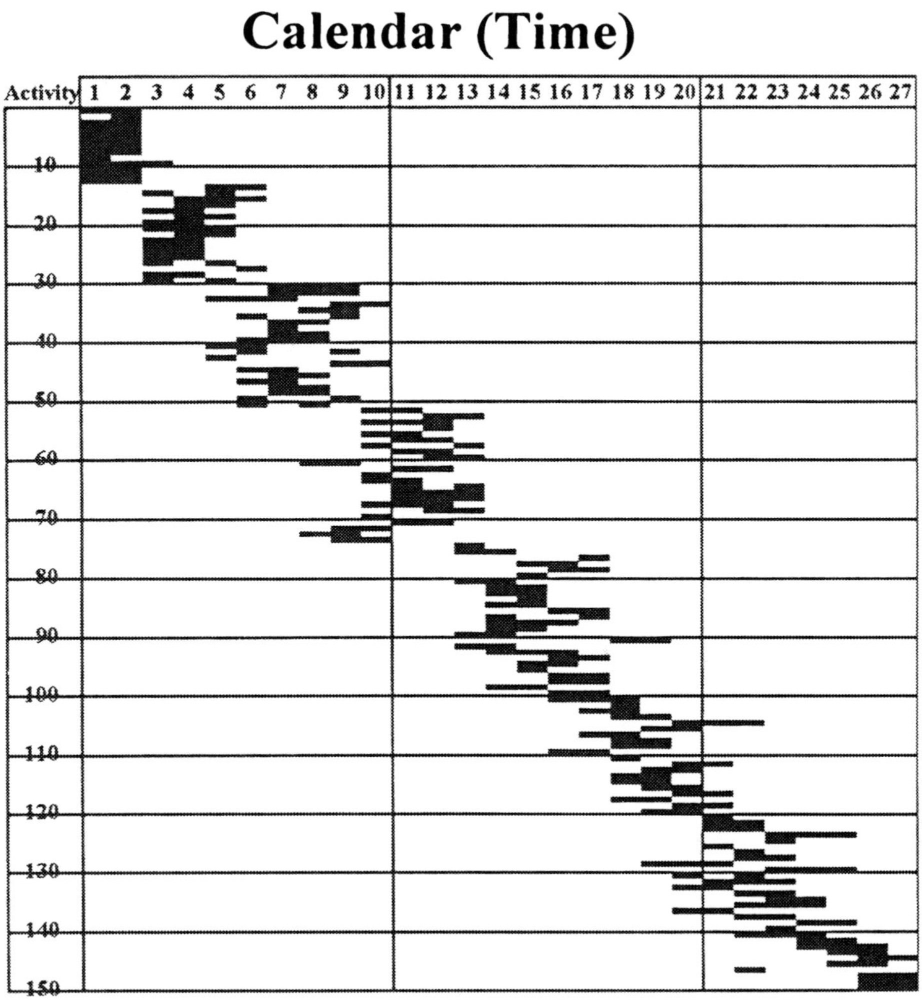


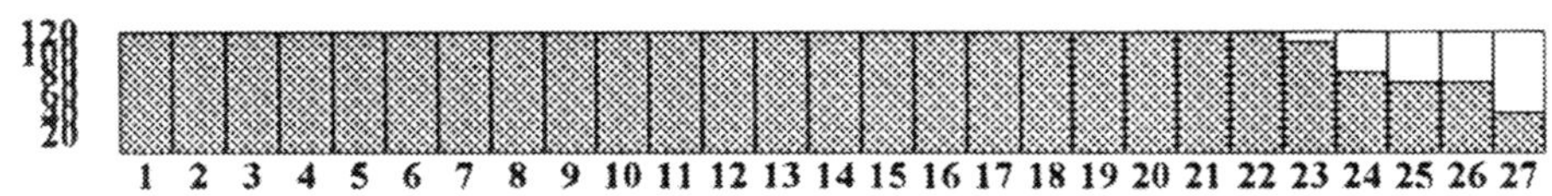


**FIG. 6.32**

**COMPUTED SCHEDULE WITH Lm = 120 (CASE B)**

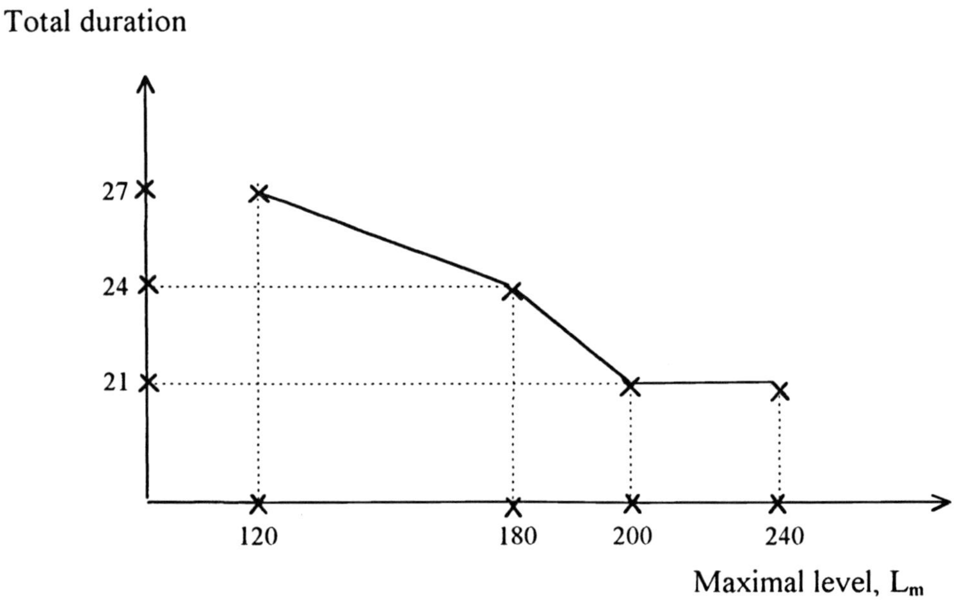


**FIG. 6.33**
**TOTAL DURATION IN TERMS OF THE MAXIMAL RESOURCE (CASE B)**

Two additional examples were studied with variable $L_m$ (t) and presented in Figs. 6.34 and 6.35.

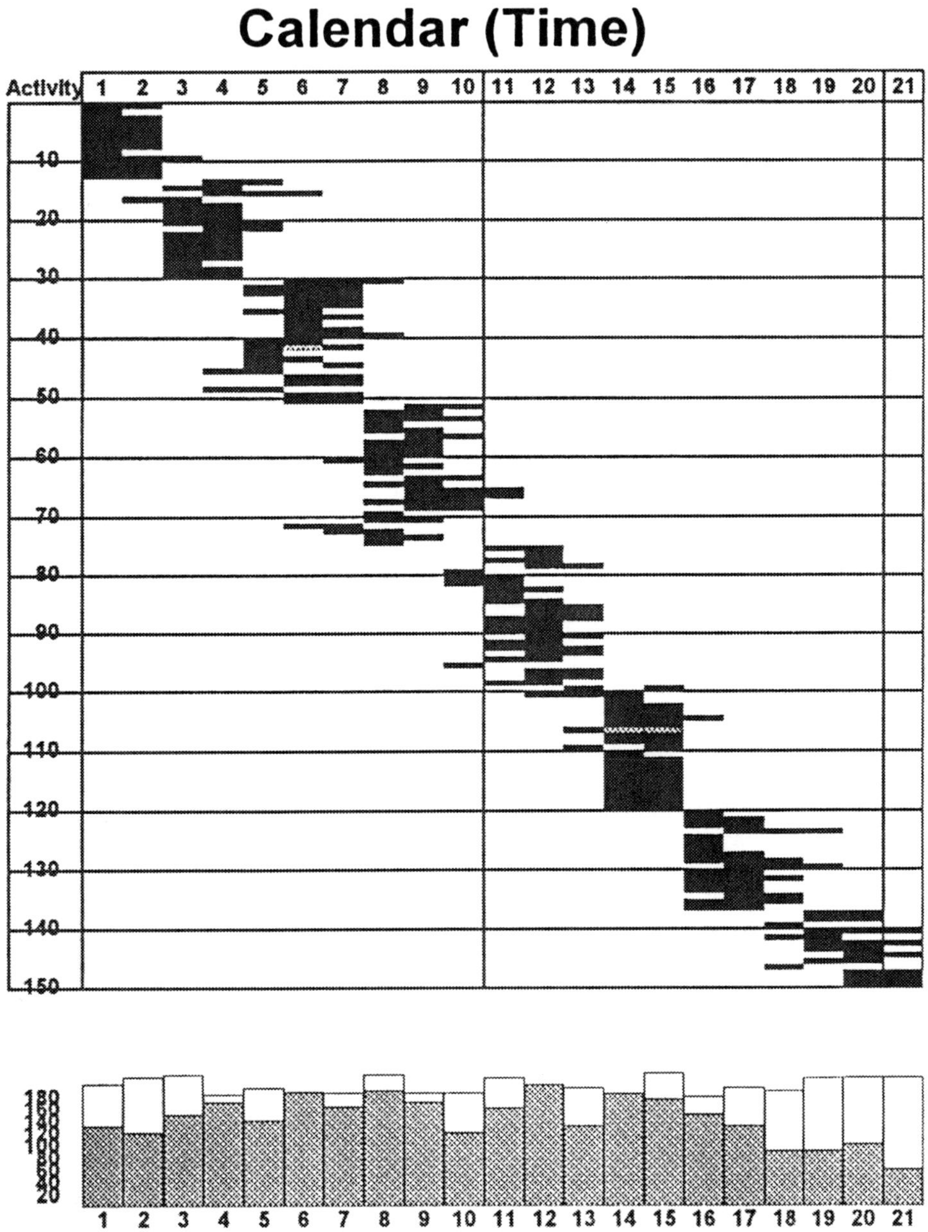


FIG. 6.34

COMPUTED SCHEDULE WITH VARIABLE Lm STARTING WITH Lm = 180

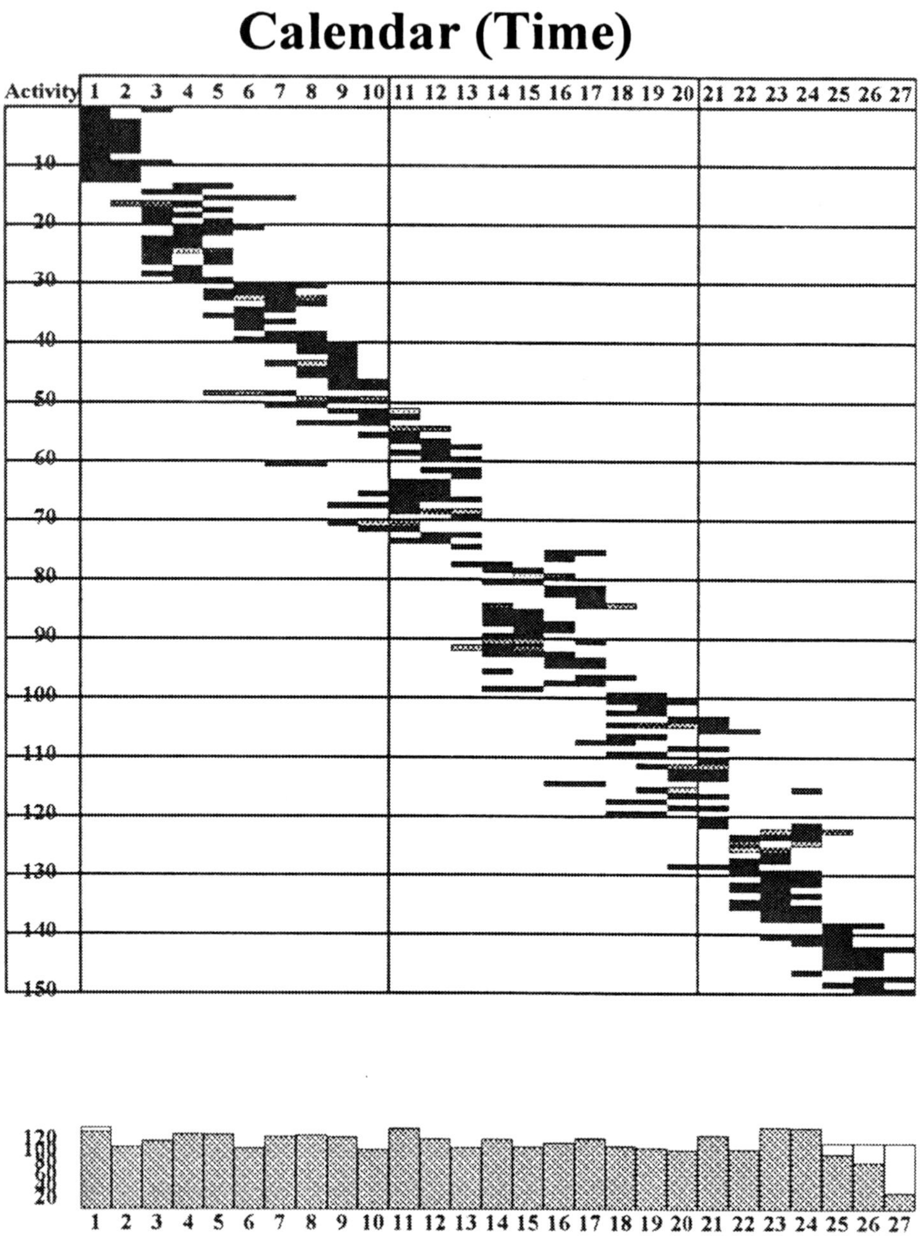


FIG. 6.35
COMPUTED SCHEDULE WITH VARIABLE Lm STARTING WITH Lm = 120

# 6. Dynamic Programming

## 6.1 General formulation

The proposed model in 4. has the advantage of being formulated in terms of the resources allocated in each time unit to each activity.

The optimization of these decisions can be also studied by **Dynamic Programming** (DP) as is shown by Tavares (1987) whenever the project will be easily formulated as a sequence of stages.

Dynamic Programming has the main advantage of reducing the number of alternative solutions to be computed as it is based on the **Bellman's principle** (Bellman et al., 1962):

A sequence of decisions from the present until the end of the studied horizon is a Bellman optimal policy if it is optimal for the present state of system, independently of the past states. Therefore, such a policy has to be determined for each state which can occur at each moment of decision.

The basic model of DP requires the formulation of:

- **Stages** of the project, j = 1, ..., J, corresponding to the moments of decision.
- **State** definition of the system at each stage, j, S(j)
- **Decision variable,** Y(j, S(j))
- **Objective function**:
  $F_j[S(j), Y(j)] = U_j[S(j), Y(j)] + f \cdot F^*{}_{j+1}[S(j+1)]$ for j = 1, ..., J-1 where f is the discount factor and being $F^*_{j+1}$ the optimal $F_{j+1}$ the for S = S(j+1) which is determined in terms of S(j) and Y(j) through the state equation.

It should be noted that $F_j$ describes the optimization criteria from stage j until the end of the studied horizon. (Fig. 6.36)

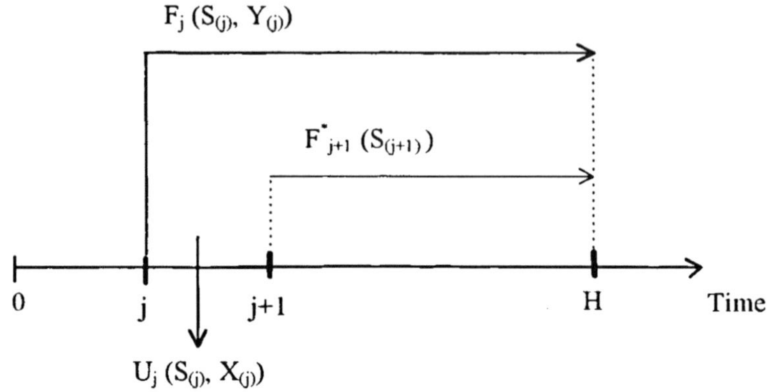


**FIG. 6.36**
**THE DP FORMULATION**

The optimization is carried out by finding the minimal or maximal $F_t(S(j), Y(j))$ given S(j):

$$F^*_j(S(j)) = \underset{Y(j)}{\text{Max or Min}} \left\{ U_j(S(j), Y(j)) + f \cdot F^*_{j+1}[S(j+1)] \right\}$$

This process can be recursively solved from j = H until j = 1, computing at each j and for each S(j), all values of $F_j[S(j), Y(j)]$ in order that the best decision, $Y^*(j)$ will be determined in terms of S(j) : $Y^*(j)$ [S(j)]. (optimal policy rule).

- **State equation**

This equation relates S(j+1) with S(j) and Y(j):

$$S(j+1) = g[S(j), Y(j)]$$

The application of this model to the scheduling problem was studied by a few authors and analytical results were presented by Tavares (1987).

The application of DP to project scheduling implies a careful discussion of how the previous concepts can be formulated.

**A) Stage j = 1, ..., J.**

The concept of stage should be associated to each moment of decision making.

In Project Scheduling, a moment of decision making occurs whenever an activity is started or a change of its intensity is introduced. This can happen at any time and so each time unit would be a stage. However, this formulation is too heavy and so a simplified model may be preferred:

a) associating j to the start of each progressive order

J is much lower than before but the decision taken at each j should concern all activities starting between j and j+1.

b) associating j to the start of each activity

The difficulty of this formulation concerns the selection of the sequence of the starting times which is unknown, a priori. Then, for each sequence of starting times, the optimal sequence of decisions can be computed. Unfortunately, the number of alternative sequences is usually too high.

c) associating j to the start of each activity of the critical path

This approach avoids the previous difficulty but then the decisions concerning the non-critical activities have to be associated to those concerning the critical activities.

d) associating j to the stage of implementation of the project which can be defined in terms of sub-projects or intermediate milestones.

This formulation is more appropriate to macro-projects including a set of interconnected sub-projects. The decisions concern the allocation of resources to the sub-projects and their allocation to the different stages.

The two last formulations may be the most appropriate ones as they correspond to a compromise between dimension and detail of the process of decision-making.

**B) States and Decisions**

The definition of the state should describe the level of implementation of the project and should give sufficient information to support the process of decision making. The definition of the decisions should be made in terms of the adopted state variables and the definition of these variables has to be consistent with the choice of the adopted stages.

Usually, the state should describe the amount of resources already spent and the level of implementation of the activities. The decisions should describe the start of the activities and the resources allocated to each stage.

**C) Objective Function**

The objective function has to verify the additivity conditions allowing the decomposition of $F_j[S(j),Y(j)]$ into $\{U[S(j),Y(j)] + f \cdot F^*_{j+1}[S(j+1)]\}$, where the term U (S(j), Y(j)) describes usually the cost or the benefit due to the implementation of stage j.

Additional terms can be included in U(S(j), Y(j)) to describe other criteria such as a penalty due to the fluctuation of the level of allocation of resources:

$$[Y(j) - Y(j-1)]^2$$

However, in this case Y (j-1) should be included in the description of the state of j and so the dimension of the state vector is increased.

## 6.2 A discounting model

A model can be proposed in terms of the concepts presented in 6.1 to optimize the management of a project including J stages and $I_j$ sub-projects at each stage j (Tavares, 1987) adopting the following definitions:

a) each sub-project is denoted by k = (i, j) where j = 1, ..., J and i = 1, ..., $I_j$.

b) the cost of k, $C_k$, discounted for its starting time, $t_k$, is given by

$$C_k = \int_0^{T_k} \left[ f^t \cdot X_k(t) \right] dt$$

assuming that the sub-project k is carried between $t_k$ and $t_k + T_k$ where $X_k$ (t) is the expenditure level at time t. The parameter f is the appropriate discounting factor.

The duration $T_k$ has to satisfy: $l_k \le T_k \le L_k$ where $(l_k, L_k)$ are the lower and upper bounds of $T_k$.

c) at the end of each stage, j, a benefit can be generated, $B_j$ and its discounted value will be given by $B_j \cdot f^{t_{j+1}}$ with j = 1, ..., J where $t_{J+1}$ is the completion time of the project.

d) the objective function can be expressed by the overall net present value:

$$\sum_{j=1}^{J} \left\{ B_j f^{t_{j+1}} - \sum_{i=1}^{Ij} C_{k=(j,i)} f^{t_k} \right\}$$

This model can be represented by the sequence shown in Fig. 6.37 and, obviously, the optimal solution implies ending each project of each stage j, at time $t_{j+1}$, due to the nature of the objective function.

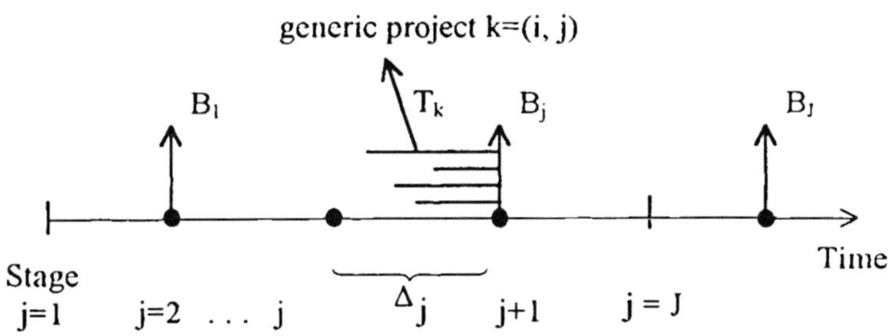


FIG. 6.37
THE STUDIED PROCESS

The objective function can be expressed by:

$$F_j(t_j, \Delta_j) = \left[ B_j f^{t_j + \Delta_j} - \sum_{i=1}^{Ij} C^*_{k=(j,i)} f^{t_j} \right] + F^*_{j+1}(t_{j+1})$$

for j = 1, ..., J with $F^*_{J+1} \equiv 0$ and with the state equation given by

$$t_{j+1} = t_j + \Delta_j$$

The variable $t_j$ is the state variable and $\Delta_j$ is the duration of stage j (decision variable):

$$\underset{k=(j,i)}{\text{Max}}\, l_k \le \Delta_j \le M\Delta_j$$

being $M\Delta_j$ the maximal duration of stage j, with $M\Delta_j \le \underset{k}{\text{Max}}\, L_k$.

The function $C_k^* = (j,i)$ is the minimal present cost of project k for the duration, $T_k$, with $l_k \le T_k \le L_k$.

It should be noted that the factor $f^{tj}$ can be eliminated by making

$$G_j\left(\Delta_j\right) = \frac{F_j\left(t_j, \Delta_j\right)}{f^{tj}} \text{ and then } G_j\left(\Delta_j\right) = \left[B_j f^{\Delta_j} - \sum_{i=1}^{Ij} C_{k=(j,i)}^*\right] + f^{\Delta_j} \cdot G_{j+1}^*$$

$$\text{and } G_j^* = \underset{\Delta_j}{\text{Max}}\left[G_j(D_j)\right]$$

This formulation is very convenient because at each stage, $G_j$ should be studied just in terms of $\Delta_j$.

Determining $C_{k=(j,i)}^*$ depends on the type of function adopted for $X_k(t)$. This function can be defined as a continuous function of time, and different shapes can be adopted. The discount factor is usually defined by an exponential expression when a continuous formulation of time is considered: $f=e^{-\gamma}$ with $\gamma>0$.

The easiest case corresponds to have $X_k$ (t) constant on time. Then:

$$C_k = e^{-\gamma(\Delta_j - D)} \int_0^D e^{-\gamma t} X\, dt \text{ with } l_k \le D \le \text{Min}\left[L_k, \Delta_j\right] \text{and } X \cdot D = M.$$

Therefore:

$$C_k = \frac{M}{\gamma D} e^{-\gamma(\Delta_j - D)}\left[1 - e^{-\gamma D}\right]$$

Obviously, the optimal solution is given by $D_k^* = \mathrm{Min}\left[L_k, \Delta_j\right]$ and $X^* = \dfrac{M}{D_k^*}$ if M is assumed to be constant.

Alternatively, one can assume that M is not constant because it may be a function of D:

a) $M = M_o e^{-\beta(D-l_k)}$ with $\beta \geq 0$ and $D > l_k$.

b) $M = M_o e^{\beta(D-l_k)}$ with $\beta \geq 0$ and $D > l_k$

In the former case, the optimal solution is the same as before $\left(D_j^* = \mathrm{Min}\left[L_k, \Delta_j\right]\right)$ but for the latter, one should minimize:

$$C_k = \frac{M_o e^{\beta(D-l_k)}}{\gamma D} e^{-\gamma(\Delta_j - D)}\left[1 - e^{\gamma D}\right]$$

and so the optimal $C_k$, $C^*_k$, can be obtained by making

$D_k = l_k$

or

$D_k = \mathrm{Min}\left[L_k, \Delta_j\right]$

or

$D_k \Leftrightarrow \dfrac{dC_k}{dD} = 0$

The presented model can be generalized by assuming that X(t) can be non constant and that a penalty is due to the variation of X with time:

$$C_k = e^{-\gamma(\Delta_j - D)} \int_0^D \left[ e^{-\gamma t} X_l(t) + \delta \cdot \dot{X}_k(t)^2 \right] dt$$

where $\delta$ is a positive penalty coefficient and where $\dot{X}_k$ is the time derivative of $X_k(t)$, satisfying $\int_0^D X_k(t)\,dt = M(D)$ and with $l_k \leq D \leq \mathrm{Min}\left[L_k, \Delta_j\right]$.

This problem was solved by Tavares (1987) obtaining the analytical solution by **Calculus of Variations** as the given problem is equivalent to:

$$\text{Min}\int_0^D \left\{ \underbrace{e^{\gamma D}\left[e^{-\gamma t}\cdot X_k(t)+\delta\cdot \dot{X}_k(t)^2\right]-\lambda\cdot X_k(t)}_{S} \right\} dt.$$

This problem is the so-called "**isoperimetric problem**" and the optimal solution is obtained by the **Euler equation** (necessary condition):

$$\frac{\partial S}{\partial X_k(t)}=\frac{d}{dt}\frac{\partial S}{\partial \dot{X}_k(t)} \quad \text{or} \quad e^{\gamma D}e^{-\gamma t}-\lambda=2e^{\gamma D}\delta\cdot\ddot{X}_k(t)$$

where $\ddot{X}_k(t)$ is the second derivative of $X_k(t)$. Finally, the application of the **Legendre inequality** (sufficient condition) produces the result:

$$\frac{\partial^2 S}{\partial \dot{X}_k(t)^2}=2e^{\gamma D}\delta>0$$

In Figure 6.38, an example of optimal function $X^*(t)$ is given assuming that $\gamma = 0.0012$ (corresponding to an annual discount rate of 6.5%, using the week as time unit), $\delta$=0.1 and $\Delta_j = 50$.

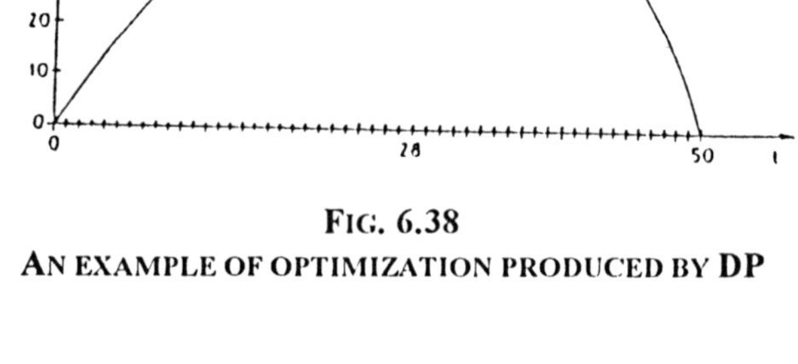


FIG. 6.38
AN EXAMPLE OF OPTIMIZATION PRODUCED BY DP

# 7. Risk

## 7.1 Risk and scheduling

The concept of **risk** can be considered as the probability of occurrence undesired events and it is a major source of concern for any project manager when making decisions about the scheduling of a project. This is due to the random or uncertain nature of the duration and of the consumption of resources by each activity. Therefore, the selection of a schedule tends to be considered as a "weapon" to fight the negative effects of such randomness or uncertainty (see, Soroush, 1993 and Tavares, 1994).

Actually, an earlier start is usually adopted to avoid the risk of failing to meet project milestones the scheduling of activities to produce a constant level of consumption of resources is a strategy to avoid the occurrence of peaks of resources requirements; and, finally, a later schedule of the start of an activity can reduce the impact of an increase of the consumption of resources on the discounted total cost of the project.

Unfortunately, most of the developments of project scheduling have ignored completely the study of the risk as they assume a full knowledge of all data about the network and their deterministic nature. This is well illustrated by the interesting survey by Elmaghraby (1995).

Therefore, priority should be given to the modelling of risk to study the optimal scheduling.

Such modelling should consider the randomness or uncertainty not just of the durations as it was studied by the classical authors (see, e. g., Kelly, Jr., 1963) but also of the consumption of resources.

The approach developed in Chapter 4 to simulate networks has adopted this formulation assuming that the cost of each activity is correlated with its duration. This correlation is particularly meaningful because a longer duration of an activity is usually due to an increase of its contents (additional works, etc.) or of its complexity, to a decrease of efficiency or to problems associated by the use of resources. In any case, the cost of the activity tends also to be higher.

Thus, one can estimate the effects on cost and on duration of alternative schedules by the estimation of the statistical distributions of the total duration and cost using the simulation model presented in Chapter 4.

This means that the probability of exceeding a specific level of cost or of duration can be immediately computed by the estimated distribution.

However, other results also can be deduced as is shown in the next two sections.

### 7.2 Risk of failing to meet a predefined schedule

A schedule can be defined as a set of starting times for all the activities of the network, $\left\{t_\bullet^s(i) : i = 1,\dots,N\right\}$ and therefore the risk of failing to meet this schedule, R, is given by

$$R = 1 - P$$

where P is the probability of being able to implement this schedule.

The probability P is determined by:

$$P = \prod_{\substack{i=1 \\ \text{with} \\ J(i)\neq\emptyset}}^{N} \prod_{j\in J(i)} P\left[t_\bullet^s(j) + D(j) \le t_\bullet^s(i)\right]$$

if an activity i which can start before $t_\bullet^S(i)$ will just start at $t_\bullet^S(i)$. This condition is essential to avoid the interdependence of the probabilities used in the previous formula. D(j) is the duration of the activity j and usual hypothesis of independence between the durations of different activities is assumed also. The activities with J(i)=Ø are excluded from this formula because these activities can start on schedule without risk. The probability $P\left[t_\bullet^s + D(j) \le t_\bullet^s(i)\right]$ can be computed easily by:

$$P\left[t_\bullet^s(j) + D(j) \le t_\bullet^s(i)\right] = P\left[D(j) \le \underbrace{t_\bullet^s(i) - t_\bullet^s(j)}_{d_{ij}}\right]$$

and this probability can be directly derived from the distribution of D(j).

If the upper tail of this distribution can be approximated by a negative exponential law given by $P(D(j) > d) = 1 - F_{D_{(j)}}(d) = e^{-\lambda_j d}$ with $\lambda_j \geq 0$, one has $P(D(j) \le d_{ij}) = 1 - e^{\lambda_j d_{ij}}$ and so:

$$\ln P = \ln\left\{\prod_{\substack{i=1\\ \text{with}\\ J(i)\neq\varnothing}}^{N} \prod_{j\in J(i)} \left[1-e^{-\lambda_j d_{ij}}\right]\right\} = \sum_{\substack{i=1\\ \text{with}\\ J(i)=\varnothing}}^{N} \sum_{j\in J(i)} \ln\left[1-e^{-\lambda_j d_{ij}}\right]$$

if the usual independence assumption is adopted.

The study of the risk of the total expenditure (or consumed resource) exceeding a specific level L also can be carried out easily under the same independence assumption. Actually, the total cost, $C_T$, is defined by $\sum_{i=1}^{N} C_i$ and for a reasonable number of activities (N > 20 ; 30) the theorem of the Central Limit applies and $C_T$ follows a gaussian distribution.

Thus:

$$C_T \approx N(\mu = \sum \mu(C_i); \sigma^2 = \sum \sigma^2(C_i))$$

where $\mu$ $(C_i)$ and $\sigma^2$ $(C_i)$ are the mean and the variance of $C_i$.

The probability of exceeding L then can be immediately determined.

### 7.3 A stochastic risk model for duration and cost

This model was proposed by Tavares (1994) and it has the objective of producing appropriate estimates of the risk of exceeding given levels of duration and cost for the studied project.

The **duration** and the **cost** of **each activity, i**, are assumed to follow a **bivariate** distribution, with marginal parameters denoted by:

$$\mu_i = E(D_i) \; ; \; \mu'_i = E(C_i)$$

$$\sigma_i^2 = VAR(D_i) \; ; \; \sigma'^2_i = VAR(C_i)$$

and $\rho_i$ being the correlation coefficient between $D_i$ and $C_i$. The usual assumption about the independence of $(D_i, C_i)$ between different activities is adopted.

This model uses AoA and adopts the traditional hypothesis of PERT defining the occurrence time of each node of the critical path (k = 0, 1, ..., K where k = 0 corresponds to the start and K to the end) just in terms of the previous critical activities assuming that the duration of each activity is equal to its mean:

$t_{k+1} = t_k + D_k$ with k =0, ..., K-1, where $t_l$ is the occurrence time of node k and $D_k$ is the duration of the critical activity linking k and k+1.

Therefore:

$$E(t_{k+1}) = ET_{k+1} = ET_k + \mu_k$$
$$V(t_{k+1}) = VT_{k+1} = VT_k + \sigma_k^2$$

The cost of the activities is allocated to each node of the critical path, k, proportionally of the fraction of each activity located between $ET_k$ and $ET_{k+1}$ and assuming the adopted schedule with the duration of each activity, i, equal to its mean, $\mu_i$.

This allocation is exemplified in Figure 6.39.

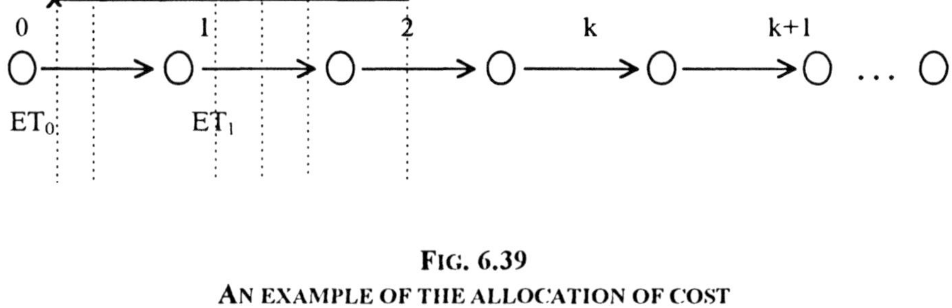


**FIG. 6.39**
**AN EXAMPLE OF THE ALLOCATION OF COST**

where:

a) $t_1^s, t_2^s, t_3^s$ ⟶ scheduled starting times of 1, 2, 3.

b) are represented assuming that $D_1 = \mu_1, D_2 = \mu_2, D_3 = \mu_3$

c) the allocation of costs is given by:

$$\text{Activity 1} \rightarrow \text{node 1} = C_1 \cdot \frac{ET_1 - t_1^s}{\mu_1}$$

$$\text{Activity 1} \rightarrow \text{node 2} = C_1 \cdot \frac{ET_2 - ET_1}{\mu 1}$$

$$\text{Activity 1} \rightarrow \text{node 3} = C_1 \cdot \frac{t_1^s + \mu_1 - ET_2}{\mu_1}$$

$$\text{Activity 2} \rightarrow \text{node 1} = C2 \cdot \frac{ET_1 - t_2^s}{\mu_2}$$

$$\text{Activity 2} \rightarrow \text{node 2} = C_2 \cdot \frac{t_2^s + \mu_2 - ET_1}{\mu_2}$$

$$\text{Activity 3} \rightarrow \text{node 2} = C_3$$

Therefore, the total cost allocated to each node with k = 1, ..., K can be obtained by adding up all the components of activities allocated to it. The mean and the variance of $R_k$ are denoted by $ER_k$ and $VR_k$.

It should be noted that one component of $R_k$ is always $C_k$ due to the activity k of the critical path which is correlated with $D_k$ but the other components of $R_k$ are independent of $D_k$.

Thus, $R_k$ is also correlated with $D_k$ and $\rho(R_k, D_k)$ is given by:

$$\rho(R_k, t_k) = \rho_1 \cdot \frac{\sigma_1'}{\sqrt{VR_1}}$$

The cumulative cost until node k is denoted by $CR_l$ and can be obtained by:

$$CR_k = CR_{k-1} + R_k$$

with l =1, ..., K and $CR_o = 0$. Therefore, the expected value and variance of $CR_k$, which are denoted by $ECR_k$ and $VCR_k$ will be given by:

$$ECR_k = ECR_{k-1} + ER_k$$
$$VCR_k = VCR_{k-1} + VR_k$$

Finally, the correlation between $CR_k$ and $t_k$, $\rho(CR_k, t_k)$ can be then deduced:

$$\rho(CR_k, t_k) = \frac{\rho(CR_{k-1}, t_{k-1}) \cdot \sqrt{VCR_{k-1}}\sqrt{VT_{k-1}} + \rho(R_k, D_k) \cdot \sqrt{VR_k} \cdot \sigma_k}{\sqrt{VCR_{k-1} + VR_k}\sqrt{VT_{k-1} + \sigma_k^2}}$$

with k =1, ..., K and having $VCR_0 = VT_0 = 0$.

Therefore, under the presented assumption, this model produces for each node k of the critical path a bivariate distribution ($CR_k$, $t_k$) of the cumulative cost allocated until k, $CR_k$, and of the time spent until k, $t_k$.

Under the gaussian assumption several results can be easily determined in terms of $ECR_k$, $VCR_k$, $ET_k$, $VT_k$ and $\rho$ ($CR_k$, $t_k$):

a) Risk, $\gamma_c$, of the cost due until node k exceeding a specific level, CR*.

$$\gamma_c = \int_{CR^*}^{\infty} f(CR_k)\, d\, CR_k$$

CR* is often made equal to $ECR_k$ plus a safety component.

Alternatively, any quantile ($CR_k$ ($\alpha$)) can be also estimated:

$$1 - \alpha = \int_{CR_k(\alpha)}^{\infty} f(CR_k) dCR_k$$

The presentation of $CR_k$ ($\alpha$) in terms of $\alpha$ and k can help the role of the manager (Fig. 6.40).

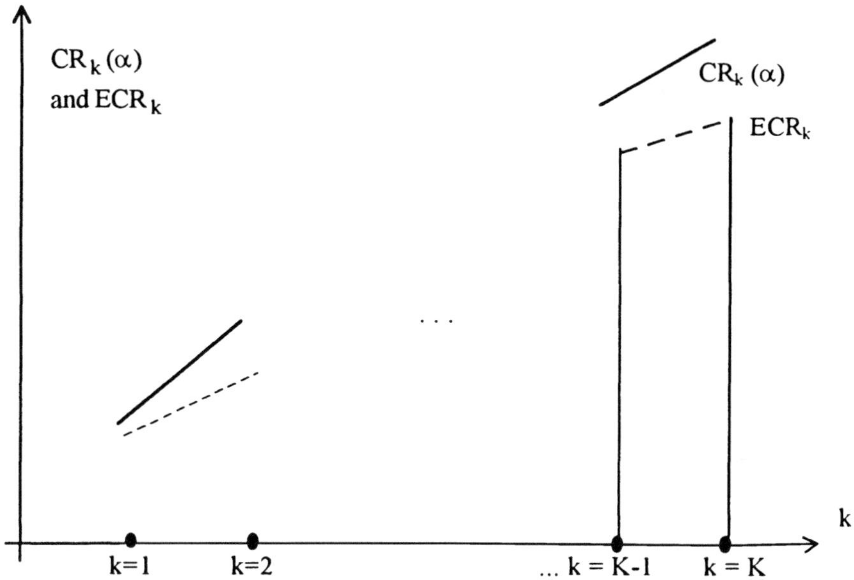


**FIG. 6.40**
**$CR_K(\alpha)$ AND $ECR_K$ IN TERMS OF k**

b) $\gamma_t$, of the time spent until node k exceeding a specified level, t*:

$$\gamma_{t^*} = \int_{t^*}^{\infty} f(t_k)\, dt_k$$

Alternatively, any quantile, $t_k(\alpha)$, can be also estimated

$$1-\alpha = \int_{t_k(\alpha)}^{\infty} f(t_k)\, dt_k$$

and presented in terms of $\alpha$ and k (Fig. 6.41).

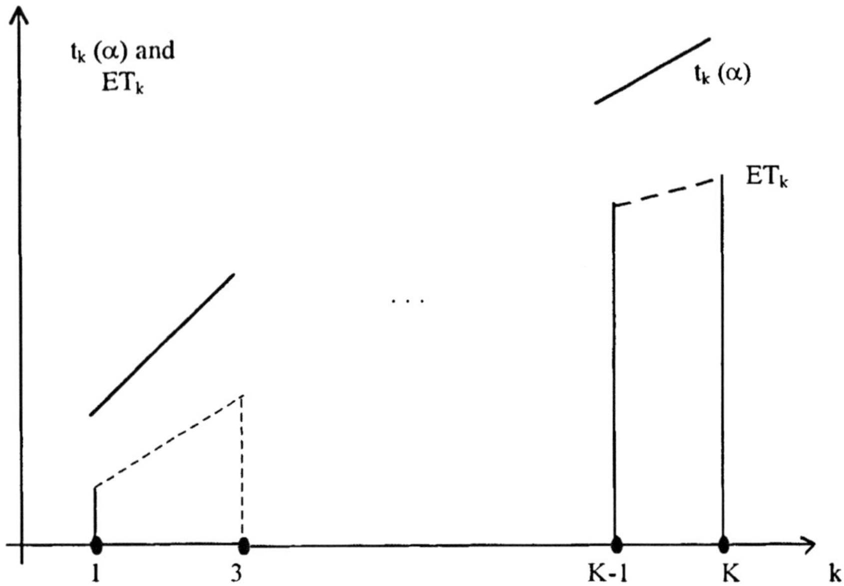


**FIG. 6.41**
**$t_K(\alpha)$ AND $ET_K$ IN TERMS OF k**

c) Risk, $\gamma_{tc}$, of the cost due until a specific time, t*, exceeding a certain level, CR*.

In this case,

$$\gamma_{ct} = \sum_{k=1}^{K} \left[ \int_{CR^*}^{\infty} f(CR_{k-1}) dCR_{k-1} \right] \cdot P(t_{k-1} < t^* \le t_k)$$

where

$$P(t_{k-1} < t^* \le t_k) = \int_0^{t^*} f(t_{k-1}) \cdot \left[ \int_{t^*-t_{k-1}}^{\infty} f(D_k)\, dD_k \right] dt_{k-1}$$

and, alternatively, the α quantile of CR due until t*, $CR(\alpha)_{t^*}$ also can be determined by allowing the presentation of $CR(\alpha)$ t* in terms of time, t* (Fig. 6.42):

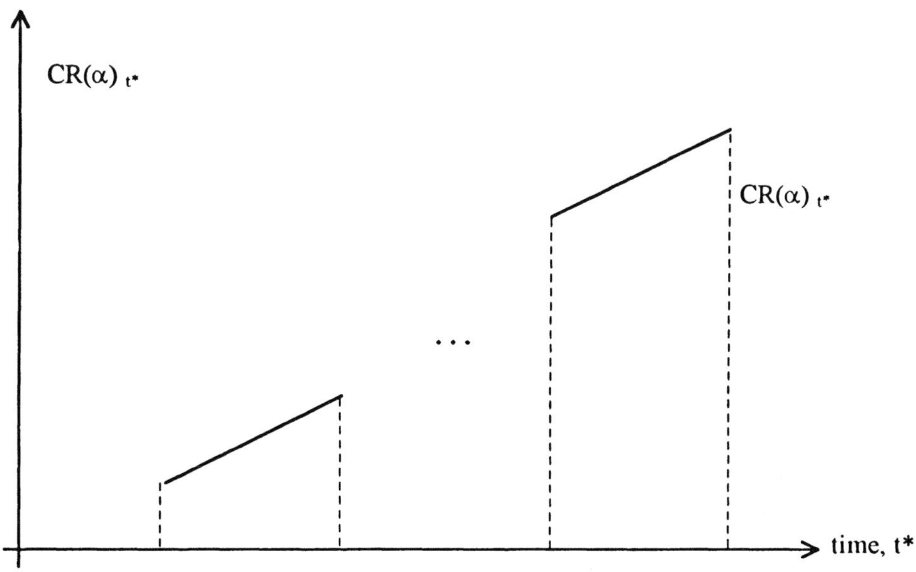


**FIG. 6.42**
**CR(α)t* IN TERMS OF t***

# 7

# THE ASSESSMENT AND EVALUATION OF PROJECTS

## *1. Typology*

The **assessment** of projects is a very important part of project management because the number of alternative candidates is usually much higher than the number of selected projects and because even after selecting a project, there is a wide range of optional alternatives. Actually, this is increasingly true in the last few years mainly for two reasons:

- the globalization of markets offers a wider range of options for the provision of supplies and for the targeting of demand segments.

  This means that a project to be carried out in US or in Europe may find the most convenient suppliers in Asia or in South American because of the progressive liberalization of international trade which is reducing the traditional barriers between countries. The selection of beneficiaries or consumers for a project is also becoming worldwide and very recent examples of important new constructions (like the bridge Vasco da Gama over the river Tagus in Lisbon or the Euro Tunnel between UK and France), new industrial products (like the aircrafts produced by the european consortium AIRBUS and new consumer products (like the famous lines of products marked by american cosmetic brands) illustrate the selection of locations for the demand market very differently from those associated to major suppliers (capital, goods, etc.)

- the technological progress is increasing the spectrum of feasible alternatives to carry out a project. Actually, the development of new materials, of new ingredients, of new digital systems to measure and to control processes as well as of new models to solve logistic problems is also offering more options to implement a project.

Summing up, recent trends are giving more freedom to conceive and to design alternative projects. Therefore, the comparison between alternative projects and between alternative solutions for a selected project is becoming a more crucial issue in Project Management.

Several types of comparisons have to be considered according to different perspectives and calendars:

a) **Project assessment**

The concept of assessment is specific to name the comparisons taking place before starting the project ("ex-ante").

The project assessment should compare the expected benefits and costs but also should give attention to other perspectives, such as social and environmental impacts, risk levels, etc.

b) **Project monitoring**

This type of analysis should be carried out after starting the project and its goal is comparing the actual implementation and the obtained results with the approved decisions and the expected results, respectively.

This comparison is called project monitoring and includes a first stage during the project development (construction of an infra-structure, launching of a new product, etc.), a second one during the operation of the developed system (use of the constructed infra-structure, production of the new product, etc.) and, eventually, a third one concerning the closing of operations (rehabilitation of the infra-structure, shutting down the production unit, etc.).

The monitoring of the first stage is based on a systematic process of comparing quantities and quality specifications between the adopted design and the implemented project. This monitoring requires a good Management Information System.

The monitoring of the second or third stages include the comparison between the expected and the achieved results (benefits and costs, receipts and expenses, etc.) and therefore focus the behavior of the developed system, the generated supply and the captured demand.

All those comparisons are oriented to assess how the plans and designs have been fulfilled or how large have been the gaps between expectation and reality.

### c) **Project evaluation**

This comparison should be carried out in similar terms to the project assessment but after completing the project. The word evaluation is reserved for this "ex-post" analysis.

## 2. Financial assessment and evaluation

Traditionally, the project assessment or evaluation has been done by using trivial financial calculus to estimate financial indicators in terms of the expected flows of income ($B_t$) and expenses ($C_t$) during the adopted horizon, H, such as:

a) Net Present Value, NPV

$NPV = \sum_{t=0}^{H} f^t (B_t - C_t)$ where $f = \frac{1}{1+\alpha}$ is the discounting factor and $\alpha$ is the discounting rate.

b) Relative NPV, RNPV

In this case, one has the NPV per unit of resource allocated to the project,

$$RNPV = \frac{NPV}{\sum_{t=0}^{H} f^t C_t}$$

c) Internal Rate of Return, r

This rate, r, is the maximal value for which the $NPV \geq 0$ with $f = \frac{1}{1+r}$

d) Return Period, $T_r$

Usually, most expenditures take place in the beginning of the project and income is generated just after a certain milestone. The return period, $T_r$, is given by the minimal H, for which $NPV \geq 0$.

These three indicators are supposed to compare the flow of consumed financial resources with the flow of generated resources but they consider different perspectives.

The NPV is an additive balance between these two flows; the internal rate of return measures the maximal rate of interest for borrowing the required financial resources which is competitive with the break even of the project; and the return period indicates the minimal period for getting back the required capital.

Obviously, different decision makers or actors may prefer different indicators. For instance, a bank may be more interested on r because this will be the upper bound of the interest rate which can be asked of this client without causing a cancellation of the project.

An investor may prefer $T_r$ as it means that after $T_r$ he can use his capital again for other purposes or RNPV which measures the profitability of each unit of capital allocated to the project. Finally, the institution (company, agency, etc.) owning the project may prefer to use NPV as it really measures the net value generated by the project.

As was discussed before, these indicators (except r) are quite sensitive to the adopted discounting factor and they tend to be responsible for myopic decisions as any benefit or receipt generated after 20 or 25 years has no significant impact on their values. The use of r or T can be heavily criticized as they measure more the "speed" of the results than their magnitudes, and so they will be responsible for the selection of "fast" projects rather than "good" projects.

Another dimension completely ignored by this financial indicators is the risk associated to a project.

Often, as it was discussed before , one type of risk receiving special importance by the project managers is the probability of having too large a delay and not meeting important milestones. Obviously, the features of a project like the NPV and the risk depend very much on the adopted schedules and so the same project can be studied for different alternative schedules and obtaining different assessments or evaluations. This means that the traditional attitude of evaluating or assessing a project before studying its schedule can be responsible for important errors. The need to consider several alternative schedules for the same project also increases the number of candidates to be studied as well as the importance of developing better models for projects assessment and evaluation.

All this criticism has justified the adoption of a more consistent and integrated approach to assess projects, considering this financial perspective expressed by any of these indicators as well as other perspectives.

This is called the Multi-Criteria Assessment or Evaluation of Projects and it will be studied in the next section.

## 3. Multi-criteria assessment or evaluation

The rational approach suggested by Decisional Sciences to the assessment or evaluation problem is based on the sequence of stages presented in Fig. 7.1 and similar to the sequence presented in Fig. 1.1, Chapter 1.

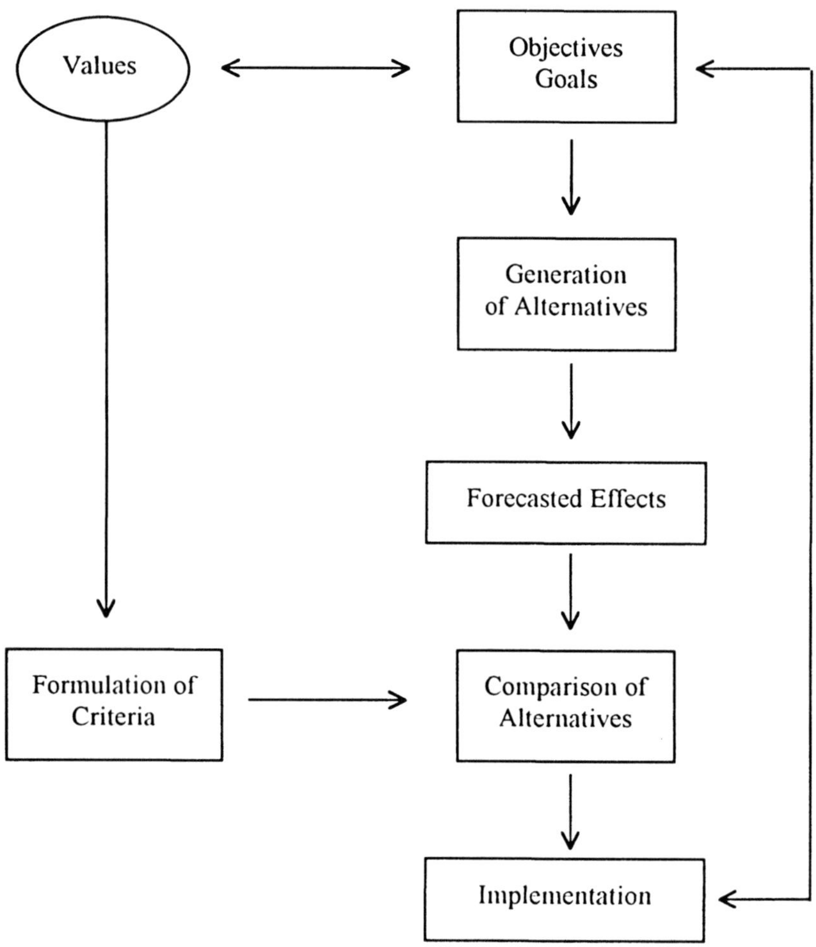


**FIG. 7.1**
**THE COMPARISON OF ALTERNATIVES**

This approach has always given particular attention to:

- the explicit definition of values as they are obviously related to the selection of objectives;

- the generation of alternatives and the estimation of their effects;

- the formulation of criteria based on the adopted values to develop the comparison stage.

Even during the early years of Operational Research, the need to consider multiple criteria has been explicitly mentioned by several authors, for example, (see Kauffman, 1968).

However, the usual approach was to formulate the multiplicity of criteria:

- as a weighted sum to be used as objective function (minimization or maximization)

and, or

- as restrictions to be fulfilled by any feasible solution.

Fortunately, the new methods of Multi Criteria Decision Analysis (See Roy, 1985) offer much more flexible models and systems to support the process of decision making.

The consideration of several criteria can be particularly well adapted to assess or to evaluate projects.

The approach recommended by the author considers a small number of criteria (namely, three) branching out into other clusters of criteria, keeping the number of criteria in each cluster small to avoid overly difficult comparisons. These trees are often called "**value trees**" (Keeney, 1992).

In Figures 7.2 and 7.3, examples are given for the analysis of tenders concerning the development of projects and the acquisition of a piece of new equipment. The concepts of integrated price and duration correspond to the "effort" associated with the project and the quality of the benefits obtained from the project. However, the concept of quality deficit defined as Maximal quality - quality is preferred to the quality to have the same direction of preference as the price or the duration.

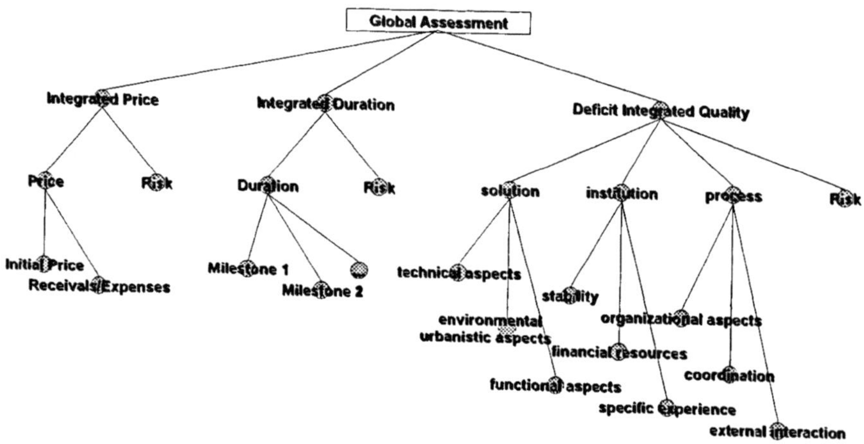


**FIG. 7.2**
**EXAMPLE OF VALUE TREE FOR THE ASSESSMENT OF PROJECTS**

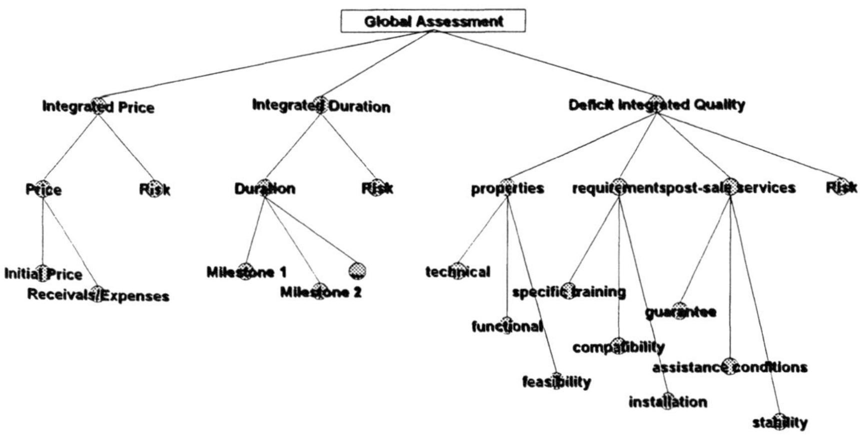


FIG. 7.3
EXAMPLE OF VALUE TREE FOR THE ASSESSMENT OF ALTERNATIVE EQUIPMENTS

In other situations, other criteria can be proposed:

a) for the development of a firm

- short term profitability
- long term prospects
- risk

b) for the choice of a technology

- economic criterion (e. g., cost-effectiveness)
- environment impact
- social-political impact

The concept of risk has to considered in any project assessment and can be formulated in terms of the statistical distribution of the studied magnitude (cost, duration or deficit of quality) and it is defined by the probability, $\alpha$, of exceeding the given value, V, by more than a certain fraction of V, denoted by $\beta V$. A common example for $\beta$ is making $\beta = \frac{1}{3}$. The negative exponential law may be assumed for the exceedence, X, and therefore the corrected expected value can be easily computed:

$$f(X) = \lambda e^{-\lambda X} \text{ with } X \geq 0$$

$$\text{and } P(X > \beta V) = e^{-\lambda \beta V} = \alpha$$

$$\text{or } \lambda = -\frac{\ln \alpha}{\beta V}$$

$$\text{and therefore } E(X) = -\frac{\beta V}{\ln \alpha}.$$

This means that the corrected magnitude will be given by:

$$V\left[1 - \frac{\beta}{\ln \alpha}\right].$$

The use of the expected value assumes a neutral attitude towards risk but a different assumption can be adopted and the corresponding result easily computed.

The development of Multi-Criteria Decision Models has opened new perspectives mainly, by the development of two approaches:

A) **Binary Comparison**

These models are based on binary comparisons between alternatives according to each criterion and adopting an appropriate **Relational System of Preference** (RSP).

This means that instead of mapping each alternative into a metric space with one or more dimensions, this analysis will aim to associate one of the alternative relations belonging to the adopted **RSP** to each pair of alternatives, (a, b).

For instance, adopting the **RSP** called **Quasi-Order**, one has the following alternative relations:

aPb - a is preferable to b
aEb - a is equivalent to b
aIb - a is indifferent to b

The preference relation satisfies the antisimmetrical property:

$aPb \Rightarrow bP'a$

where P′ is the inverse relation of P.

The relation E is a reflexive (aEa) and symmetrical ($aEb \Rightarrow bEa$).

The relations E and P are transitive:

If aEb and bEc then aEc;

The relation I is introduced to represent the comparison between alternatives which are not equivalent but which have differences below the threshold of sensitivity by the decision maker. Obviously, this relation is also symmetric.

However, a major distinction concerns the transitivity property because I is not transitive.

Actually, the relation I should satisfy:

- If aIb then bIa (symmetry)

- If aIb and bIc then aRc can be R = P or E or P′

- If aPb and bIc then aRc can be with R=P or I or E.

This Relational System of Preference can be formally defined by the following conditions:

a) E → symmetrical, reflexive and transitive
b) P → asymmetrical and transitive
c) I → symmetrical and reflexive (but no transitive)
d) $P^2 \cap I^2 = \emptyset$
e) P. I. P = P

The notation (P. I. P) means the relation between a and b if

$$aPc_1 \cap c_1\ I\ c_2 \cap c_2\ Pb$$

The notation $P^2$ means the relation between a and b if aPc and cPb.

Consider the following example: 5 alternatives projects {a, b, c, d, e} are being compared in terms of the profitability criterion. The decision maker then gives the following comparative matrix (Table 7.1).

Three types of analysis now can be performed on this matrix:

| | a | b | c | d | e | f |
|---|---|---|---|---|---|---|
| a | | P′ | P′ | P′ | P′ | P′ |
| b | P | | I | P | P′ | P′ |
| c | P | I | | P | I | I |
| d | P | P′ | P′ | | P′ | P′ |
| e | | P | I | P | | P′ |
| f | P | P | I | P | P | |

**TABLE 7.1**
**AN EXAMPLE OF A BINARY COMPARATIVE MATRIX USING QUASI-ORDER RSP**

a) filling up missing comparisons

For instance, the missing case eRa easily can be determined because ePb and bPa. Therefore, one should have ePa.

b) checking up the consistency of the comparisons. For instance, the case eIa is inconsistent with ePb and bPa.

c) selecting those alternatives which are candidates to the optimal choice. Obviously, the candidates to the optimal decision are those which do not "suffer" a preference relation. In this case, just c and f can be candidates to the optimal choice (Fig. 7.4).

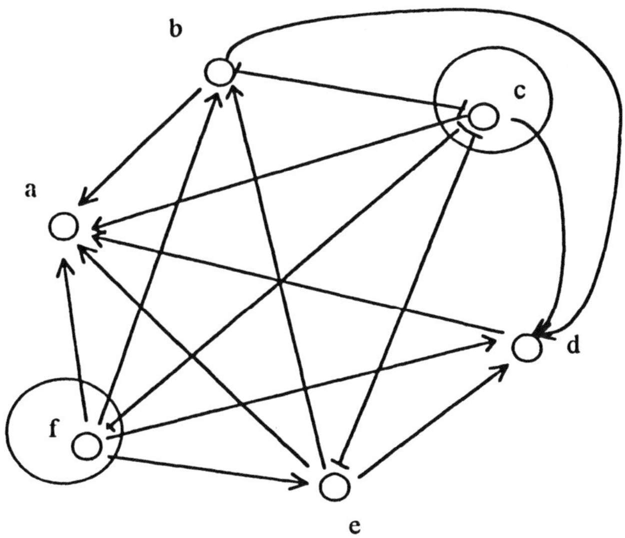


**A graphical representation of the matrix presented in Table 7.1**

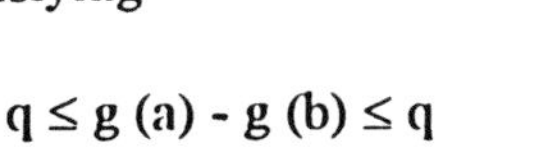


FIG. 7.4

It can be shown that if a Quasi-Order RSP is adopted, then a real function g can be defined for the set of alternatives satisfying

$$aIb \leftrightarrow -q \leq g(a) - g(b) \leq q$$
$$aPb \leftrightarrow g(a) > g(b) + q$$

where q is a non-negative constant called an "**indifference threshold**".

This function, $g(\bullet)$, corresponds to a concept of value defined in terms of q.

The presented analysis can be extended to the case of having multiple criteria and so several comparative matrices have been built.

The construction of a multi-criteria RSP implies the adoption of composition rules to relate the multi-criteria relation (aRb) with the single criteria ones $\left(aR_j b\right)$ where $j = 1,...,N$.

For instance, with N = 3 the following rules can be adopted:

a) aIb if and only if any Rj belongs to $I \cup E$ and where there is at least one j with $R_j = I$.

b) aEb if and only if $R_j = E$ with $j = 1,2,3$;

c) aPb if and only if $R_j$ belongs to $E \cup I \cup P$ being P at least for one j.

B - **Metric Approach**

This approach is based on the construction of real functions representing the values associated to each alternative according to each criterion.

Several models have been proposed to construct the value function for each criterion and to estimate the coefficient representing the relative importance of each criterion (**weight**), allowing the integration of these multiple value functions into a multi-criterion value function.

For each alternative i, the **value function** concerning each criterion, j, $v_{ij}$, is usually defined in terms of a magnitude, $m_{ij}$, measuring for each alternative i, a relevant property expressing such criterion (cost, time, etc.).

These value functions should have the same domain and scale. Graphical representations or numerical transformations can help the decision maker to construct $v_{ij}$.

In Figs. 7.5 and 7.6, two usual shapes are presented and the following linear transformation is also a function often adopted to obtain $v_{ij}$:

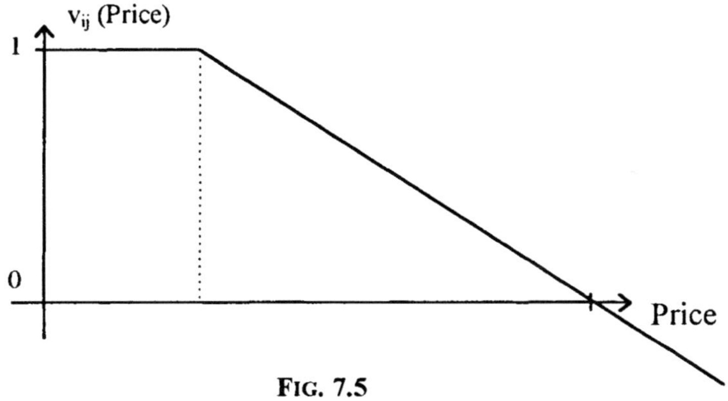


**FIG. 7.5**
**FIRST EXAMPLE OF VALUE FUNCTION**

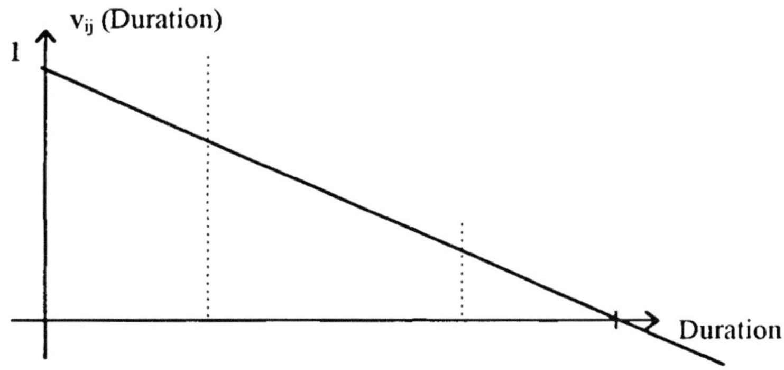


**FIG. 7.6**
**SECOND EXAMPLE OF VALUE FUNCTION**

$$v_{ij} = \frac{\max_i(m_{ij}) - m_{ij}}{\max_i(m_{ij}) - \min_i(m_{ij})} \text{ or } v_{ij} = \frac{m_{ij} - \min_i(m_{ij})}{\max_i(m_{ij}) - \min_i(m_{ij})}$$

if the preference decreases or increases with $m_{ij}$, respectively, and assuming that a maximal and a minimal bounds can be define to describe the domain of choice.

The aggregation of the value functions can be studied more easily if the linear assumption is adopted and if the number of criteria does not exceed three.

Then, one has:

$$v_i = \sum_{j=1}^{3} \lambda_j v_j \text{ with } 0 \le \lambda_j \le 1 \text{ and } \sum_{v=1}^{3} \lambda_j = 1$$

where $\lambda_j$ is the weight for j = 1, 2, 3.

Obviously, the meaning of any weight is defining a trade-off. For instance, if $v_{i3}$ is kept constant, the variation $dv_{i2}$ which is required to compensate $dv_{i1}$ can be determined easily.

Then,

$$dv_i = 0 = \lambda_1 \, dv_{i1} + \lambda_2 \, dv_{i2}$$

or

$$dv_{i2} = -\frac{\lambda_1}{\lambda_2} dv_{i1}$$

and so

$$dv_{i2} = -\frac{\lambda_1}{\lambda_2} \text{ if } dv_{i1} = 1.$$

The selection of $\{\lambda_1, \lambda_2, \lambda_3\}$ corresponds to the selection of a point within the triangular space $\{0,0;0,1;1,0\}$ as is shown in Fig. 7.7.

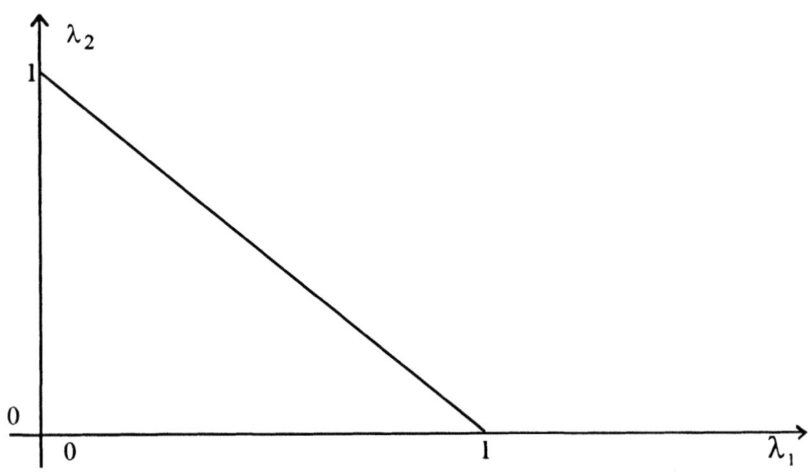


FIG. 7.7
THE WEIGHTS SPACE

The model **TRIDENT** (Tavares, 1984) can help select such a point by developing the following sensitivity analyses:

a) two alternatives, i, i', have the same utility if $v_i = v_i'$ and this is the equation of a straight line separating the sub-space where i⊐i' (i is preferable to i') and the sub-space where i'⊐i;

b) a set of $\frac{N(N-1)}{2}$ lines can be produced where N is the number of alternatives;

c) each sub-space not crossed by any line is the domain of a specific ranking of alternatives.

This model automatically produces these domains and the winning alternative for each sub-set of domains.

Let be given a simple example to compare five alternatives in terms of three criteria ($v_{ij} \leq i$):

| | j = 1 | j = 2 | j = 3 |
|---|---|---|---|
| $A_1$ | 1 | 2/3 | 0 |
| $A_2$ | 0 | 1 | 1/2 |
| $A_3$ | 1/3 | 1/3 | 1 |
| $A_4$ | 2/3 | 0 | 4/5 |
| $A_5$ | 2/3 | 1/2 | 4/5 |

Obviously, $A_4$ is dominated by $A_5$ and therefore it will be excluded from the analysis.

The weighted average for i = 1, ..., 5, $v_i$, will be then given by:

$$v_1 = \lambda_1 + \frac{2}{3}\lambda_2$$

$$v_2 = \lambda_2 + \frac{1}{2}(1 - \lambda_1 - \lambda_2)$$

$$v_3 = \frac{1}{3}\lambda_1 + \frac{1}{3}\lambda_2 + 1(1 - \lambda_1 - \lambda_2)$$

$$v_5 = \frac{2}{3}\lambda_1 + \frac{1}{2}\lambda_2 + \frac{4}{5}(1 - \lambda_1 - \lambda_2)$$

The lines defining the equality $v_i = v_k$ with i, k = 1, ..., 5 and i ≠ k are presented in Figure 7.8. The sub - domains for each winning alternative are presented in Figure 7.9.

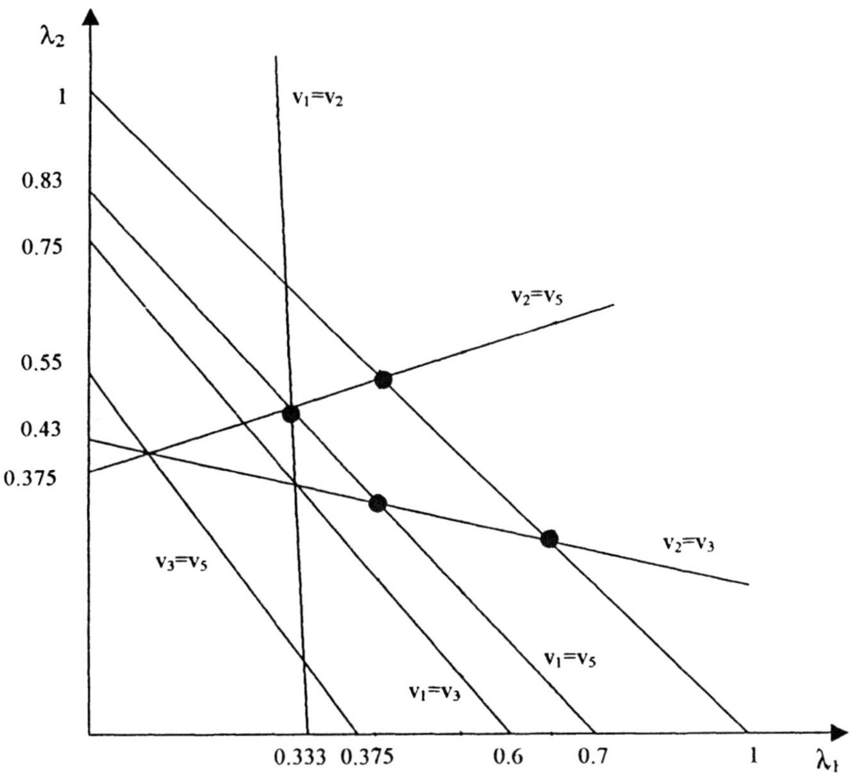


**Fig. 7.8**
**The TRIDENT model**

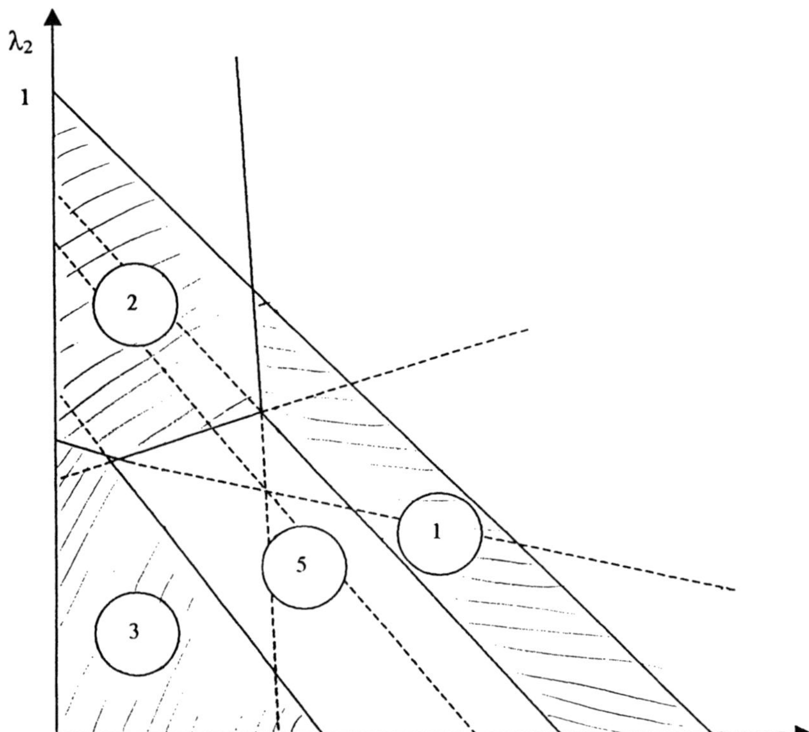


**FIG. 7.9**
**THE SUB - DOMAINS FOR THE WINNING ALTERNATIVE**

These outputs help the decision maker to assess how robust each ranking is and the corresponding choice of the winning alternative.

## 4. The MACMODEL

A model was developed to assess and to evaluate alternative projects or solutions for any specific project: MACMODEL.

This model incorporates the presented analyses and it includes five blocks (Fig. 7.10) which are illustrated in the following figures:

**A - Construction of a value-tree.**

The basic values-trees for tenders and for acquisitions already have been presented but the user is given the possibility of building up a new one.

**B - Construction of the value functions and selection of their weights**

The user can adopt two alternatives approaches:

- building a value function, $v_{ij}$, for each criterion j using a graphical display which allows the construction of a linear piece-wise function (Fig. 7.11)

- adopting the measure, $m_{ij}$, of each criterion (price, time, quality) as its value function, $v_{ij}$.

The weights for the criteria also have to be given.

**C - Introduction of data about alternatives**

This data includes their features about the adopted criteria (duration, milestones, proce, quality, risk, etc.).

**D - Comparative analysis**

The ranking of alternatives is produced as a table (Fig. 7.12) and as a graphical display (Fig. 7.13).

**E - Sensitivity Analyses**

This block includes three different types of sensitivity analyses:

a) An one - dimension analysis on weights

The thresholds for the weights allowing any alternative i coming first are computed assuming that the ratio between the two weights associated to criteria where i is better (or worse) than the winning alternative keeps constant. An example is given in Fig. 7.14.

b) An one-dimension analyses on $v_{ij}$

This analysis is similar to the previous one but the thresholds are computed for $v_{ij}$ rather than for $\lambda_j$ (Fig. 7.15).

c) A two-dimension analysis on weights

This is the TRIDENT analysis. The studied example is shown in Fig. 7.16.

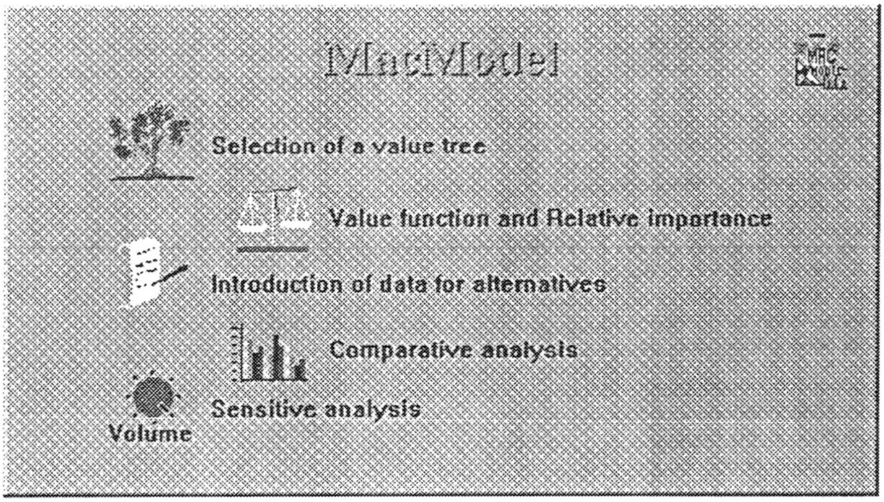


FIG. 7.10
GENERAL MENU OF MACMODEL

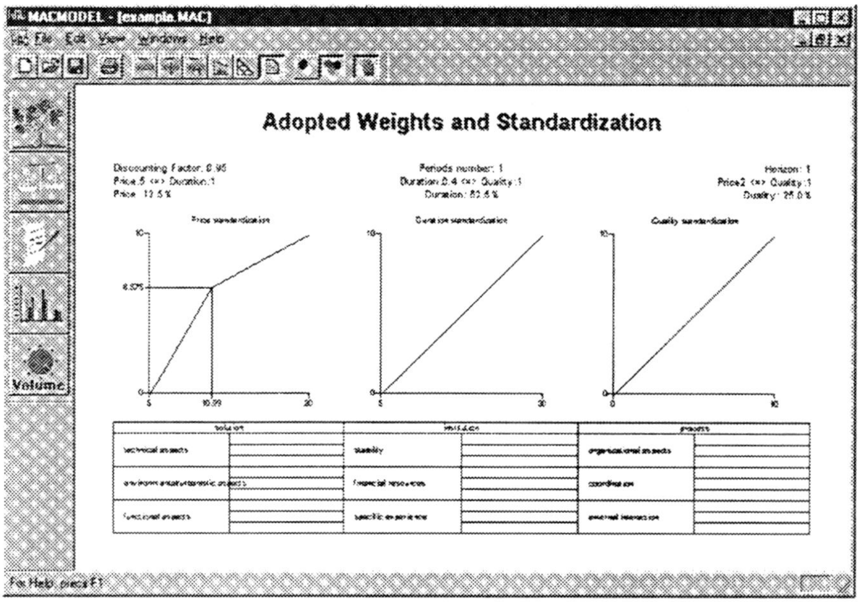


**FIG. 7.11**

**CONSTRUCTION OF A VALUE FUNCTION FOR EACH CRITERION**

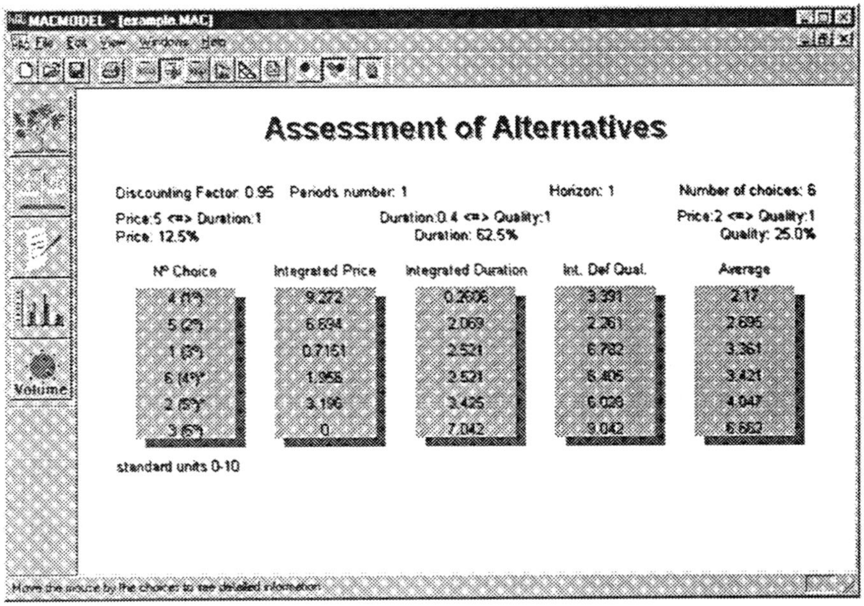


**FIG. 7.12**
**RANKING OF ALTERNATIVES**

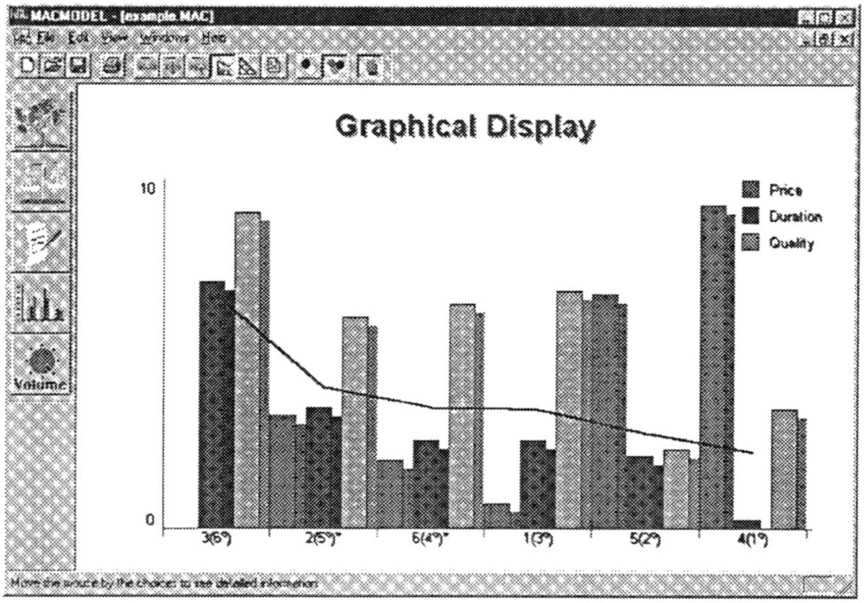


**FIG. 7.13**
**GRAPHICAL COMPARISONS OF ALTERNATIVES**

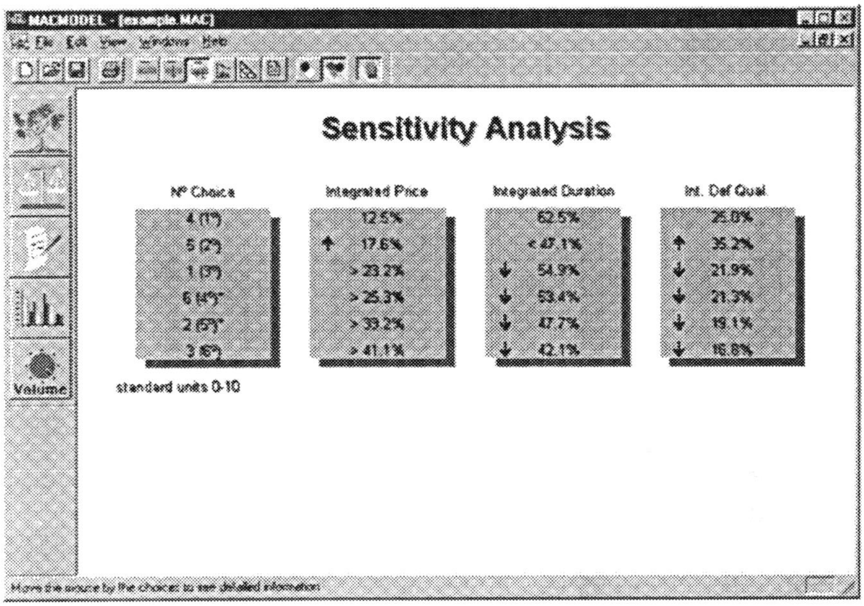


**FIG. 7.14**

**ONE-DIMENSION SENSITIVITY ANALYSIS ON WEIGHTS**

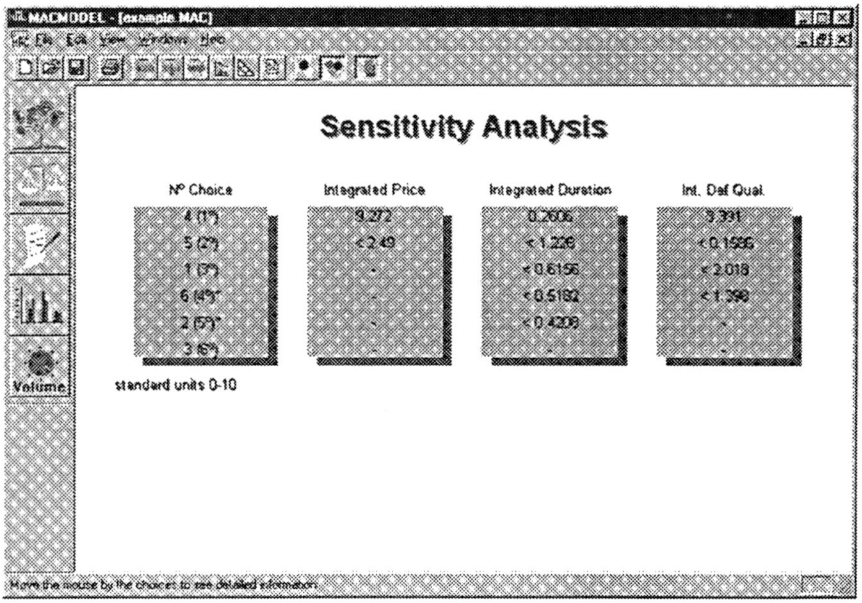


**FIG. 7.15**
**ONE-DIMENSION SENSITIVITY ANALYSIS ON $V_{ij}$**

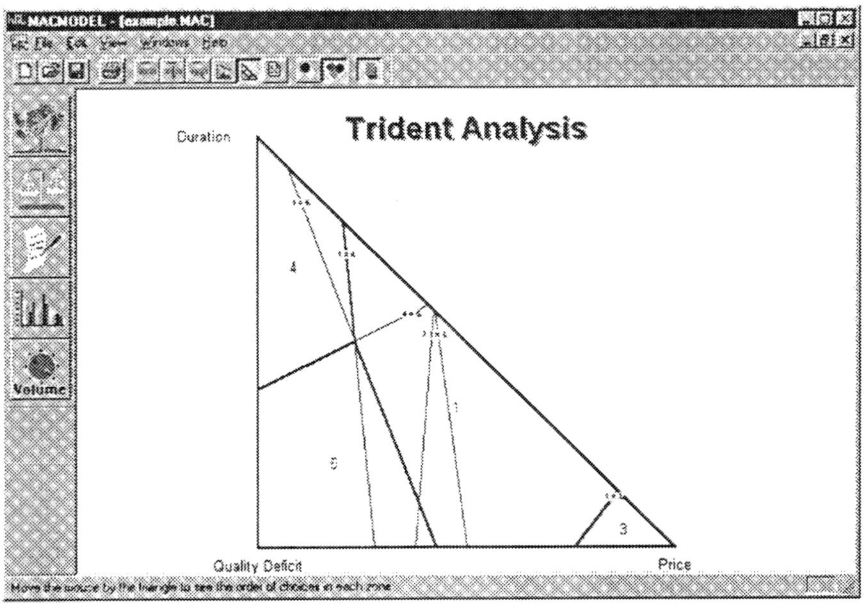


FIG. 7.16
THE TRIDENT ANALYSIS

# 8

# THE OPTIMAL SCHEDULING OF A PROJECT IN TERMS OF ITS DURATION

## *1. The synthesis of decision variables*

In the previous section, the scheduling of a projects was studied for several types of restrictions and objective functions. The scheduling decision was defined in terms of the set of starting times of the activities, $t_i^s$, with i=1, ..., N and therefore, for any reasonably large project, this is a very large set of decision variables.

Therefore, solving this optimization problem is a complex task not allowing the deduction of any general conclusions but, at most, the computation of a numerical solution for each specific problem.

However, such conclusions might be derived if a synthetic description of such decisions set will be built.

The selection of a starting time corresponds always to a decision about how much of the available **activity's float** is going to be used because:

$$t_i^s(E) \leq t_i^s \leq t_i^s(L)$$

or

$$t_i^s = t_i^s(E) + \alpha\left[t_i^s(L) - t_i^s(E)\right]$$

with $0 \leq \alpha \leq 1$

where $\alpha$ expresses such decision. The adoption of a very low $\alpha$ means that an earlier schedule is being produced or that such float will be used for activities coming later on and a higher $\alpha$ means precisely the opposite choice.

Therefore, the selection of $\alpha$ to synthetize the large number of starting times seems a very intuitive choice although it is hardly mentioned in the litterature.

However, a very basic question has to be answered:

Can we adopt the same $\alpha$ for all the activities?

Obviously, if the answer is negative, there would be no significant gain from using $\alpha$ as major decision variable because each activity would require a specific $\alpha$, $\alpha_i$. Fortunately, the answer is yes as is proved in the next section (see Tavares, Ferreira and Coelho, 1998).

As was discussed in the last chapter, the assessment and the evaluation of a project should consider more than one criteria, and quite often there is a crucial compromise to be achieved between too early a schedule, which would reduce the risk of not meeting a milestone, but it is responsible for a higher discounted cost and too late a schedule, which is better for the discounted cost, but it would increase such risk. Therefore, this means that the two presented objectives, Discounted Cost and Risk of Delay, can be selected as major criteria to be optimized in terms of $\alpha$ with $0 \leq \alpha \leq 1$.

The optimization study will adopt a project network model similar to that one already presented in Chapter 5: the duration of each activity is lognormally distributed and its cost is related to the duration by a linear and gaussian regression. The following sections are based on the quoted paper by Tavares, Ferreira and Coelho (1998).

## 2. A theorem on the propagation of floats

The proposed synthesis is described by the factor $\alpha$ (**float factor**) with:

$$t_i^s(\alpha) = t_i^s(E) + \alpha.\Delta_i$$

where $0 \leq \alpha \leq 1$

The latest (earliest) schedule is obtained with $\alpha$=1 ($\alpha$=0). In order to use $t_i^s(\alpha)$ with the same $\alpha$ for the whole network it is necessary to prove the finishing time of i assuming $t_i^s(\alpha)$ is not incompatible with adopting $t_k^s(\alpha)$ for any activity $k \in K(i)$ where K(i) is the set of immediate successors of i, as usual.

Therefore the following theorem should be proved:

**Theorem on the propagation of floats:**

The finishing time of i assuming that i starts at $t_i^s(\alpha)$ is not greater than the starting time of any k using the same $\alpha$, $t_k^s(\alpha)$, with $k \in K(i)$.

This means that:

$$t_i^s(\alpha) + D(i) \leq t_k^s(\alpha) \text{ with } k \in K(i) \text{ and } 0 \leq \alpha \leq 1.$$

Proof:

The earliest start time of any activity k, with $k \in K(i)$, $t_k^s(E)$ is given by $t_k^s(E) = t_i^s(E) + D(i) + \delta_i$ with $\delta_i \geq 0$ in order to account for other paths converging at the starting node of k.

The latest start time of i, $t_i^s(L)$ is given by $t_i^s(L) = \left[\min_{k \in K(i)} t_k^s(L)\right] - D(i)$ and the finishing time of i adopting $t_i^s(\alpha)$ is given by $\left[t_i^s(\alpha) + D(i)\right]$.

Then:

$$t_k^s(\alpha) = t_k^s(E) + \alpha\left[t_k^s(L) - t_k^s(E)\right]$$
$$= (1-\alpha)\, t_k^s(E) + \alpha\, t_k^s(L)$$
$$= (1-\alpha)\left[t_i^s(E) + D(i) + \delta_i\right] + \alpha\, t_k^s\,(L)$$
$$= (1-\alpha).\, t_i^s(E) + (1-\alpha)\, D(i) + \alpha\, t_k^s\,(L) + (1-\alpha)\delta_i$$

and

$$t_i^s(\alpha) + D(i) = t_i^s(E) + \alpha\left[t_i^s(L) - t_i^s(E)\right] + D(i)$$
$$= (1-\alpha)\, t_i^s(E) + \alpha\, t_i^s(L) + D(i)$$
$$= (1-\alpha)\, t_i^s(E) + \alpha\left[\min_{k\in K(i)} t_k^s(L) - D(i)\right] + D(i)$$
$$= (1-\alpha)\, t_i^s(E) + (1-\alpha)\, D(i) + \alpha \min_{k\in K(i)} t_k^s(L).$$

Finally, $t_i^s(\alpha) + D(i) \le t_k^s(\alpha)$ because

$$\alpha. \min_{k\in K(i)} t_k^s(L) \le \alpha\, t_k^s(L) + (1-\alpha)\,\delta_i.$$

The deduced result allows the project manager to select the same float factor $\alpha$ for the whole set of activities.

## *3. The developed stochastic model of the project network*

The adopted model assumes that the duration of each activity is lognormally distributed because of the reasons already discussed (Chapter 5).

The cost of each activity is assumed to be linearly related to the duration by a gaussian linear regression and the correlation coefficient, $\rho$, is given.

The project manager has to make a decision about the starting time of each activity and such time can be expressed in terms of $\alpha$ multiplied by the float of $i, \Delta_i$ determined by assuming that the durations of the activities are equal to their expected values. These scheduled starting times are denoted by $t_j^s(\alpha)$ and given by: $t_j^s(\alpha) = t_j^s(E) + \alpha . \Delta_j$.

Therefore, a simulation model easily can generate a random set of realizations (k = 1,..., K) of the studied project network defining each realization by a duration and a cost for each activity $\left[D_i(k);\ C_j(k)\right]$ with i = 1,..., N.

After selecting the set of values $\left\{t_j^s(\alpha) \text{ with } j = 1,\ldots, N\right\}$, the project manager can start the project and the real set of durations and costs $\left\{D_j(k),\ C_j(k), \text{with } j = 1,\ldots, N\right\}$ will be considered as a realization of the adopted model.

The occurred durations $\left\{D_j(k),\ j = 1,\ldots,N \text{ and } k = 1,\ldots,K\right\}$, imply that for each j and k, one may have $t_i^s(\alpha)$ with $j \in J(i)$ not feasible for one or more activities. Actually, the earliest starting time of any activity i for such a set of occurred durations in realization k can be determined by:

$t_i^s(E)_k = \max_{j \in J(i)} \left\{t_j^s(\alpha)_k + D_j(k)\right\}$ where J (i) just includes all the set of activities directly precedent to i and being $t_j^s(\alpha)_k$ the real starting time of j which is given by:

$t_j^s(\alpha)_k = t_j^s(E)_k$ if $t_j^s(E)_k > t_j^s(\alpha)$

and

$t_j^s(\alpha)_k = t_j^s(\alpha)$ if $t_j^s(E)_k \leq t_j^s(\alpha)$

Thus, a progressive iterative procedure starting with the activities, i, having an empty J (i), will determine the whole set of $\left\{t_i^s(\alpha)_k,\ i = 1,\ldots,N \text{ and } k = 1,\ldots,K\right\}$.

Therefore, for each realization k, the total duration and the discounted cost can be computed.

The support to the process of decision making concerning the selection of $\alpha$ can be developed then in terms of the risk estimated from the K generated realizations of the network being K a sufficiently large number. This **risk function**, R, is the probability of falling outside a target domain defined for the total duration and the discounted cost.

For each target domain, the optimal $\alpha$ is its value minimizing R.

The study of the total duration is presented in terms of

$$\tau_k = \frac{T_k}{T_0}$$

where $T_k$ is the total duration correspondent to the realization k and $T_0$ is the total deterministic network duration assuming that the duration of each activity is equal to its mean.

The analysis of the discounted cost can be made for each k, in terms of the Present Cost, PC, which is given by:

$$PC_k = \sum_i C_i(k).f^{t_i^s(\alpha)k}$$

where f is the discount factor and assuming that the cost of each activity, $C_i$ (k), is allocated to its starting time, $t_i^s(\alpha)_k$. Then, PC will be studied in relative terms by:

$$v_k = \frac{\sum C_i(k) f^{t_i^s(\alpha)k}}{\sum_i \mu_{(C_i)} f^{t_i^s(\alpha_0)}}$$

where $\mu_{(C_i)}$ is the average of $C_i(k)$, and $\alpha_0$ is the reference value adopted for $\alpha$ (usually, $\alpha_0 = 0$) .

Therefore, the target domain will be defined by an upper bound for $\tau$, $L_\tau$, and an upper bound for v, $L_v$.

This model is implemented as a decision aid to help the project manager to select the most appropriate $\alpha$ in terms of $\tau$ and v, using the structure presented in Fig. 8.1

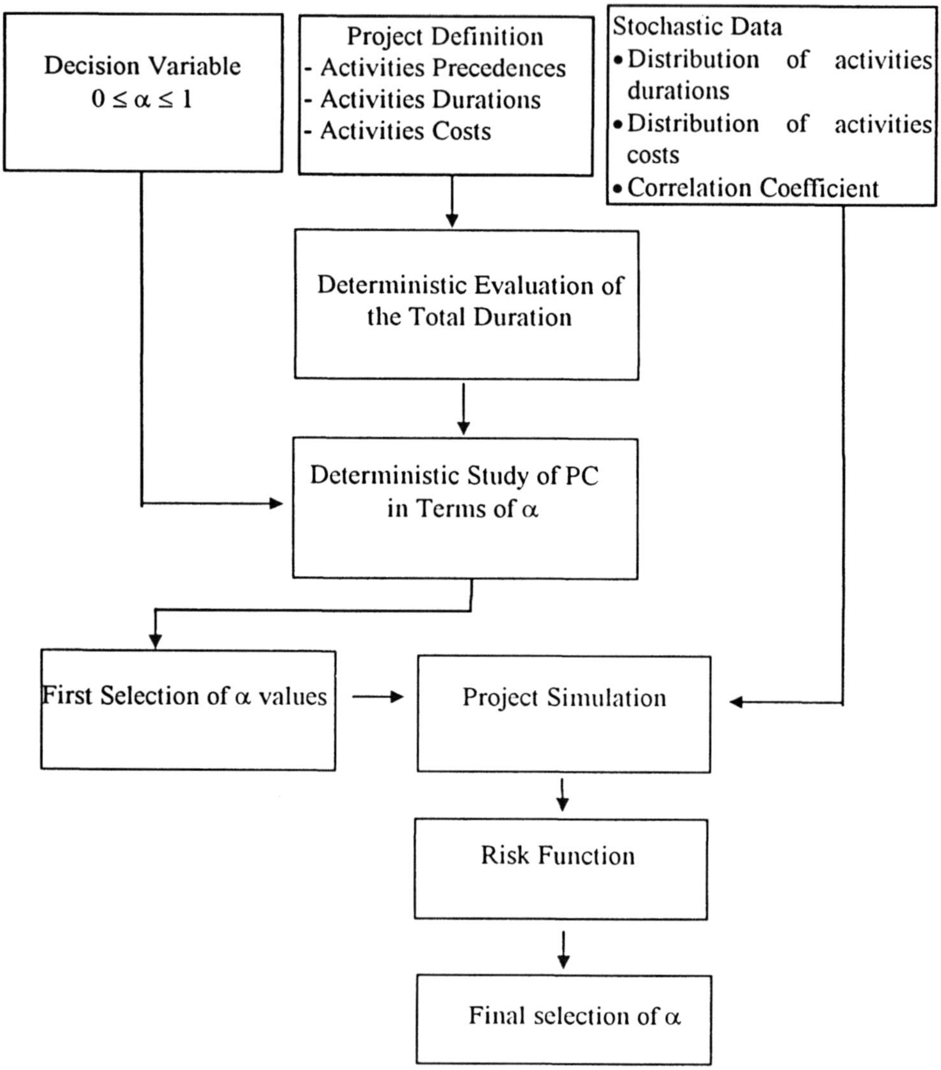


**FIGURE 8.1**
**STRUCTURE OF THE DECISION-AID MODEL**

## 4. The results

The developed model was applied to an illustrative project network with the data presented in Figure 8.2 and in Table 8.1 (Case A). The parameter ρ is assumed to be equal to 0.7.

**Table 8.1**
**Case A**

| | Duration | | | | | Cost | |
|---|---|---|---|---|---|---|---|
| Activity | Mode | μ | Lower Limit | σ | γ (skewness) | μ | σ |
| 1 | 58.26 | 67.00 | 50.66 | 3.65 | 2.99 | 151.00 | 7.07 |
| 2 | 59.13 | 75.00 | 51.42 | 4.98 | 4.32 | 34.00 | 4.47 |
| 3 | 129.57 | 149.00 | 112.67 | 5.44 | 2.99 | 219.00 | 5.92 |
| 4 | 81.74 | 94.00 | 71.08 | 4.32 | 2.99 | 126.00 | 5.48 |
| 5 | 41.74 | 48.00 | 36.29 | 3.09 | 2.99 | 220.00 | 9.49 |
| 6 | 69.57 | 86.00 | 60.49 | 5.04 | 3.97 | 97.00 | 8.37 |
| 7 | 8.70 | 10.00 | 7.56 | 1.41 | 2.98 | 34.00 | 3.87 |
| 8 | 20.87 | 24.00 | 18.15 | 2.19 | 2.99 | 103.00 | 6.32 |
| 9 | 43.48 | 50.00 | 37.81 | 3.15 | 2.99 | 170.00 | 7.94 |
| 10 | 41.74 | 54.00 | 36.29 | 4.40 | 4.58 | 51.00 | 5.57 |
| 11 | 6.09 | 7.00 | 5.29 | 1.18 | 2.97 | 20.00 | 3.16 |
| 12 | 32.17 | 37.00 | 27.98 | 2.71 | 3.00 | 207.00 | 8.94 |
| 13 | 77.39 | 89.00 | 67.30 | 4.21 | 2.99 | 154.00 | 10.00 |
| 14 | 67.83 | 85.00 | 58.98 | 5.17 | 4.16 | 194.00 | 10.95 |
| 15 | 56.52 | 65.00 | 49.15 | 3.60 | 2.99 | 292.00 | 10.00 |
| 16 | 39.13 | 45.00 | 34.03 | 2.99 | 2.99 | 69.00 | 8.94 |

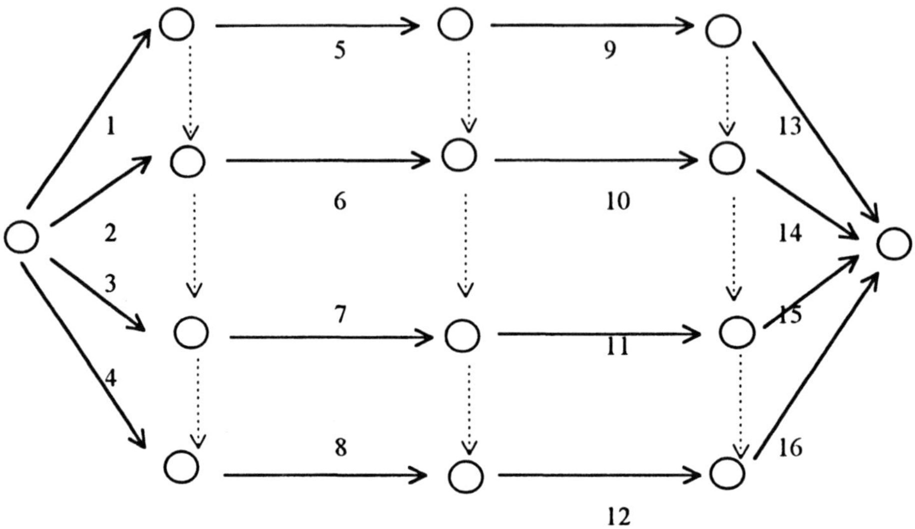


**FIG. 8.2**
**THE NETWORK UNDER STUDY USING AOA**
**(ACTIVITIES I=1, ..., 16)**

Alternatively, a network with a higher uncertainty is also studied (Case B) multiplying the previous standard deviation of the duration and of the cost by a factor of 2 (Table 8.2).

**Table 8.2**
**Case B**

| | Duration | | | | | Cost | |
|---|---|---|---|---|---|---|---|
| Activity | Mode | μ | Lower Limit | σ | γ | μ | σ |
| 1 | 51.07 | 67.00 | 50.66 | 7.31 | 44.64 | 151.00 | 14.14 |
| 2 | 51.71 | 75.00 | 51.42 | 9.97 | 87.54 | 34.00 | 8.94 |
| 3 | 113.58 | 149.00 | 112.67 | 10.90 | 44.71 | 219.00 | 11.83 |
| 4 | 71.66 | 94.00 | 71.08 | 8.64 | 44.29 | 126.00 | 10.95 |
| 5 | 36.58 | 48.00 | 36.29 | 6.20 | 45.19 | 220.00 | 18.97 |
| 6 | 60.86 | 86.00 | 60.49 | 10.07 | 74.82 | 97.00 | 16.73 |
| 7 | 7.62 | 10.00 | 7.56 | 2.83 | 45.49 | 34.00 | 7.75 |
| 8 | 18.30 | 24.00 | 18.15 | 4.35 | 43.75 | 103.00 | 12.65 |
| 9 | 38.12 | 50.00 | 37.81 | 6.29 | 44.08 | 170.00 | 15.87 |
| 10 | 36.49 | 54.00 | 36.29 | 8.77 | 94.98 | 51.00 | 11.14 |
| 11 | 5.33 | 7.00 | 5.29 | 2.39 | 47.67 | 20.00 | 6.32 |
| 12 | 28.21 | 37.00 | 27.98 | 5.41 | 43.97 | 207.00 | 17.89 |
| 13 | 67.85 | 89.00 | 67.30 | 8.40 | 44.22 | 154.00 | 20.00 |
| 14 | 59.33 | 85.00 | 58.98 | 10.31 | 80.38 | 194.00 | 21.91 |
| 15 | 49.55 | 65.00 | 49.15 | 7.19 | 44.40 | 292.00 | 20.00 |
| 16 | 34.31 | 45.00 | 34.03 | 5.97 | 43.93 | 69.00 | 17.89 |

The major decision variable is $\alpha$ which varies between 0 and 1 and the risk function, R, is studied in terms of the limits of the target domain ($L_v$, $L_\tau$), as it was previously presented.

The study of this network is initially performed assuming deterministic data equal to the means of durations and costs. The result for the start and finish times are presented in Table 8.3. The calendar is presented in Figure 8.3 and 8.4 for $\alpha = 0$ and $\alpha = 1$, obtaining a total duration of 300 units.

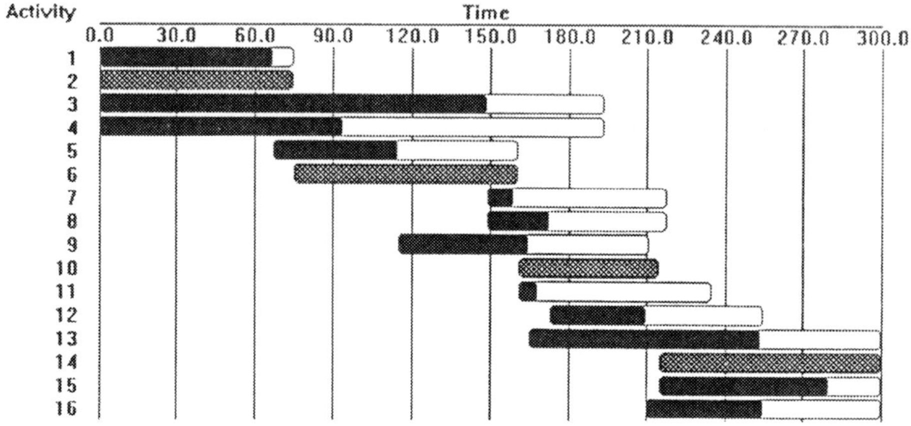


FIGURE 8.3
CALENDAR FOR $\alpha = 0$

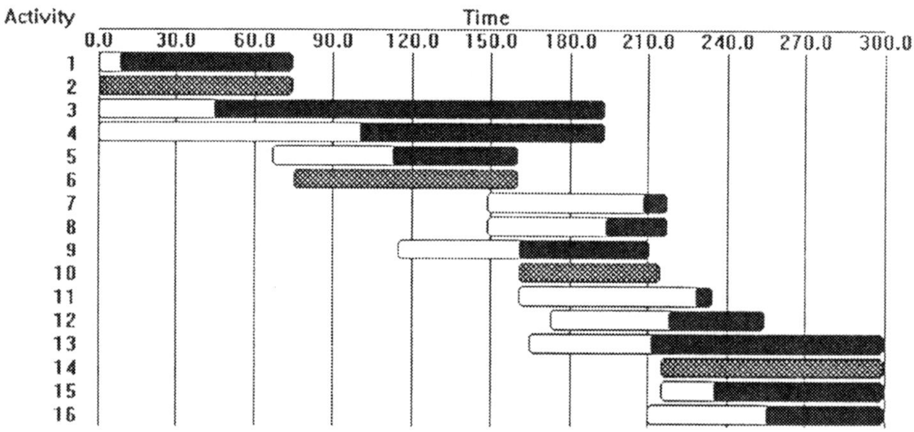


FIGURE 8.4
CALENDAR FOR $\alpha = 1$

TABLE 8.3
SCHEDULE FOR THE STUDIED NETWORK

| Activity | Precedences | ESTime | EFTime | LSTime | LFTime | Start Time | Finish Time |
|---|---|---|---|---|---|---|---|
| 1 | - | 0.00 | 67.00 | 8.00 | 75.00 | 0.00 | 67.00 |
| 2 | - | 0.00 | 75.00 | 0.00 | 75.00 | 0.00 | 75.00 |
| 3 | - | 0.00 | 149.00 | 45.00 | 194.00 | 0.00 | 149.00 |
| 4 | - | 0.00 | 94.00 | 100.00 | 194.00 | 0.00 | 94.00 |
| 5 | 1 | 67.00 | 115.00 | 113.00 | 161.00 | 67.00 | 115.00 |
| 6 | 1; 2 | 75.00 | 161.00 | 75.00 | 161.00 | 75.00 | 161.00 |
| 7 | 2: 3 | 149.00 | 159.00 | 208.00 | 218.00 | 149.00 | 159.00 |
| 8 | 3; 4 | 149.00 | 173.00 | 194.00 | 218.00 | 149.00 | 173.00 |
| 9 | 5 | 115.00 | 165.00 | 161.00 | 211.00 | 115.00 | 165.00 |
| 10 | 5; 6 | 161.00 | 215.00 | 161.00 | 215.00 | 161.00 | 215.00 |
| 11 | 6; 7 | 161.00 | 168.00 | 228.00 | 235.00 | 161.00 | 168.00 |
| 12 | 7; 8 | 173.00 | 210.00 | 218.00 | 255.00 | 173.00 | 210.00 |
| 13 | 9 | 165.00 | 254.00 | 211.00 | 300.00 | 165.00 | 254.00 |
| 14 | 9; 10 | 215.00 | 300.00 | 215.00 | 300.00 | 215.00 | 300.00 |
| 15 | 10; 11 | 215.00 | 280.00 | 235.00 | 300.00 | 215.00 | 280.00 |
| 16 | 11; 12 | 210.00 | 255.00 | 255.00 | 300.00 | 210.00 | 255.00 |

The stochastic simulation is carried out using the presented model. A sample of 10 x 1000 realizations is generated and, for Case A and Case B, the estimated distribution of $\tau$ and $v$ (where $p[\tau]$ or $p[v]$ denotes the estimated probability of occurrence of $\tau$ or $v$ within the corresponding interval) is presented in Figures 8.5 to 8.10 for $\alpha = 0$; $\alpha = 0.5$ and $\alpha = 1.0$.

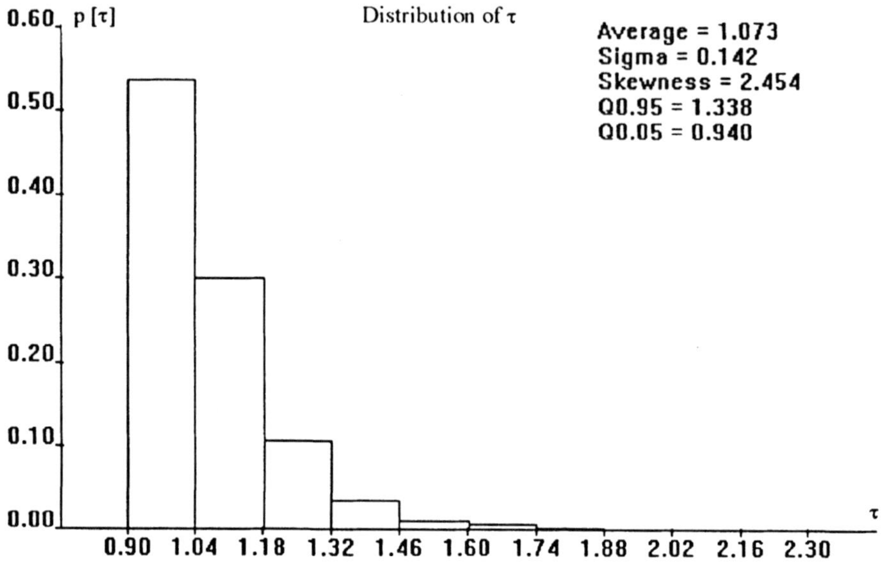


**Figure 8.5 A - Distribution of τ for Case A (α = 0)**

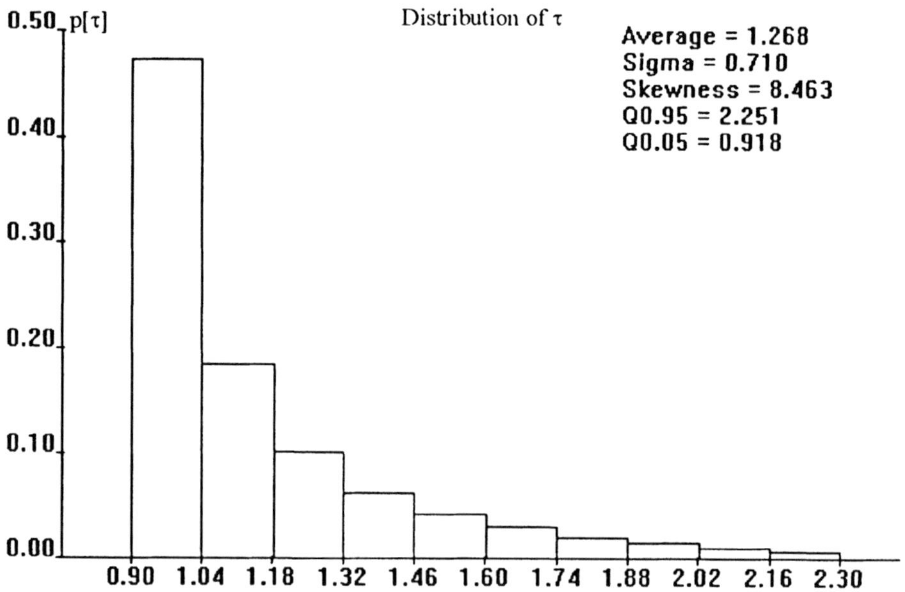


**Figure 8.5B- Distribution of τ for Case B (α = 0)**

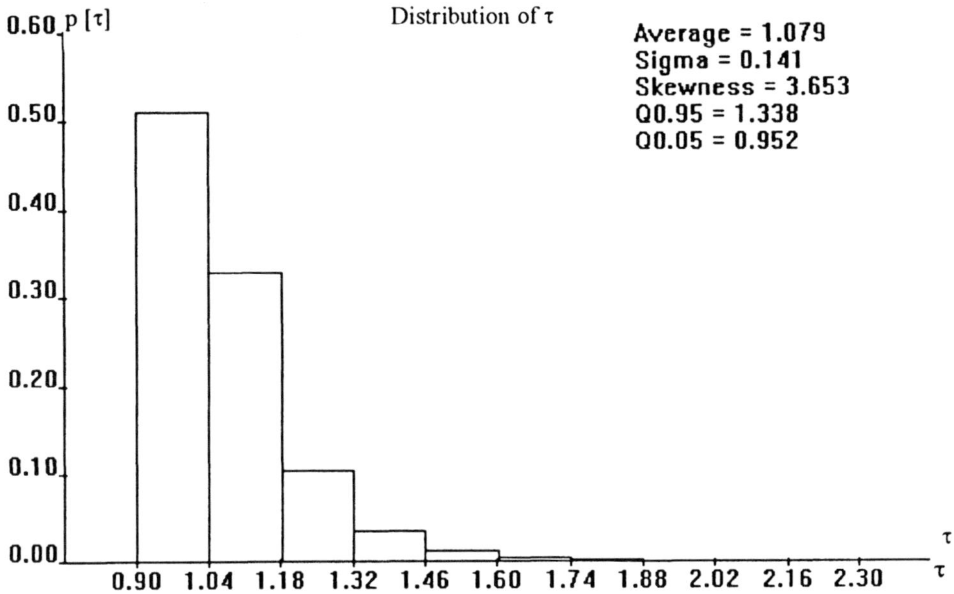


**Figure 8.6 A - Distribution of τ for Case A (α = 0.5)**

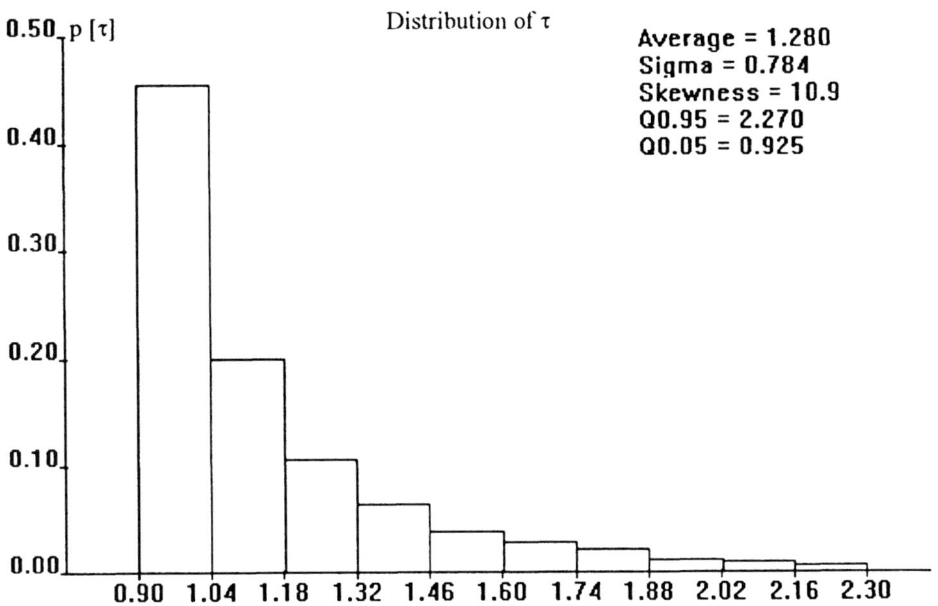


**Figure 8.6 B- Distribution of τ for Case B (α = 0.5)**

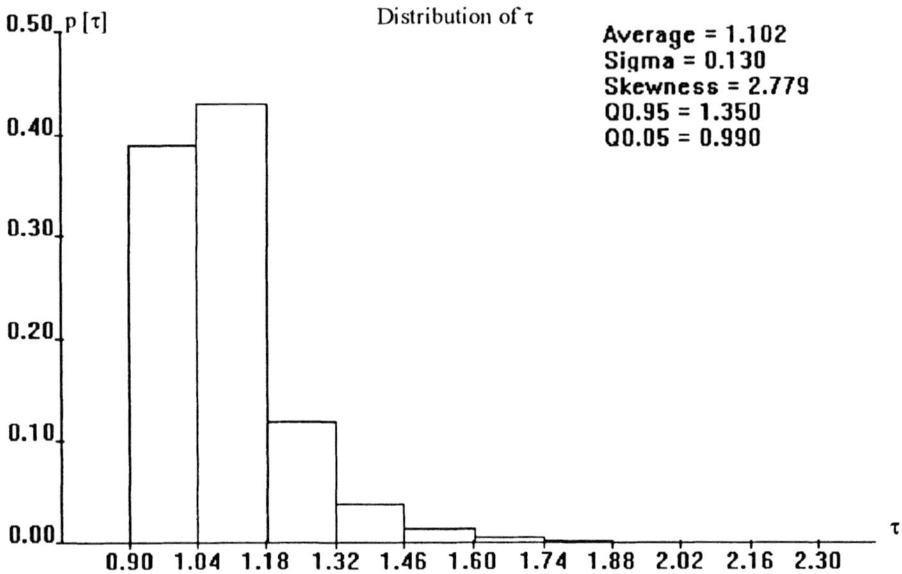


**Figure 8.7 A - Distribution of τ for Case A (α = 1)**

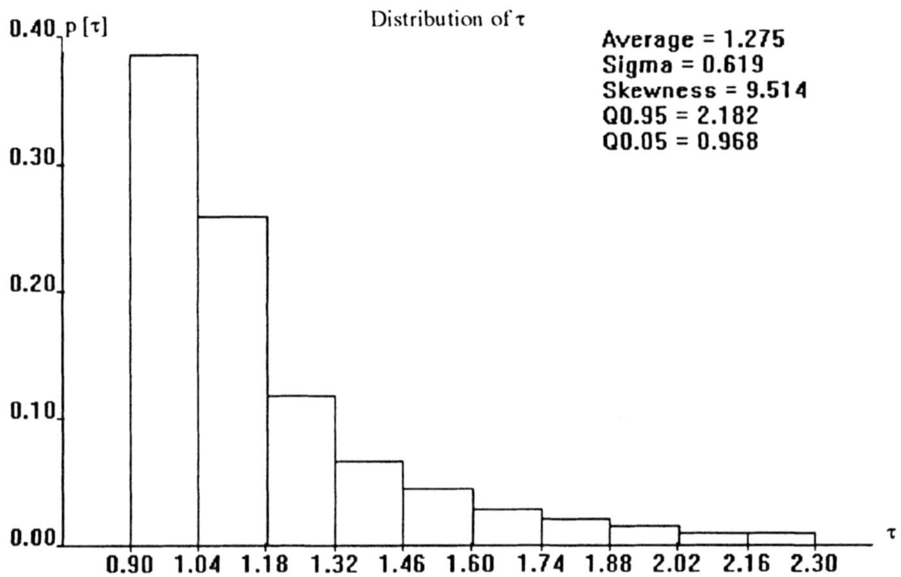


**Figure 8.7 B - Distribution of τ for Case B (α = 1)**

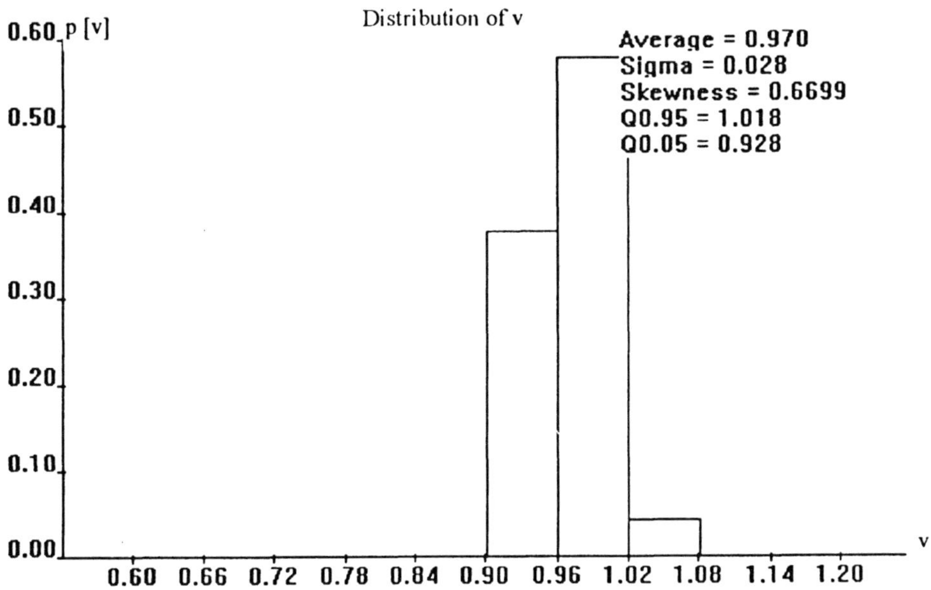


**Figure 8.8 A - Distribution of v for Case A ($\alpha = 0$)**

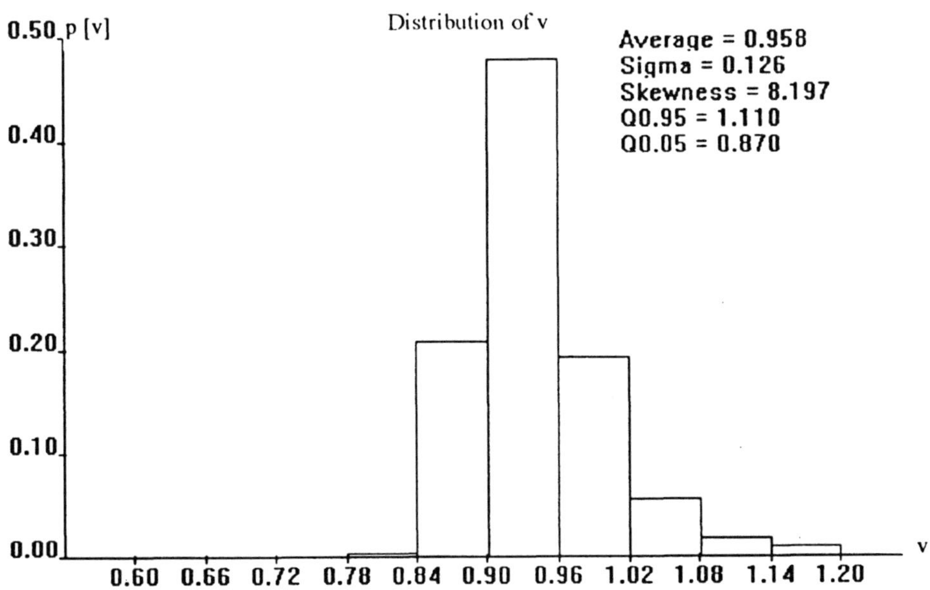


**Figure 8.8 B - Distribution of v for Case B ($\alpha = 0$)**

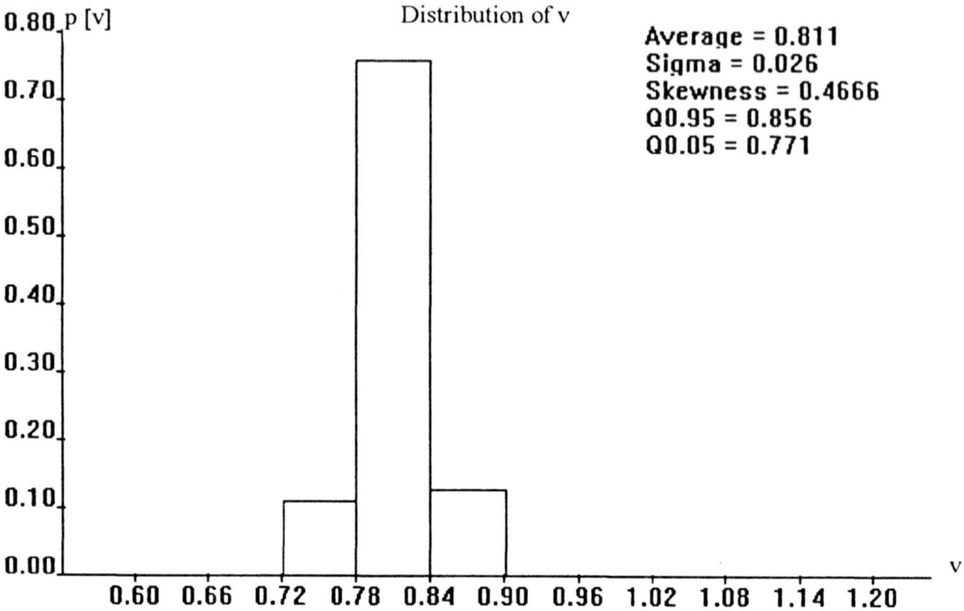


**Figure 8.9 A - Distribution of v for Case A (α = 0.5)**

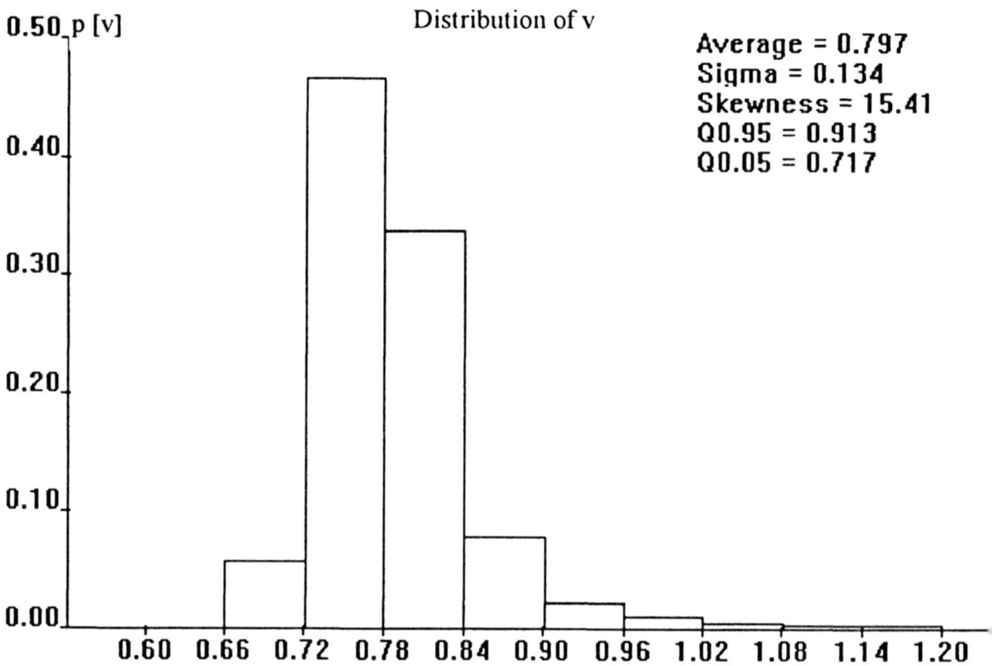


**Figure 8.9 B - Distribution of v for Case B (α = 0.5)**

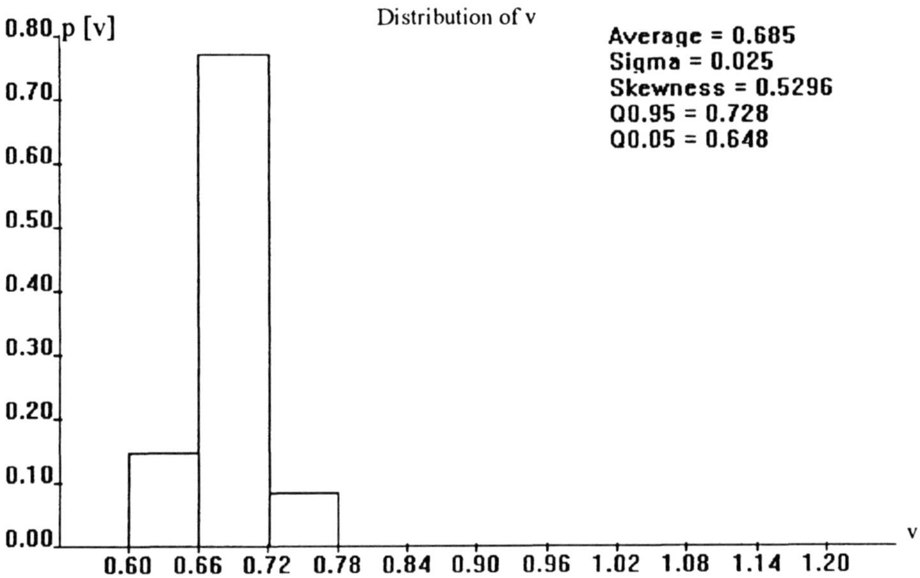


**Figure 8.10 A - Distribution of v for Case A ($\alpha$ = 1)**

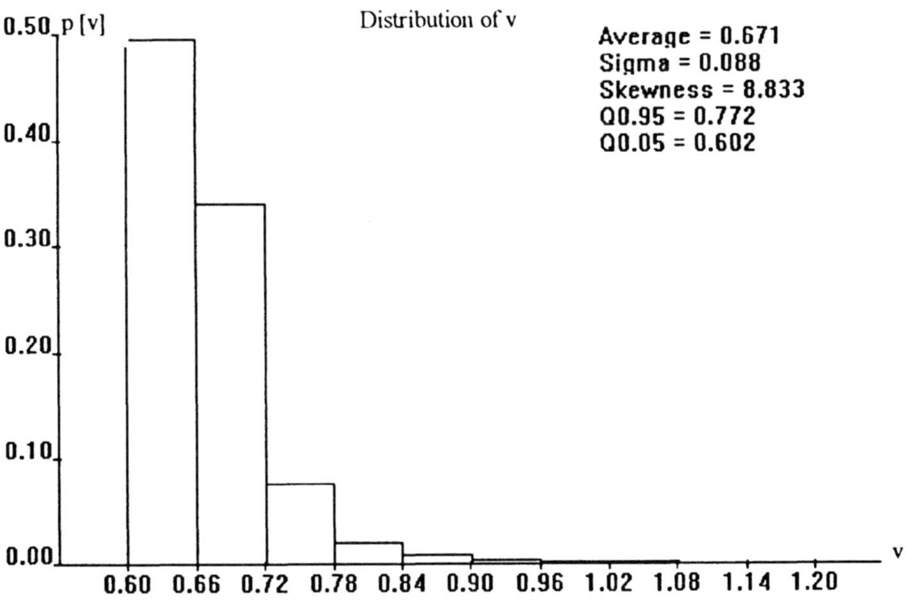


**Figure 8.10 B - Distribution of v for Case B ($\alpha$ = 1)**

The analysis of $\tau$ and v in terms of $\alpha$ is also studied in Figures 8.11 and 8.12 where the estimated mean$(\mu)$ and the 5% and 95% quantiles ($Q_{0.05}$, $Q_{0.95}$) are plotted.

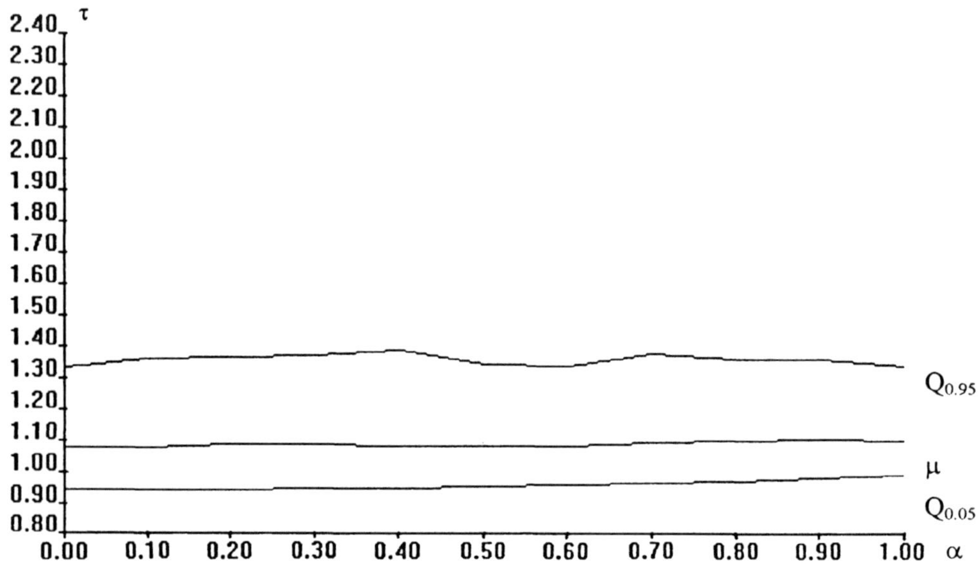


**Figure 8.11 A - Estimated mean and quantiles of $\tau$ in terms of $\alpha$ for Case A**

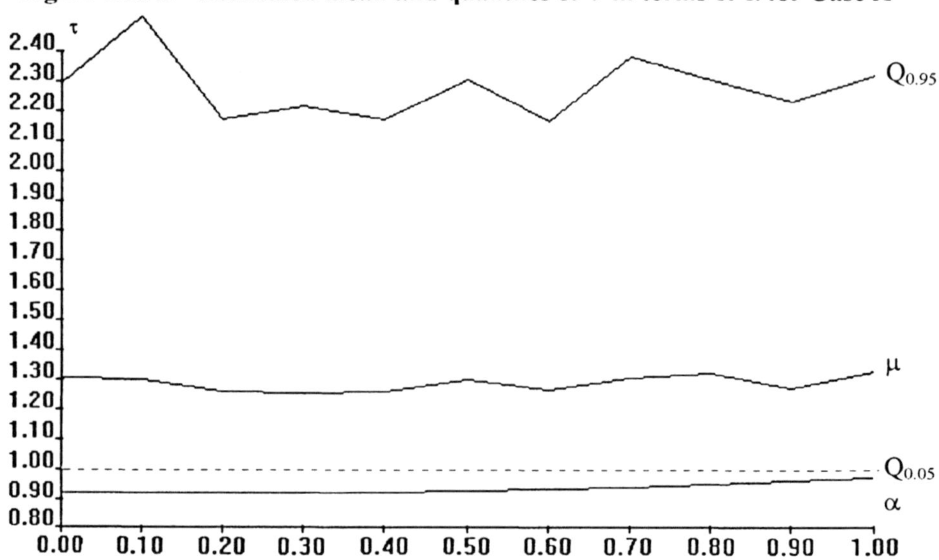


**Figure 8.11 B - Estimated mean and quantiles of $\tau$ in terms of $\alpha$ for Case B**

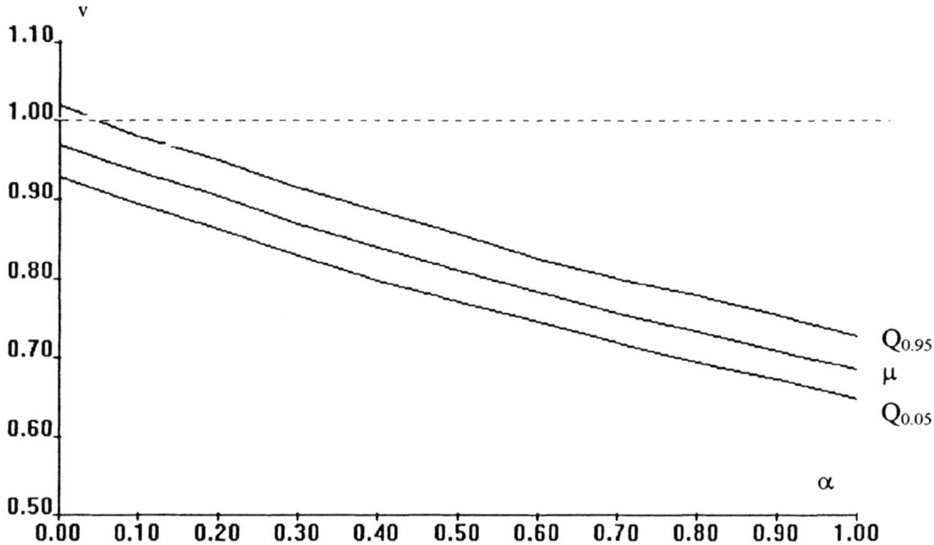


**Figure 8.12 A - Estimated mean and quantiles of v in terms of α for Case A**

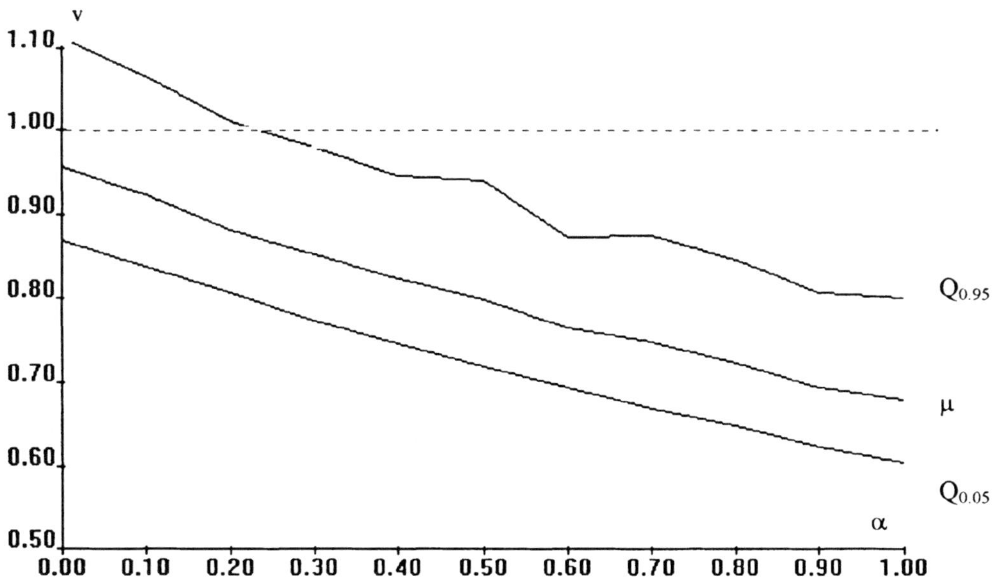


**Figure 8.12 B - Estimated mean and quantiles of v in terms of α for Case B**

The estimated results for the average, standard deviation and quantiles of $\tau, v$ in terms of $\alpha$ and of the adopted Case are presented in Table 8.4 and 8.5. In these tables, the standard error of such parameters (SE (.) ) is estimated using 10 samples of 1 000 realizations.

**Table 8.4**
**Estimated parameters of the distribution of $\tau$**

| $\alpha$ | 0,0 | | 0,5 | | 1,0 | |
|---|---|---|---|---|---|---|
| Variance | LV | HV | LV | HV | LV | HL |
| $\mu$ | 1.074 | 1.266 | 1.081 | 1.275 | 1.102 | 1.306 |
| SE($\mu$) | 0.001 | 0.011 | 0.001 | 0.010 | 0.001 | 0.007 |
| $\sigma$ | 0.146 | 0.833 | 0.143 | 0.870 | 0.132 | 0.782 |
| SE($\sigma$) | 0.003 | 0.134 | 0.003 | 0.240 | 0.004 | 0.099 |
| $Q_{0.95}$ | 1.345 | 2.226 | 1.350 | 2.237 | 1.348 | 2.279 |
| SE($Q_{0.95}$) | 0.007 | 0.037 | 0.005 | 0.035 | 0.006 | 0.030 |
| $Q_{0.05}$ | 0.940 | 0.917 | 0.952 | 0.925 | 0.990 | 0.968 |
| SE($Q_{0.05}$) | 0.001 | 0.000 | 0.000 | 0.000 | 0.001 | 0.000 |

LV - Low variance (Case A)
HV - High Variance (Case B)

**Table 8.5**
**Estimated parameters of the distribution of $v$**

| $\alpha$ | 0,0 | | 0,5 | | 1,0 | |
|---|---|---|---|---|---|---|
| Variance | LV | HL | LV | HV | LV | HL |
| $\mu$ | 0.969 | 0.957 | 0.811 | 0.797 | 0.686 | 0.675 |
| SE( $\mu$) | 0.001 | 0.002 | 0.000 | 0.001 | 0.001 | 0.001 |
| $\sigma$ | 0.028 | 0.148 | 0.026 | 0.128 | 0.025 | 0.110 |
| SE($\sigma$) | 0.001 | 0.023 | 0.000 | 0.012 | 0.001 | 0.012 |
| $Q_{0.95}$ | 1.018 | 1.101 | 0.855 | 0.913 | 0.728 | 0.780 |
| SE($Q_{0.95}$) | 0.001 | 0.006 | 0.001 | 0.005 | 0.001 | 0.004 |
| $Q_{0.05}$ | 0.928 | 0.819 | 0.765 | 0.718 | 0.648 | 0.603 |
| SE($Q_{0.05}$) | 0.001 | 0.001 | 0.019 | 0.001 | 0.000 | 0.001 |

LV - Low variance (Case A)
HV - High Variance (Case B)

Finally, the risk function, R, was computed (Cases A and B) in terms of $\alpha$ and using different bounds for $L_V$ and $L_\tau$ :

| | V | $\tau$ |
|---|---|---|
| Target domain 1: | $L_v = 0.6$ | $L_\tau = 1.0$ |
| Target domain 2: | $L_v = 0.8$ | $L_\tau = 1.20$ |
| Target domain 3: | $L_v = 1.0$ | $L_\tau = 1.40$ |

This set of results is presented in Figures 8.13 to 8.15 allowing the estimation of the optimal $\alpha$ which is given by the $\alpha$ minimizing R.

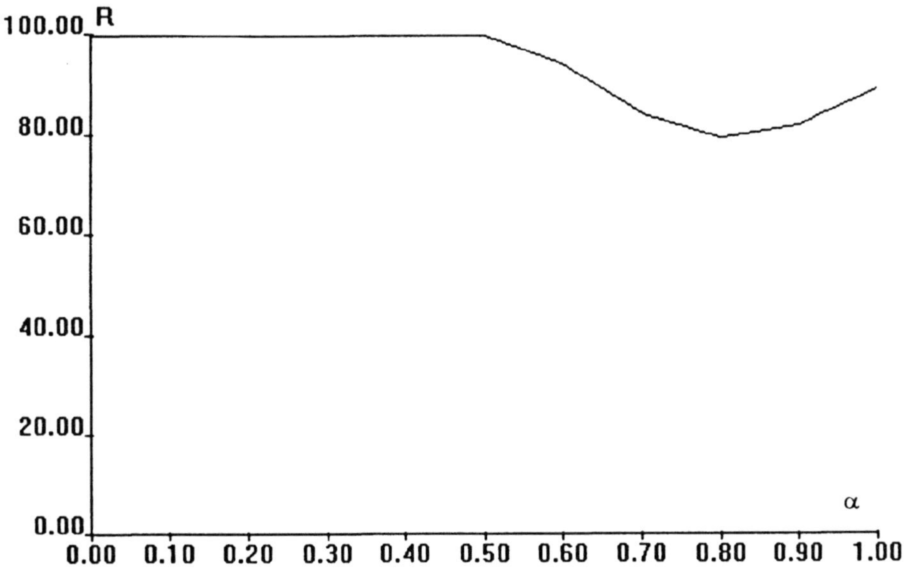


**Figure 8.13 A - Estimated risk for domain 1 in terms of $\alpha$ in Case A**

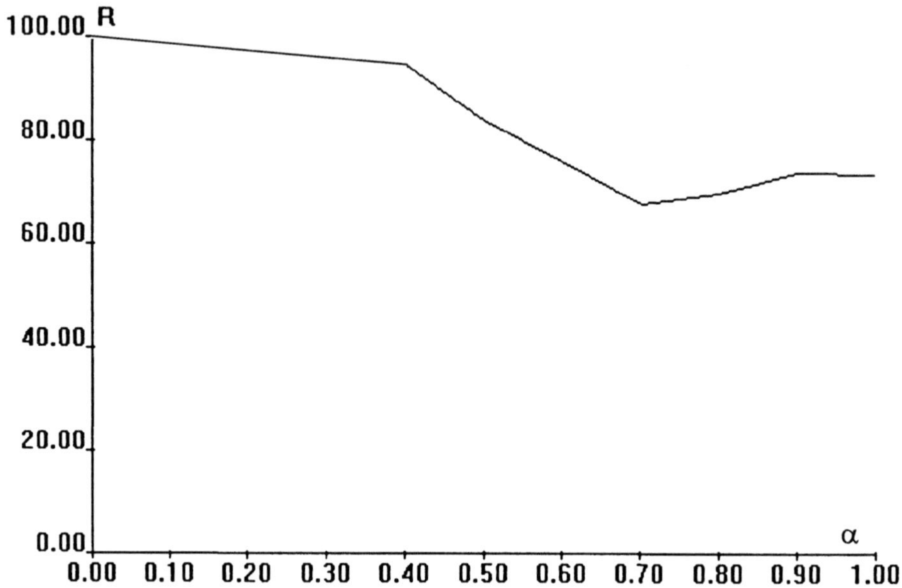


**Figure 8.13 B - Estimated risk for domain 1 in terms of α in Case B**

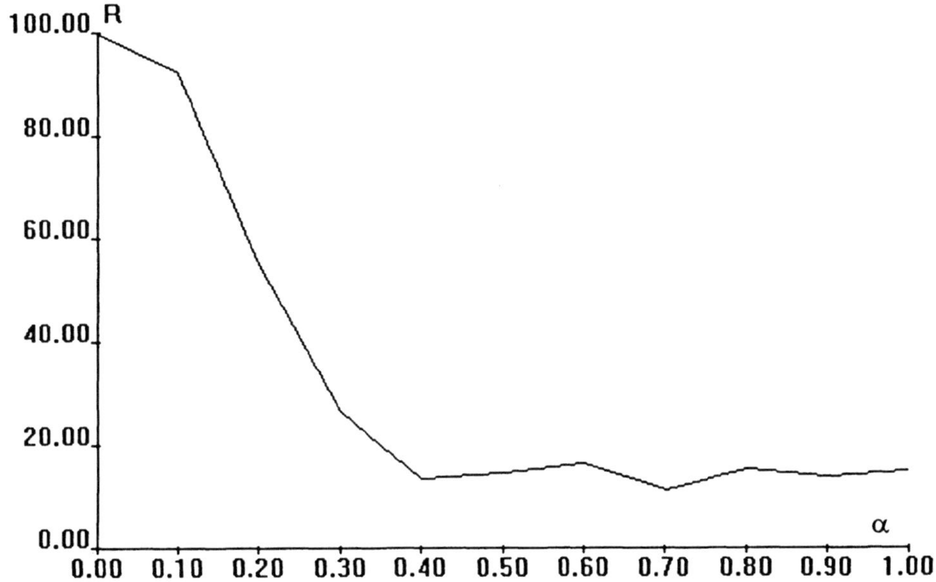


**Figure 8.14 A - Estimated risk for domain 2 in terms of α in Case A**

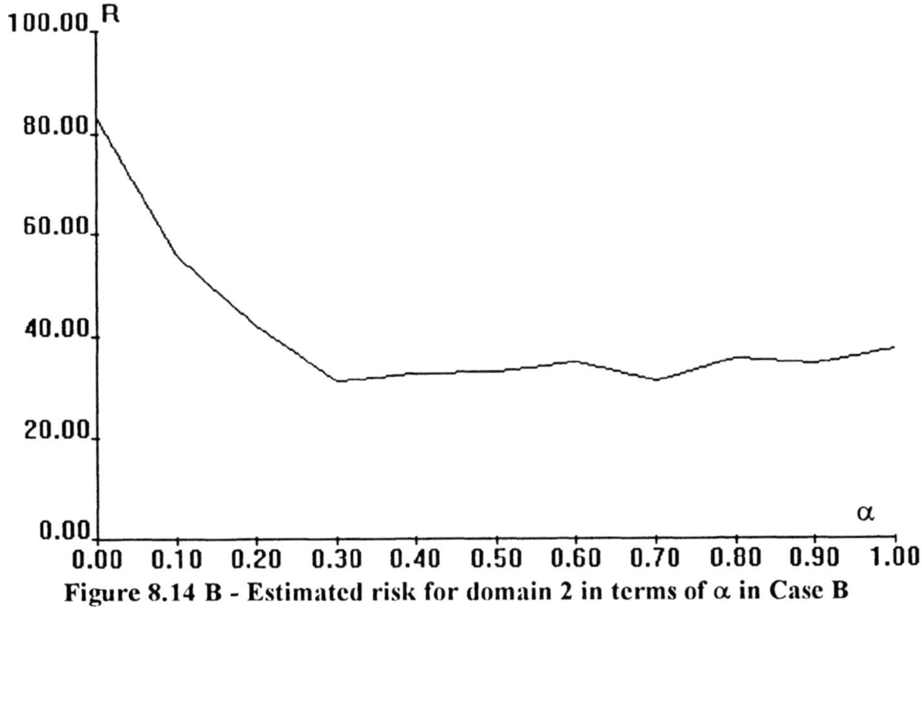


**Figure 8.14 B - Estimated risk for domain 2 in terms of α in Case B**

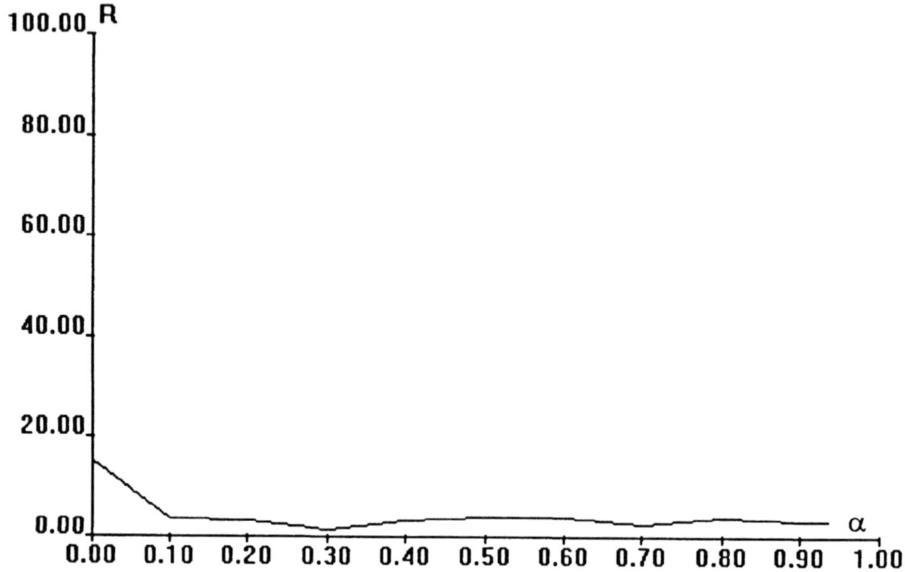


**Figure 8.15 A - Estimated risk for domain 3 in terms of α in Case A**

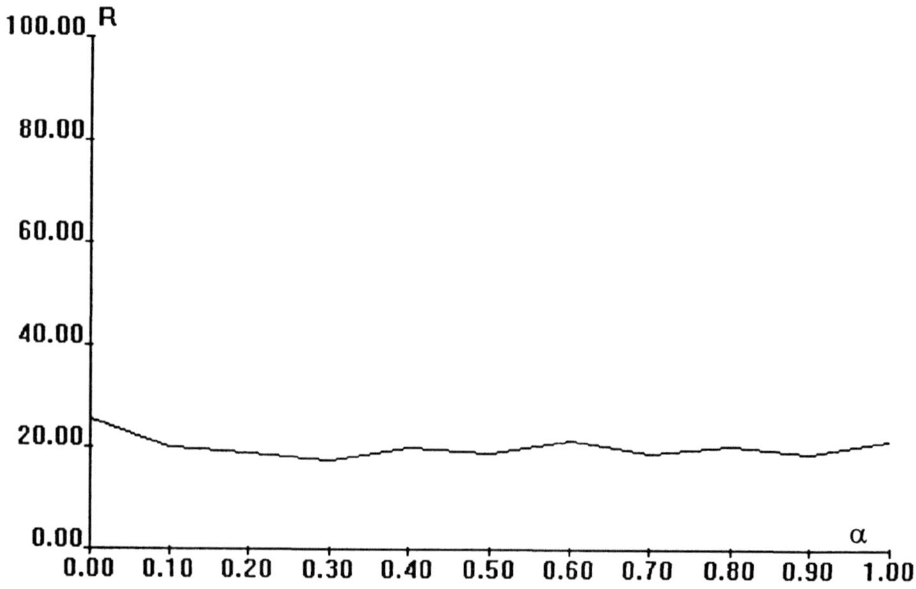


**Figure 8.15 B - Estimated risk for domain 3 in terms of α in Case B**

## 5. Conclusions

The presented results deserve the following comments:

- the distribution of $\tau$ always has a positive skewness showing that the gaussian assumption of PERT is far from acceptable for this type of example.

  However, the distribution of v is reasonably symmetrical.

  It should be noted that these results were obtained assuming a lognormal law for the duration of the activities and a linear gaussian regression for the cost in terms of the duration.

- the level of variance adopted is crucial as the assumption of a higher variance for the durations of the activities increases very significantly the effect of $\alpha$. Actually, the increase of the duration and the decrease of the present cost is much clearer when $\alpha$ increases for case B than for case A.

  This effect is particularly strong for the skewness and the upper quantile.

- the presented results for the risk, R, show that the concept of optimal $\alpha$ can be estimated through the presented model. The risk gain due to adopting the optimal $\alpha$ grows with the uncertainty of the data as would be expected and its value also depends significantly on the bounds adopted and on the level of variance (case A or B).

# Annex 1 - Developed Software*

## *1. RISKNET*

The purpose of this software (developed in Microsoft Visual C++ 4.0) is carrying out advanced analyses and graphical presentation of project networks.

**Module 1: Tables**

A set of tables including the list of activities, precedent activities, successor activities and their data is produced.

**Module 2: Adjacency and Level Matrices**

This mode produces the adjacency matrix and the level matrix.

The direct precedence links are considered.

The level matrix is presented including:

a) the number of activities of each level within the diagonal elements;
b) the other elements are presented in black if all possible direct links exist and a clearer shade is adopted if just a lower percentage exists.

**Module 3: Schedule**

The schedule of the project is presented using a bar chart representation for the earliest and latest starting times. The floats are also represented.

---

* for further information, please contact the author:
L. Valadares Tavares, CESUR - IST, Av. Rovisco Pais, - 1000 Lisboa, Portugal
(lavt@alfa.ist.utl.pt)

The user can increase or decrease the duration of each activity and can schedule its start between its earliest and latest starting time.

**Module 4: Graphical display**

The graphical display of the network is produced using this mode.

This software is based on the following procedures:

- the activities are represented by nodes and they are located in terms of their progressive level being equally spaced in each level along a central strip (S) of the plane. Every node is represented by a circle except the starting node and those nodes not preceding any other activity which are represented by a small square. The final node is not represented.

- a specific indicator, $\pi_I$, measuring for each activity, i, the degree of difficulty for drawing its link is computed by: $\pi_i = \sum_{L=1}^{V} n_i (L).2^{2\,(L-1)}$ i where $n_i(L)$ is the number of links including i with length L=1,2,...V. (excepting those connecting the activity to the final node). This function is adopted because a longer link is more difficult to be drawn without confusion with other links than a shorter one.

- at each level, the nodes are ordered in terms of $\pi_i$ allocating an activity to each node from bottom up. This procedure guarantees that the most difficult nodes will be located near the upper border which is convenient to reduce the number of long links crossing most of the network.

- the links with length equal to one are represented by bold full straight lines;

- the links longer than one are represented by curves distributed on strips of space outside S. A link with length L'>L will go through a strip more peripheral than the curve representing the link with length equal to L. These curves are described by B - splines (Foley and Van Dams., 1982) and the following conventions are used:

  - thick full line (L=1)
  - full line (L=2)
  - thin full line (L≥3)

- the links connecting each activity with the final node are not represented as they are not required to understand the network.

Graphical representations of the resource requirements are also produced for each schedule.

**Module 5: Network generation**

The purpose of this module is the generation of project networks in terms of the proposed six morphologic indicators: $I_1$, $I_2$, $I_3$*, $I_4$, $I_5$, $I_6$.

The user should specify a value for each indicator of the set $I_1$, ..., $I_6$ and then the following procedure to generate a network is adopted.

1- Generation of N and M given $I_1$, $I_2$.

2- Allocation of the activities to the levels, a = 1, ..., M, in terms of $I_3$ (a) and introduction of precedences with level length equal to one between arbitrarily chosen activities, one from each level.

3- Generation of the population of possible links with level length L, P (L), starting with L=1.
Generation of the population of activities of level a = 2, ..., M not yet receiving a precedence link from level (a - 1) (population, $\overline{P}$ ).

4 - Random selection from P (L) of a link and eventual updating of $\overline{P}$. Repetition of this step until the number prescribed by $I_4$ is reached.

After generating the network, the duration and the cost of each activity can be generated using a normal, a lognormal or a negative exponential law.

**Module 6: Risk Analysis**

This module receives as input a set of instances of a specific network and is devoted to the estimation of all the statistical features of the total duration and corresponding risk measures.

**Module 7: Optimization**

The purpose of this module is contributing to the optimization of the schedule of a specific network.

The inputs include:

– the features of the network;

– the restrictions concerning the resources ($LM_t$ and $Lm_t$).

This model can produce the following outputs:

a) the earliest schedule
b) the latest schedule
c) an intermediate schedule defined by $\alpha$ (Chapter 8).
d) the files to be received by GAMS-MINOS to optimize the schedule using as objective function the minimal total duration or the Net Present Value.

## 2. MACMODEL

This software is a decision aid to help project managers assessing or evaluating alternative solutions.

This model includes five components:

- Introduction of data about alternatives
- Definition of the value tree
- Construction of the value functions and assignment of the weights
- Comparative analysis
- Sensitivity analysis

A value-tree based on (price, duration and quality) can be used or a specific value tree can be built.

The risk analysis is developed assuming that the distribution of each relevant magnitude (for instance, the price) follows a negative exponential law and that the risk of exceeding a certain threshold is given.

The value functions can be graphically built using a piecewise linear relationship.

The outputs include a table with the comparative analysis of alternatives and their ranking as well as three types of sensitivity analysis.

# Annex 2 - Main Notation

$a(\cdot), A(\cdot), La(\cdot), LA(\cdot), M(\cdot), LM(\cdot), m(\cdot), Lm(\cdot), V(\cdot), Q(\cdot), q(\cdot), S(\cdot), s(\cdot), v(\cdot),$
$\overline{S}(\cdot), \overline{X}(\cdot), \overline{Y}(\cdot), \overline{Z}(\cdot), \overline{I}(\cdot), R(\cdot), C(\cdot), p(\cdot), q(\cdot), \tau(\cdot), \tau'(\cdot), N(\cdot), B.\text{or } B(\cdot), I(\cdot), O(\cdot),$
$W(\cdot), D(\cdot), C(\cdot), S(\cdot), r(\cdot), R(\cdot) \rightarrow$ function of $(\cdot)$

a, b, c, u, v, w → activities

$A_i$ → activity i.

d → devaluation rate

D → decision maker

$E(\cdot)$ → expected value of $(\cdot)$

f → discount factor

F → objective function

$f(\cdot)$ → probability density function

$F(\cdot)$ → cumulative distribution function

g → state function

G → goal

i, j, k, m, p, q, $\tau$, $\delta$, s, s' → integer variables

$I_1, \ldots, I_6$ → indicators

IRR → internal rate of return

L(•, *) → path from • to *

N, K, M, P, Q, H, MW, V, D, W → constants

NPV → net present value

PB → present benefit

PC → present cost

P, p → probability

RNPV → relative net present value

RPV → relative present value

t, $\tau$ → time

$t^s(\cdot), t^f(\cdot)$ → starting, finishing time of •

T, $T_r$, T*, V, R, Q, X, Y, Z, W, k, e, h, r → variables

U → utility function

$X_{it}, a_{ik}, A_{ij}, m_{it}, AX_{it}, \Delta_{it}$ → real matrix

VAR(·), $V(\cdot)$ → variance of $(\cdot)$

Ⓐ $(\alpha)$ (circled) → system

$\alpha$ → float factor

$\alpha$, $\alpha'$ → discount rate

$\beta_\bullet, I_\bullet, J_\bullet, K_\bullet, \Omega_\bullet$ → set defined in terms of •

$\textcircled{\beta}$ → border

$\mu_\bullet, \sigma_\bullet$ → mean, standard deviation of •

$\gamma, \tau, \pi, \lambda$ → coefficients

$\rho(\cdot)$ → correlation

$\Omega$, $\Sigma$ → domain

# References

Ackoff, R., 1967, "Fundamentals of Operations Research", Wiley, N.Y., USA.

Agrawal, M. K., Elmaghraby, S. E. and W. S. Herroelen, 1996, "DAGEN a generator of testsets for project activity nets", European Journal of Operational Research, 90, 376-382.

Ahn, T. and S. S. Erengue, 1998, "The resource constrained project scheduling problem with multiple crashable modes: a heuristic procedure", European Journal of Operational Research, 107, 2, 250-259.

Ashby, W. R., 1956, "An introduction to Cybernetics", University Paperbacks, õxford, U.K.

Battersby, A., 1967, "Network analysis for planning and scheduling", Mac Millan, 2nd Edition, London, UK.

Bein, W. W., Kamburowski, J. E. and M. M. Stallmann, 1992, "Optimal reduction of two-terminal directed acyclic graph", SIAM Journal Computing, 21, 1112 - 1129.

Bellman, R. and S. Dreyfus, 1962, Applied Dynamic Programming, Princeton University Press, NJ.

Bianco, L. P. dell'Olmo and M. G. Speranza, 1998, "Heuristics for multimode schedulling problems with dedicated resources", European Journal of Operational Research, 107, 2, 260-271.

Blazewicz, J., K. H. Ecker, G. Schmidt and J. Weglarz, 1994, "Scheduling in computer and manufacturing systems", 2nd ed., Springer - Verlag, Berlin.

Boctor, F., 1990, "Some efficient multi-heuristic procedures for resource constrained project scheduling", European Journal of Operational Research 49, 3-13.

Brooke, A. D. Kendrick and A. Meeruns, 1996, "GAMS, release 2.25. A user's guide", GAMS Development Corporation, Washington, USA.

Brucker, P., S. Knust, A. Schoo and O. Thiele, 1998, "Abranch and bound algorithm for the resource-constrained project scheduling problem" European Journal of Operational Research, 107, 2, 272-288.

Clark, C. E., 1962, "The PERT model for the distribution of an activity time", Operations Research, 10, 405-406.

Crauwels, HAJ, Poth CN, L. N. Van Wassenhove, 1996, "Local search heuristics for single machine scheduling with batching to minimize the number of late jobs, European Journal of Operational Research, 200-213.

Davies, 1973, "Project scheduling under resource constraints: historical review and categorization of procedures", AIIE Transactions, 5, 297-313

Davies, 1974, "An experimental investigation of resource allocation in multiactivity projects", Operations Research, 24, 587-591.

Davis, E. W., and G. E. Heidorn, 1971, "An algorithm for optimal project scheduling under multiple resource constraints", Management Science, 17/12, 803-816.

De, P., E. J. Dunne, J. B., Ghosh and C. E. Wells, 1994, "The discrete time-cost trade-off problem revisited", Dept. of MIS and Decision Sciences, University of Dayton, Dayton, OH, USA.

De, P., E. James Dunne, J. Ghosh and C. E. Wells, 1994, "Complexity of the discrete time-cost trade-off problems for project networks", Operations Research, 45, 2, 302-306.

Dean, Burton, S. Mertel Jr., L. Roepke, 1969, "Research project cost distribution and budget forecasting", IEEE Transactions on Engineering Management. Vol. E.M. 16, n°4, 176-191.

Demeulemeester, E., B. Dodin and W. Herroelen, 1993, "A random activity network generator", Operations Research, 41, 5, Sept-Oct., 972-980.

Demeulemeester, Erik L., and W. S., Herroelen, 1997, "A branch-and-bound procedure for the generalized resource-constrained project scheduling problem", Operations Research, 45, 2, 201-212.

Doersch, R. H. And J. H. Patterson, 1977, "Scheduling a project to maximize its present value: a zero-one programming approach", Management Science, 23/8, 882-889.

Doersch, R. H. And J. H. Patterson, 1977, "Scheduling a project to maximize its present value: a zero-one programming approach", Management Science, 23/8, 882-889.

Drucker, P., 1954, "The practice of management", Harper and Row, N.Y., USA.

Drucker, P., 1970 "Management for results", Pan Books, N.Y., USA.

Eglese, R. W., 1990, "Simulated annealing: a tool for OR", European Journal of Operational Research, 46, 271-281.

Elmaghraby, S. E., 1977, "Activity networks: project planning and control by network models", Wiley, N.Y., USA.

Elmaghraby, S. E. and W. S. Herroelen, 1980, "On the measurement of complexity in activity networks", European Journal of Operational Research, 5, 223-234.

Elmaghraby, S. E., 1995, "Activity nets: A guided tour through some recent developments", European Journal of Operational Research, 82, 383-408.

Fayol, H., 1949, "General industrial management", Pitman, London, UK.

Ferreira, J. A., 1989, "Modelos para a gestão de empreendimentos de Engenharia Civil", (in portuguese), Ph.D dissertation, Instituto Superior Técnico, Lisboa.

Foley, J. D. and A. Van Dams, 1982, "Fundamentals of Interactive Computer Graphics", Addison Wesley.

Glover, 1989, "Tabu search, Part I, ORSA Journal on Computing, 190-206.

Goldberg, 1989, "Genetic algorithms in search, optimization and machine learning", Addison Wesley.

Hammer, M. and S. Stanton, 1995, "The re-engineering revolution: the handbook", Harper Collins, New York.

Handly, 1979, "Understanding organizations", Penguin.

Herroelen, Willy, S., P. Van Dommelen and Eric L. Demeulemeester, 1997, "Project network models with discounted cash flows: a guided tour through recent developments", European Journal of Operational Research, 100, 97-121.

Johnson, T. J. R., 1967, "An algorithm for the resource-constrained project scheduling problem", PhD dissertation (Unpublished), MIT.

Jones, Robert E., 1990, "Managing the political context in PMS organizations", European Journal of Operational Research, 49, 60-67.

Kaimann, R. A., 1974 "Coefficient of Network Complexity", Management Science, 21, 172-177.

Kamburowski, J., 1985, "An upper bound on the expected completion of PERT networks", European Journal of Operational Research, 21, 206-212.

Kaufman, A., 1968, "The science of decision-making", World University Library, London, UK.

Kearney, A. T. Inc., 1993, "Activity-based management", Monograph n° 7, A. T. Kearney, Chicago, USA.

Keeney, R. L., 1992, "Value - focused thinking: a path to creative decision making", Harvard University Press, Boston, Massachusetts, USA.

Kelley, J. E. and M. R. Walker, 1959, "Critical path planning and scheduling", Proc Eastern Joint Computer Conf., 16, 160-172.

Kelly, J. E., 1961, "Critical path planning and scheduling mathematical bases", Operations Research, 9, 246-320.

Kelly, Jr., J. E., 1963, "The critical-path method: resource planning and scheduling", in Math and Thompson (ed.), 1963, Industrial scheduling, Prentice-Hall, N. Y., USA.

Kolish, Rainer, Arno Sprecher and Andreas Drexl, 1995, "Characterization and generation of a general class of resource - constrained project scheduling problems", Management Science, Vol. 41, 1693 - 1703.

Maggot, J. and K. Skudlarski, 1993, "Estimating the mean completion time of PERT networks with exponentially distributed durations of activities", European Journal of Operational Research, 71, 70-79.

Malcolm, D. G., J. M., Rosebloom, C. E. Clark and W. Fazar, 1958, "Application of a technique for research and development program evaluation", Operational Research, 7, 649-669.

Muller, J. H. and J. Spinrad, 1989, "Incremental modular decomposition", Journal of the ACM, 36, 1-19.

Pascoe, T. L., 1966, "Allocation resources - CPM, Revue Française de Recherche Opérationelle, 38, 31-38.

Patterson, J. H., 1973, "Alternative methods of projects scheduling with limited resources", Naval Research Logistics Quarterly, 20/4, 767-784.

Patterson, J. H., and G. Roth, 1976, "Scheduling a project under multiple resource constraints: A zero-one programming approach", IIE Trans. 8, 449-455.

Patterson, J. H., 1984, "A comparison of exact approaches for solving the multiple constrained resource project scheduling problem", Management Science, 30, 845-867.

Patterson, J. H., R. Slowinski, B. Talbot and J. Weglarz, 1990, "Computational experience with a backtracking algorithm for solving a general class of precedence and resource - constrained scheduling problems", European Journal of Operational Research, 49, 68 - 79.

Pritsker, A. A. B., L. J. Watters and P. M. Wolfe, 1969, "Project scheduling with limited resources: a zero one programming approach", Management Science, 16, 93-108.

Roy, B., 1985, "Methodologie Multicritère d'Aide à la Decision", Economica, Paris, France.

Simon, H. A., 1960, "The new science of management decision", Harper and Row.

Sprecher, A. and A. Drexl, 1998, "Multi-mode resource-constrained project scheduling by a simple, general and powerful sequencing algorithm", European Journal of Operational Research, 107, 2, 431-450.

Soroush, H., 1993, "Risk taking in stochastic PERT networks", European Journal of Operational Research, 67, 221-241.

Talbot, F. B., 1982, "Resource constrained project scheduling with time-resource trade-offs: the nonpreemptive case", Management Science, 28, 1197-1210.

Tavares, L. V., 1984, "The TRIDENT approach to rank alternative tenders for large engineering projects", Foundations of Control Engineering, 9/4, 181-191.

Tavares, L. V., 1987, "Optimal resource profile for program scheduling", European Journal of Operational Research, 29, 83-90.

Tavares, L. V., 1989, "A multi-stage model for project scheduling under resource constraints", in R. Slowinski and J. Weglarz (ed.), 1989, "Advances in Project Scheduling", Elsevier, Amsterdam, Netherlands.

Tavares, L. V., 1990, "A multistage non-deterministic model for project scheduling under resources constraints", European Journal of Operational Research, 49, 92-101.

Tavares, L. V., 1994, "A stochastic model to control project duration and expenditure" European Journal of Operations Research, 78, 262-266.

Tavares, L. V., 1995, "A review on the contributions of Operational Research to Project Management", EURO XIV, 67-82.

Tavares, L. V., J. A. Antunes Ferreira and J. Silva Coelho, 1997, "The risk of delay of a project in terms of the morphology of its network", EURO XV - INFORMS XXXIV, Barcelona, Spain, to appear in EJOR.

Tavares. L. V., J. A. Antunes Ferreira and J. S. Coelho, 1998 "On the optimal management of project risk", European Journal of Operational Research, 107, 2, 451 - 469.

Taylor, F., 1947, "Scientific management", Harper and Row, N.Y., USA.

Wiest J. D., 1967, "A heuristic model for scheduling large projects with limited resources", Management Science, 13/6, B359-B377.

# INDEX

Breinigsville, PA USA
26 May 2010
238652BV00005B/122/A

9 780792 382225